International Association of Geodesy Symposia

Jeffrey T. Freymueller, Series Editor
Laura Sánchez, Series Assistant Editor

Series Editor
Jeffrey T. Freymueller
Endowed Chair for Geology of the Solid Earth,
Department of Earth and Environmental Sciences
Michigan State University
East Lansing, MI, USA

Assistant Editor
Laura Sánchez
Deutsches Geodätisches Forschungsinstitut
Technische Universität München
München, Germany

International Association of Geodesy Symposia

Jeffrey T. Freymueller, Series Editor
Laura Sánchez, Series Assistant Editor

Gravity, Positioning and Reference Frames

Proceedings of the IAG Symposia - GGHS2022: Gravity, Geoid, and Height Systems 2022, Austin, TX, United States of America, September 12 – 14, 2022; IAG Commission 4: Positioning and Applications, Potsdam, Germany, September 5 – 8, 2022; REFAG2022: Reference Frames for Applications in Geosciences, Thessaloniki, Greece, October 17 – 20, 2022

Edited by

Jeffrey T. Freymueller, Laura Sánchez

 Springer

Series Editor
Jeffrey T. Freymueller
Endowed Chair for Geology of the Solid Earth,
Department of Earth and Environmental Sciences
Michigan State University
East Lansing, MI, USA

Assistant Editor
Laura Sánchez
Deutsches Geodätisches Forschungsinstitut
Technische Universität München
München, Germany

Associate Editors
Milad Asgarimehr
GFZ: Deutsches Geoforschungszentrum Potsdam
Potsdam, Germany

Riccardo Barzaghi
Dipartimento di Ingegneria Civile e Ambientale
Politecnico di Milano
Milan, Italy

Jean-Paul Boy
Universite de Strasbourg Ecole et Observatoire des
Sciences de la Terre
Strasbourg, France

Xavier Collilieux
Institut national de l'information geographique et forestiere
Saint-Mandé, France

Michael R. Craymer
Canadian Geodetic Survey
Natural Resources Canada
Ottawa, Canada

Jianliang Huang
Canadian Geodetic Survey
Natural Resources Canada
Ottawa, Canada

Errico C. Pavlis
Goddard Earth Science and Technology Center
University of Maryland, Baltimore County
Baltimore, USA

Benedikt Soja
Institute of Geodesy and Photogrammetry
ETH Zurich
Zurich, Switzerland

Krzysztof Sośnica
Institute of Geodesy and Geoinformatics
Wroclaw University of Environmental and Life Sciences
Wroclaw, Poland

Derek VanWestrum
National Geodetic Survey
National Oceanic and Atmospheric Administration
Boulder, USA

Pawel Wielgosz
Institute of Geodesy
University of Warmia and Mazury
Olsztyn, Poland

ISSN 0939-9585   ISSN 2197-9359   (electronic)
International Association of Geodesy Symposia
ISBN 978-3-031-63854-1   ISBN 978-3-031-63855-8   (eBook)
https://doi.org/10.1007/978-3-031-63855-8

This Springer imprint is published by the registered company Springer Nature Switzerland AG
The registered company address is: Gewerbestrasse 11, 6330 Cham, Switzerland

If disposing of this product, please recycle the paper.

Preface

This volume contains selected papers presented at three different IAG Symposia held in Fall 2022, within a few weeks of each other:

- GGHS2022: Gravity, Geoid, and Height Systems 2022, Austin, TX, United States of America, September 12–14, 2022,
- IAG Commission 4: Positioning and Applications, Potsdam, Germany, September 5–8, 2022, and
- REFAG2022: Reference Frames for Applications in Geosciences, Thessaloniki, Greece, October 17–20, 2022.

The **IAG Commission 2 and the International Gravity Field Service (IGFS) joint symposium GGHS2022** (https://www.csr.utexas.edu/gghs2022/) was hosted and organized by the University of Texas Center for Space Research (UTCSR – https://www.csr.utexas.edu/), with in-person and remote participation and was held at the Thompson Conference Center of UT Austin. This event was originally planned for Fall 2020 but was postponed due to the COVID19 pandemic. Therefore, GGHS2022 became an important opportunity for the global geodesy community to rebuild professional networks and to resume in-person interaction. Accordingly, the conference venue selection, physical layout, and the program organization deliberately emphasized providing ample time and space for relaxed conversations and safe person-to-person interactions.

The GGHS2022 followed the successful symposia GGHS2018 in Copenhagen, Denmark and GGHS2016 in Thessaloniki, Greece, and a long prior sequence of biennial IAG/IGFS meetings since 2000.

The conference was conducted against a backdrop of rapid developments in the science of the Earth's gravity field, its time variability, in the technologies available for sensing these, and the data analytic methods for extracting insights from observations. From the classical disciplines of geoid determination, geodetic reference systems, navigation and satellite orbit determination, and geophysics and interior Earth structure, the gravity field science has in the past decades also provided unique data on cryosphere and hydrological changes, and general mass transport within the earth system, primarily from the US/German GRACE and GRACE Follow-On satellite missions. At the same time, global knowledge of details of the gravity field has improved significantly due to the GOCE mission, large-scale airborne gravity campaigns, and the coverage of the oceans by satellite altimetry. New technologies such as cold atom interferometry, miniature gravity sensors, strapdown IMU gravity sensors, and new satellite mission concepts are on the verge of further advancing gravity field science.

The GGHS2022 symposium brought together geodesists, geophysicists, and space scientists who work with gravity field observations from space, airborne and surface, novel gravity field observation technologies, gravity field modelling, fundamental height systems, gravity networks, and gravity field change observations for climate change and hydrology. Contributions were classified according to the following sessions:

- S1: Current and Future Satellite Gravity Missions,
  Chairs: David Wiese, Frank Flechtner, Adrian Jäggi
- S2: Global Gravity Field Modelling
  Chairs: Jianliang Huang, Yan Wang

- S3: Local/Regional Gravity Field Modelling
  Chairs: Riccardo Barzaghi, Hussein Abd-Elmotaal, Georgios Vergos
- S4: Absolute, Relative, and Airborne Gravity – Instrumentation, Analysis, and Applications
  Chairs: Derek van Westrum, Przemyslaw Dykowski
- S5: Height Systems and Vertical Datum Unification
  Chairs: David Avalos, Davey Edwards, Laura Sanchez
- S6: Satellite Altimetry and Applications
  Chairs: Don Chambers, Ole Andersen
- S7: Gravity for Climate & Natural Hazards: Inversion, Modeling, and Processes
  Chairs: Mark Tamisiea, Annette Eicker, Carmen Blackwood

The event was carried out as a hybrid symposium, with a robust in-person attendance complemented with a substantial remote attendance. The remote attendees could fully engage in the Symposium via Zoom, with active participation in both the scientific sessions and in splinter meetings. The audio-visual facilities allowed presentations by the remote attendees, and direct Q&A engagement between the in-person and remote attendees.

GGHS2022 was composed by seven scientific sessions in plenary format, with no concurrent sessions. With a total of 87 attendees from around the globe (62 in-person, 25 remote) – good attendance for a meeting that was still relatively soon after the COVID19 pandemic – the program was completed over 3 days. The program was anchored around moderated scientific oral sessions and posters, which remained up for the duration of the conference, and extended poster sessions were scheduled to encourage networking. A daily freeform "spotlight" session was organized to promote conversation on daily topics of common interest: "Geodesy in the time of COVID"; "Upcoming reference systems (NSRS2022 and others)"; and "EGM2022." The scientific sessions, session chairs, and the detailed presentation and poster program shall remain available at the conference website https://www.csr.utexas.edu/gghs2022/.

The scientific program of the symposium was complemented by an informal social dinner event at a well-known local Texas barbecue joint.

The GGHS2022 would not have been possible with invaluable hard work of the staff at the Center for Space Research (led by Jason Peck), at the TxEEE (led by Gayle Hight), and at the Thompson Center (led by Amy Davis). The work of the Scientific Organizing Committee (https://www.csr.utexas.edu/gghs2022/scientific-organizing-commitee/) is gratefully acknowledged. Their contributions spanned over two full years, including the programming for what was to have been GGHS2020, and involved extended Zoom meetings across multiple time zones, including some at very odd hours. The Symposium was principally funded by conference fees. Partial funding support by NASA (grant 80NSSC23K0001), by the office of Associate Dean of Research at the UT Cockrell School of Engineering, by the IAG for student travel and participation support, and effort contribution by UTCSR are gratefully acknowledged.

The **IAG Commission 4 Symposium on Positioning and Applications** was originally scheduled for September 2020 and due to the COVID19 pandemic was postponed by two years. It took place at the exhibition floor of science ("*Wissenschaftsetage*") of the incorporated society "*proWissen e.V.*" within the "*WIS Bildungsforum*" Potsdam. As the COVID19 pandemic was still not over everywhere in September 2022, the symposium was organized in a hybrid format, i.e. participation was possible both on site and online. A total of 74 participants from 22 countries were registered, 28 of whom took part remotely and 46 met on site.

This symposium was carried out in close cooperation with the International GNSS Service (IGS), the IAG Global Geodetic Observing System (GGOS) Focus Area on Geodetic Space Weather Research FA-GSWR), as well as via linkages with relevant entities within scientific and professional sister organizations. The Symposium was co-sponsored by the International

Association of Geomagnetism and Aeronomy (IAGA) Inter-Division Commission (IDC) on Space Weather. Further partners have been the Institute of Navigation (ION) as well as "*Technische Universität Berlin*" and GFZ Potsdam.

The scientific program of the symposium was divided into nine regular sessions:

- S1: Symposium Opening Session
  Chairs: Allison Kealy, Vassilis Gikas
- S2: Emerging Positioning Technologies and GNSS Augmentations
  Chairs: L. M. Ruotsalainen, Ruizhi Chen
- S3: GNSS Integrity and Quality Control
  Chairs: Pawel Wielgosz, Jianghui Geng, Grzegorz Krzan
- S4: Multi-frequency Multi-constellation GNSS
  Chairs: Sunil Bisnath
- S5: Symposium Special Session
  Chairs: Robert Heinkelmann, Harald Schuh
- S6: Atmospheric Remote Sensing: GNSS-Reflectometry
  Chairs: Milad Asgarimehr, Michael Schmidt
- S7: Atmospheric Remote Sensing: Troposphere
  Chairs: Marcelo C. Santos, Michael Schmidt
- S8: Atmospheric Remote Sensing: Ionosphere
  Chairs: Michael Schmidt, M. Mahdi Alizadeh
- S9: GGOS Focus Area Geodetic Space Weather Research
  Chairs: Ehsan Forootan, Michael Schmidt, Stefan Lotz

Participants of the 2nd IAG Commission 4 Symposium on the roof of the "Wissenschaftsetage" in the center of Potsdam; photo taken from https://www.iag-commission4-symposium2022.net/symposium-foto.jpg

The scientific program of the symposium was complemented by a great social program, which included a guided tour through the GFZ and the historic park on the *Telegrafenberg*, the Ice Breaker Party and a boat trip on the lakes around Potsdam with a conference dinner.

The symposium was financially supported by the IUGG Grants Program. The funded amount was shared partly to invited speakers and experts, who received a waiver of the registration fee, and to participants from long distances, who received additional support for travel costs. It was also used for travel awards for female scientists and for young scientists, preferably from developing countries.

We would like to thank all those who contributed to the success of the 2nd Symposium of the IAG Commission 4, especially the entire Scientific Organizing Committee (SOC) consisting of Allison Kealy, Christina Arras, Sharyl Byram, Suelynn Choy, Ehsan Forootan, Vassilis Gikas, Robert Heinkelmann, Ana Paula Larocca, Jiyun Lee, Laure Lefevre, Stefan Lotz, Laura Ruotsalainen, Marcelo Santos, Michael Schmidt, Harald Schuh, PawełWielgosz, and M. Mahdi Alizadeh. Most of the gratitude goes to the Local Organizing Committee (LOC) of the Symposium, in particular Robert Heinkelmann, Harald Schuh, Anja Böhmer, and M. Mahdi Alizadeh.

The **IAG International symposium "Reference Frames for Applications in Geosciences 2022" (REFAG2022)** was organized by IAG Commission 1 "Reference Frames" with the assistance of the Department of Geodesy and Surveying of the Aristotle University of Thessaloniki. The symposium was attended by 96 participants from 22 countries. The list of participants and other relevant information can be downloaded from the symposium's website www.refag2022.org.

REFAG2022 was the fifth event in the traditional series of IAG dedicated symposia on Reference Frames that were previously held in Munich (2006), Marne-la-Vallée (2010), Luxembourg (2014), and Pasadena (2018). The primary scope of the symposium was to address current theoretical concepts, advancements, and open problems related to reference systems and their practical implementation by space geodetic techniques and their combinations, along with underlying limiting factors, systematic errors, infrastructure-related aspects, and novel approaches for future improvements. After a hard period of necessary prohibitions on face-to-face scientific meetings due to the COVID19 pandemic, REFAG2022 managed to bring together again in a traditional way leading experts from academia, public authorities, and private sector, along with a large number of young scientists and graduate students, to discuss in-person current achievements and future challenges of geodetic reference frames and their scientific and societal impact. The scientific program of REFAG2022 covered all main topics in relation to the activities of IAG Commission 1 and its sub-commissions, including also other initiatives and ongoing projects which endorse the role of geodetic reference frames in Earth science, geospatial applications, and global change studies.

An important part of the symposium was the presentation of the results for the new augmented realizations of the International Terrestrial Reference System (ITRF2020, DTRF2020, JTRF2020) and their application for Earth science and precise orbit determination. Special attention was given to the analysis and modelling of surface loading effects in terrestrial reference frames, to new approaches for co-location ties toward the rigorous combination of space geodetic techniques, and to the testing of novel space-based methods for realizing global reference frames through observations to low Earth orbiting satellites. Another important theme of the symposium addressed the ongoing efforts and the future challenges to advance the geodetic infrastructure in regional and global scale, in support of maintaining high-quality terrestrial reference frames and their operational capabilities for scientific users. It seems that the strengthening of geodetic infrastructure is perhaps the most important factor to ensure the sustainability of the International Terrestrial Reference Frame and its contributing role for the continuous monitoring of the changing Earth.

A total of 88 papers were presented during the four days of the symposium, which were organized into five thematic sessions as follows:
- S1: Global Reference Frame Theory, Concepts, and Computations
  Chairs: Xavier Collilieux, Erricos C. Pavlis
- S2: Space Geodetic Measurement Techniques
  Chairs: Urs Hugentobler, Krzysztof Sośnica

- S3: Regional Reference Frames and their Applications
  Chairs: Carine Bruyninx, Michael Craymer
- S4: Celestial Reference Frames and Earth Orientation Parameters
  Chairs: Benedikt Soja
- S5: Usage and Challenges of Reference Frames for Earth Science Applications
  Chairs: Jean-Paul Boy, Susanne Glaser

Participants of the REFAG2022 symposium in front of the conference venue at Thessaloniki's famous "Aristotle Square"

Many thanks go to all the conveners who devoted valuable time in the compilation of the scientific program of the symposium and helped to make it successful. The Local Organizing Committee was led by Ms. Niki Bai and her great team of NbEvents Co., whose help was invaluable in arranging a very memorable event with an exceptional social program and providing essential support before, during, and after the conference. Lastly, sincere thanks go to all the participating scientists and graduate students who made the REFAG2022 symposium and these proceedings a success.

Members of the scientific committees of all three symposia served as the associated editors in a peer-review process lead by Jeffrey Freymueller and Laura Sánchez, the IAG Symposia Series editors. Their support is highly appreciated. Although most of the reviewers remain anonymous for the authors, a complete list of reviewers is printed in this volume to express our gratitude for their dedication.

Austin, TX, USA   Srinivas Bettadpur
Munich, Germany   Michael Schmidt
Thessaloniki, Greece   Christopher Kotsakis
November 2023

Contents

A Comparison of Pointwise and Levelling Assisted Regional Realisations of IHRS with a Case Study over Sweden

Anders Alfredsson, Jonas Ågren, and Per-Anders Olsson

Abstract

The International Height Reference System (IHRS) was defined by the International Association of Geodesy (IAG) in 2015. The global International Height Reference Frame (IHRF) should provide access to the IHRS in a broad sense. To provide high accuracy local access, regional (or national) realisations will also be needed. This study aims at evaluating different approaches to compute a denser regional realisation of IHRS in case a high accuracy levelling network is available. Using Sweden as a case study region, a GNSS (Global Navigation Satellite System) and geoid based pointwise realisation is compared with three types of levelling assisted realisations. The latter are made by applying least squares adjustments of the precise levelling observations with fixed potential value(s) from either the global IHRF station in Sweden or the pointwise potentials of a larger number of stations. It is concluded that making a minimum constraint adjustment with one station fixed is not the best option. It is favourable to fix a reasonable number of pointwise stations at an internal distance over which the relative uncertainty of levelling is significantly lower than the relative uncertainty of the pointwise solution. The investigation is made using levelling data from the third precise levelling of Sweden, the NKG2015 quasigeoid model and the NKG2016LU postglacial land uplift model.

Keywords

GNSS · Height datum unification · IHRF densification · International height reference frame · Precise levelling

1 Introduction

The International Height Reference System, IHRS, was defined in 2015 by the International Association of Geodesy, IAG; see IAG Resolution No. 1 in Drewes et al. (2016).

A. Alfredsson (✉) · J. Ågren
Faculty of Engineering and Sustainable Development, University of Gävle, Gävle, Sweden

Department of Geodetic Infrastructure, Geodata Division, Lantmäteriet, Gävle, Sweden
e-mail: anders.alfredsson@hig.se

P.-A. Olsson
Department of Geodetic Infrastructure, Geodata Division, Lantmäteriet, Gävle, Sweden

A common global vertical reference is needed for many applications, for instance to investigate and monitor climate related changes in the Earth system (Ihde et al. 2017). The vertical coordinates in IHRS are given by the geopotential numbers, C_P, which are defined as the difference between the conventional value $W_0 = 62,636,853.4$ m^{2}s^{-2} (Sánchez et al. 2016) and the geopotential value at point P, W_P. C_P can be converted to different types of physical heights, but the preferred type is not specified in the IAG resolution. Ihde et al. (2017) point out that the computation of orthometric heights introduce discrepancies caused by dissimilarities in the hypotheses and recommend the use of normal heights.

The specification and establishment of the first IHRS realisation, the International Height Reference Frame (IHRF), is now one of the highest priorities for the international

© The Author(s) 2023
J. T. Freymueller, L. Sánchez (eds.), *Gravity, Positioning and Reference Frames*,
International Association of Geodesy Symposia 156, https://doi.org/10.1007/1345_2023_225

geodetic community. The global IHRF reference network will realise the IHRS at the highest level. One possibility to determine geopotential values referring to the IHRF is the combination of a gravity field model and ellipsoidal heights determined by a space geodetic technique, most often GNSS (Global Navigation Satellite System). The underlying gravimetric model is crucial in the realisation process (Tocho et al. 2022). Regional high resolution gravity field modelling will be used when available. Otherwise, a suitable combined global Earth Gravitational Model (EGM) of high resolution will be used instead.

The global realisation is to be supplemented by regional and national realisations to provide local access to IHRF and to enable the best possible unification of height datums. In the strategy paper of Sánchez et al. (2021b), it is outlined how the IHRS may be realised on the regional/national level. It is specified that the pointwise realisation may be densified by precise levelling to provide local accessibility to the frame with low uncertainty at short distances up to about 100 km.

1.1   Purpose and Delimitations

The main purpose of the paper is to investigate a selection of methods (Table 1) to make use of a precise levelling network when computing a regional or national IHRS realisation. The paper presents a case study for Sweden using the best GNSS dataset, gravimetric quasigeoid model and precise levelling network currently available. The pointwise IHRS realisation is made following the guidelines of Sánchez et al. (2021a, b) based on the latest Nordic/Baltic gravimetric quasigeoid model NKG2015 (Ågren et al. 2016), the third precise levelling of Sweden (Ågren and Svensson 2011) and the NKG2016LU postglacial land uplift model (Vestøl et al. 2019). Only levelling assisted methods that use potential numbers of the pointwise realisation as fixed in the adjustment are investigated. In a forthcoming study, the plan is to find out how the pointwise geopotential numbers and levelling network should be properly weighted relative to each other.

Table 1 Investigated IHRS realisations/solutions

#	Realisation/Solution	Fixed stations
1	Pointwise realisation	–
2	Minimum constraint adjustment with one global IHRF station fixed	The Swedish global IHRF station ONSA0
3	Mean of minimum constraint adjustments	One station at a time from the pointwise realisation fixed
4	Constrained adjustment	A selection of stations from the pointwise realisation with approximately 200 km distance (see Fig. 1)

The study is a part of a larger project aiming for the best possible realisation of IHRS for Sweden including the transformation to the national height frame RH 2000. Later, the project can hopefully be extended to the Nordic/Baltic level within the Nordic Geodetic Commission (NKG).

2   Method

For a levelling assisted realisation of IHRS, a pointwise realisation is needed to provide fixed (or weighted) potential numbers for the height network adjustments. Sections 2.1 and 2.2 presents the input data and conversions made prior to the levelling network adjustments. After that, in Sect. 2.3, we briefly describe the different height network adjustments. Finally, we outline how the levelling assisted IHRS realisations were compared to the pointwise realisation (Sect. 2.4).

2.1   Gravimetric Geoid Model and Ellipsoidal GNSS Heights Used for the Pointwise Realisation

As mentioned in the introduction, the study is limited to using the current official Nordic gravimetric NKG2015 quasigeoid model (Ågren et al. 2016) for the pointwise IHRS realisation. The model was computed using the Least Squares Modification of Stokes' formula with Additive corrections (LSMSA) method, also named the KTH method (Sjöberg 1991, 2003). The global satellite-only geopotential model GO_CONS_GCF_2_DIR_R5 (Bruinsma et al. 2013) with maximum degree 300 and regional gravity data from the NKG gravity database were used. The NKG2015 version we use here utilises the W_0 value of IHRS, the zero permanent tide concept, and the land uplift epoch 2000.0. It should be mentioned that the officially released version of NKG2015 includes a correction for the permanent tide and a zero-level shift to approximately adapt the model to the Nordic/Baltic height systems, but the pure gravimetric model specified above is used in this paper.

The pointwise solution is based on a dataset of 187 evenly distributed high-quality GNSS stations over Sweden that includes the Swedish global IHRF station ONSA0. The dataset has one station every 35–50 km and the coordinates are given in the official Swedish ETRS89 realisation SWEREF 99 (Jivall et al. 2022). The location of the GNSS stations can be seen in the figures in the result chapter. At least 48 hours of GNSS observations with Dorne Margolin antennas and processing with the Bernese software (Dach et al. 2015) have been used to determine the coordinates of the stations. A list of the used versions of the Bernese software can be found in Jivall et al. (2022). The dataset is also well connected to the precise levelling network; see Sect. 2.3. Like in the ITRF2014, spatial positions in SWEREF 99 are given in the tide-free concept.

2.2 Transformations and Epoch Unification

The postglacial land uplift in the Nordic area makes it crucial to be consistent regarding reference epochs for any kind of geodetic data, models or reference systems (Ekman 1996). All computations and comparisons were thus made in the reference epoch 2021.04 as this epoch was agreed for the first IHRF computation (Sánchez et al. 2021a). All input data were thus converted to this epoch prior to the computations.

The quasigeoid model was converted from the reference epoch 2000.0 to 2021.04 using the geoid change model of NKG2016LU (Vestøl et al. 2019). The GNSS dataset was converted from SWEREF 99 to ITRF2014 epoch 2021.04 applying the NKG transformation method according to Häkli et al. (2016). This method contains a seven parameter Helmert transformation together with an epoch conversion based on the velocity field model NKG_RF17vel (Lantmäteriet 2021).

Corrections to align the permanent tide from the tide-free and zero tide concept in the input data sources to the mean tide concept in the IHRF was applied as specified in Mäkinen (2021).

2.3 Height Network Adjustments

The Swedish precise levelling network is part of the Baltic Levelling Ring (BLR) and is the basis for the national Swedish realisation of the European Vertical Reference System, RH 2000 (Ågren and Svensson 2011). The Swedish levelling observations were measured during approximately 30 years, between 1975 and 2003. The adjustment of RH 2000 was made in Nordic cooperation and was finalised in 2005. In total, the Swedish part of the network consists of around 50,000 height benchmarks, of which 5108 are classified as nodal benchmarks. In the current study, the Swedish precise levelling network was extended by selected lines from other countries in BLR, see Fig. 1, and reduced to include only the measured height differences between nodal benchmarks. The resulting network includes 3380 nodal benchmarks, of which 187 are common to the pointwise IHRS realisation.

The precise levelling observations, which are the geopotential differences between nodal benchmarks, were converted to the epoch 2021.04 using the postglacial land uplift model NKG2016LU prior to the adjustment. The least squares adjustment was performed using a standard

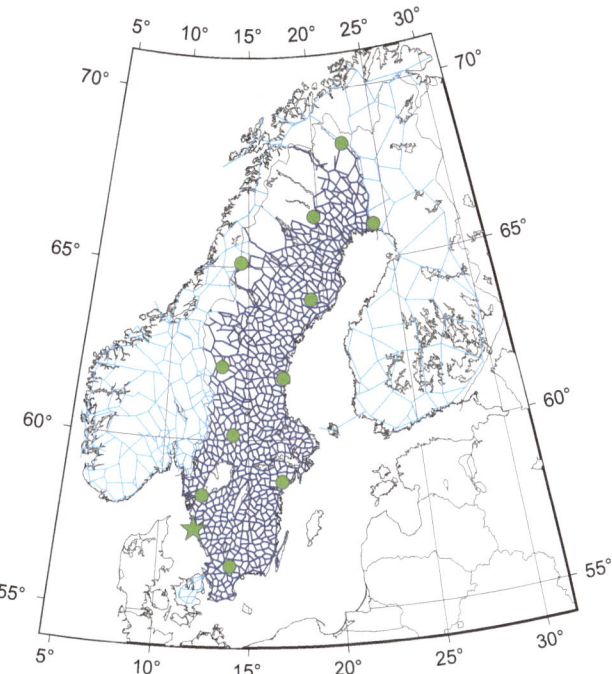

Fig. 1 The Swedish precise levelling network (dark blue) extended with selected parts of the Baltic Levelling Ring (light blue). Green markers represent the fixed pointwise IHRF stations in the adjustments

Gauss-Markoff model (Koch 1999), which is also referred to as adjustment by elements in geodesy (Fan 1997). The levelling observations were weighted using the standard model for levelling assuming weights proportional to the inverse of the length of the levelling lines. Besides this, the variance components for data from different countries presented in Mäkinen et al. (2006) were introduced to change the relative weighting between the countries.

This study includes three different levelling assisted realisations (solutions 2 to 4 in Table 1). Solution 2 was made using a minimum constraint adjustment with one station fixed, namely the Swedish station in the global IHRF network, ONSA0, which is marked by a star in Fig. 1. For reference, 187 similar adjustments were made with respect to each of the 187 stations of the pointwise realisation, one at a time. Solution 3 is the mean of all these solutions.

The relative a posteriori standard uncertainties of the adjusted heights of the Swedish precise levelling network are less than about 8–10 mm over 200 km, i.e. relative to a fixed station 200 km away (Ågren and Svensson 2011). According to Sánchez et al. (2021b), high-quality precise levelling can be used in combination with pointwise IHRF stations up to about 100 km to acquire higher resolution and high accuracy locally. Solution 4 is the result of a constrained adjustment

Table 2 Statistics for the differences between the four solutions in the study. Unit: gpu

Solution A	Solution B	Min	Max	Mean	StDev
2	1	−0.1015	0.0004	−0.0394	0.0165
3	1	−0.0621	0.0398	0.0000	0.0165
4	1	−0.0642	0.0346	0.0148	0.0150
4	3	−0.0149	0.0284	0.0025	0.0094

with a random selection of fixed stations from the pointwise realisation under the condition to get as closely as possible to 200 km between the fixed stations, cf. the black dots in Fig. 4. The 200 km distance was chosen as it corresponds to a relative standard uncertainty of 10 mm in the levelling (cf. the beginning of this paragraph) and for all levelling stations to be closer than 100 km to the nearest selected pointwise IHRF station anywhere in the network (cf. Sánchez et al. 2021b).

2.4   Comparisons

The four solutions from Table 1 were finally compared with each other. The adjusted geopotential numbers for the height network (solutions 2 to 4) were compared with the pointwise IHRS realisation at the 187 IHRF stations. Statistics of the differences between the solutions were computed as minimum, maximum, mean, and standard deviation. The constrained adjustment solution 4 was compared with both the pointwise realisation and the mean of the minimum constraint adjustments, solution 3.

3   Results

The geopotential numbers at the IHRF stations (same as GNSS stations) were compared according to Sect. 2.4. Statistics for the differences between the solutions are presented in Table 2 and illustrated in the corresponding Figs. 2, 3, 4 and 5.

4   Discussion

With the minimum constraint adjustment with respect to one station, the absolute reference level of the network is relying on one single fixed station. Using the Swedish global IHRF station, ONSA0, as fixed (solution 2), the mean difference and standard deviation compared to the pure pointwise solution (solution 1) are −0.039 gpu and 0.016 gpu, respectively, see Fig. 2. To use only the global IHRF station as fixed is clearly not very representative for the whole of Sweden. Using another station from the pointwise IHRS realisation as fixed, the mean difference will be in the interval from

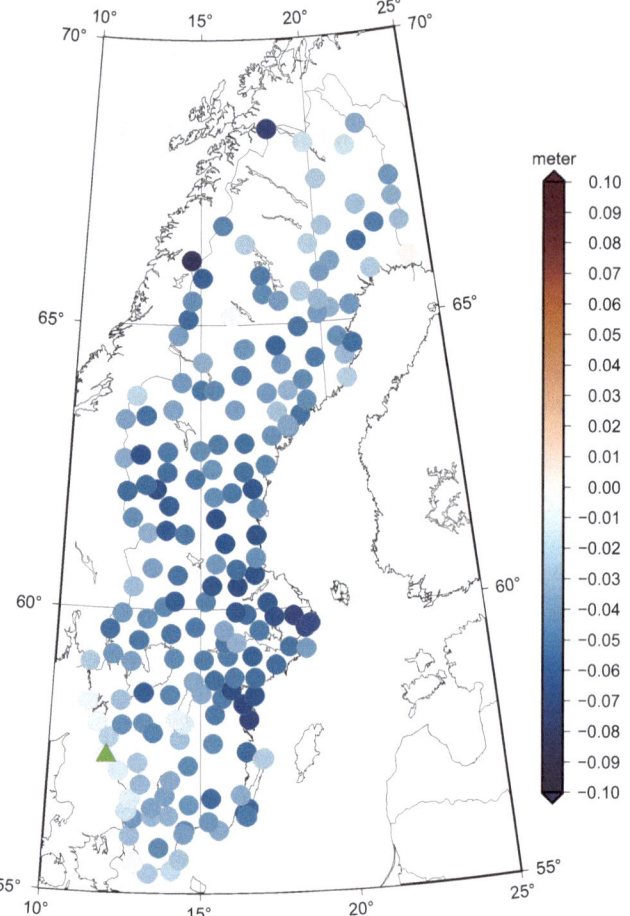

Fig. 2 Differences between the minimum constraint adjustment with respect to the global Swedish IHRF station (ONSA0 in green) and the pointwise IHRS realisation. Unit: gpu

−0.040 gpu to 0.062 gpu. Discrepancies of solutions with one fixed station are closely related to the quality of the pointwise IHRS realisation of the fixed station as the levelling observations remain the same in all compared adjustments. Uncertainties in the gravimetric model and GNSS heights are important factors for the quality of the pointwise realisation, but the uncertainty of other geodynamic modelling required in the realisation process is also crucial. In this case study, the post glacial land uplift was handled by the NKG2016LU model (Vestøl et al. 2019).

The mean of the minimum constraint adjustments, solution 3, is basically a free adjustment solution fitted to the pointwise IHRS realisation with a one-dimensional shift. The minimum and maximum differences compared to solution 1 are −0.062 gpu and 0.040 gpu, respectively, and the standard deviation is 0.016 gpu, see Fig. 3. The shape of the solution relies on the levelling observations only. In one way, solution 3 is not a realistic way to realise IHRS as it demands a pointwise realisation to compute the mean difference. However, assuming that suitable pointwise realisation

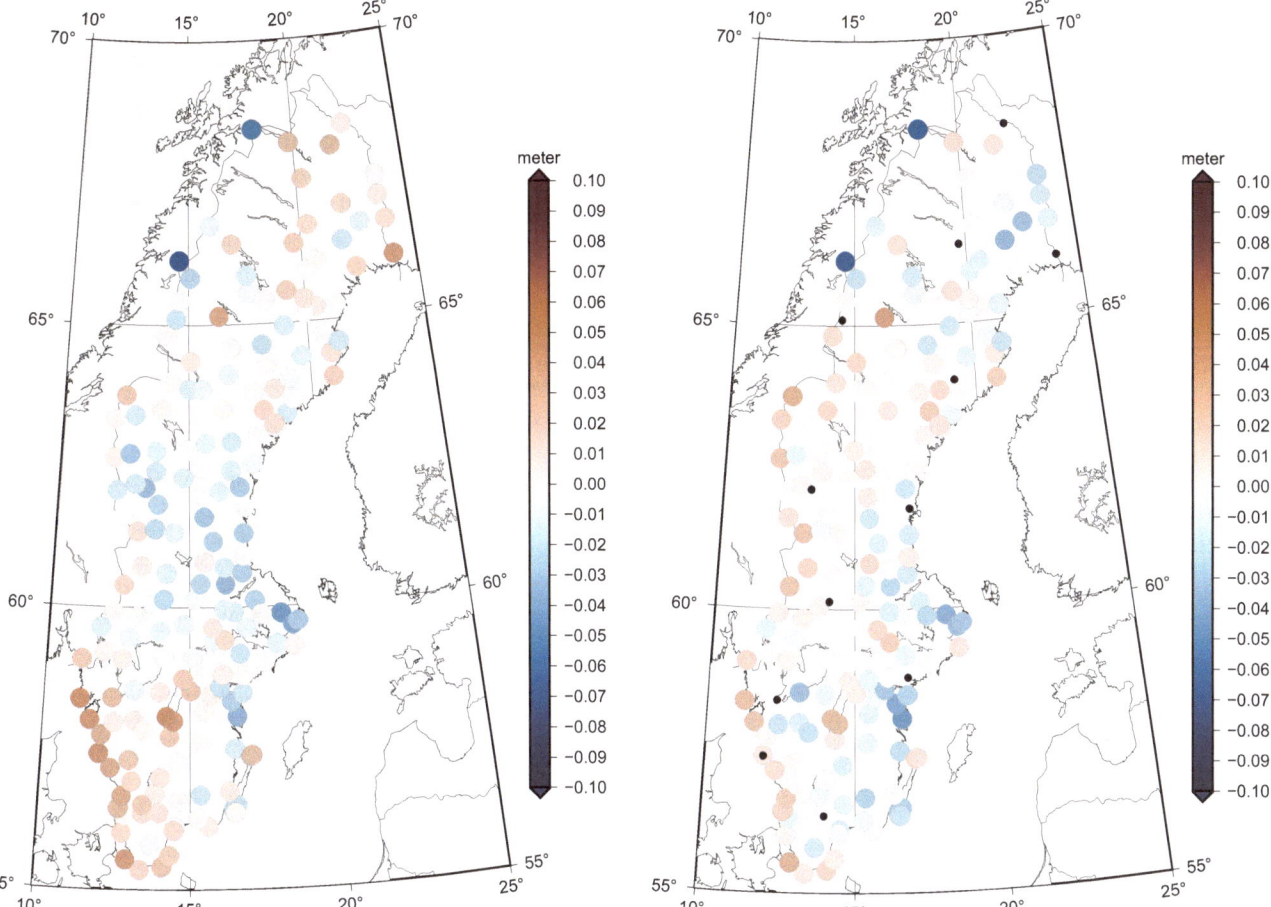

Fig. 3 Differences between the mean of the minimum constraint adjustments solution and the pointwise IHRS realisation. Unit: gpu

Fig. 4 Differences between the solution with a selection of fixed IHRF stations at an internal distance of 200 km (solution 4) and the pointwise IHRS realisation (black dots). Unit: gpu

is available, this kind of solution might be a good option in case one considers the relative uncertainty of the precise levelling to be significantly lower than the relative uncertainty of the pointwise IHRF solution over the whole target area.

For the constrained adjustment with fixed stations every 200 km, solution 4, the minimum and maximum differences to the pointwise realisation are in the same range as solution 3. The standard deviation is slightly lower, 0.015 gpu, and the mean difference is 0.015 gpu, see Fig. 4. The selection of 200 km distance between the fixed stations are based on the motivation in Sect. 2.3. With the constrained adjustment with fixed IHRF stations every 200 km, the shape of solution 4 mainly follows the NKG2015 model over longer distances than 200 km and the levelling over shorter distances. This is considered as a good solution since the accumulated relative standard uncertainty for the levelling network over 200 km (8–10 mm; see Sect. 2.3) is of about the same magnitude as the relative standard uncertainty of the NKG2015 model, which has a very small distance dependence. This means that

levelling is better than NKG2015 for shorter distances and NKG2015 is better over longer distances.

The long wavelength systematic pattern in the difference between solutions 4 and 3, see Fig. 5, is either caused by accumulated errors in the levelling network at longer distances or uncertainties in the lower degrees of the gravimetric model. Accumulated long wavelength errors in the levelling network will result in this kind of pattern with a different shape for solution 3 (mean of the minimum constraint adjustments) compared to solution 4. On the other hand, uncertainties in the lower degrees of the gravimetric model will affect the pointwise realisation over longer distances instead, but for very long distances it is well known that gravity field modelling is much better than precise levelling. The errors of the ellipsoidal GNSS heights are considered to be almost uncorrelated and will not produce this kind of long wavelength systematic pattern. The mean difference between solutions 4 and 3 is 0.002 gpu, the minimum and maximum differences are −0.015 gpu and 0.028 gpu, respectively, and

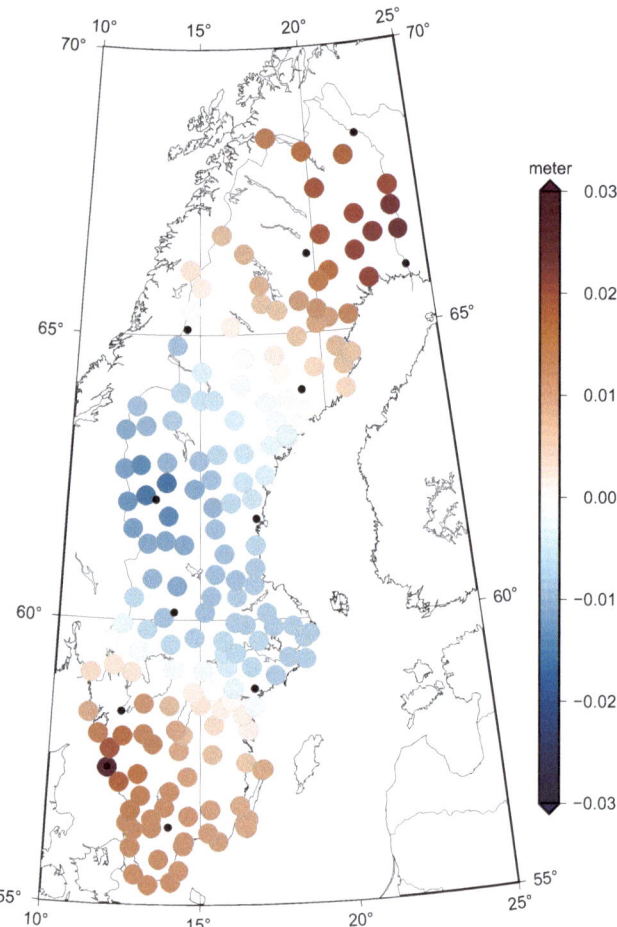

Fig. 5 Differences between the solution with a selection of fixed IHRF stations (solution 4) and the mean of the minimum constraint adjustments (solution 3). Note the different scale compared to Figs. 2, 3 and 4. Unit: gpu

the standard deviation is 0.009 gpu. The deviation represents mainly the difference in long wavelength shape between the levelling network and the gravimetric model.

5   Conclusions

The result of this paper shows that a minimum constraint adjustment with respect to one station in the height network is not optimum for a levelling assisted realisation of IHRS in the whole of Sweden.

It can be concluded that a constrained adjustment of the Swedish height network can be used to densify a sparse pointwise realisation. It is shown that a constrained adjustment with 200 km between the fixed stations performs about as well as the mean of the minimum constrained adjustments of the levelling network. It should be noted that a careful consideration of the uncertainty of the levelling network and

the pointwise realisation should form the basis of the choice of distance between fixed stations.

A densified IHRS realisation based on the adjustment of a levelling network will provide IHRF potential numbers for a large number of height benchmarks. The Swedish network consists of 3380 nodal benchmarks and about 50,000 benchmarks in total, which will provide the basis for the work on height datum unification and for computing transformation surfaces between the national height reference frame and the IHRF.

References

Ågren J, Svensson R (2011) The height system RH 2000 and the land uplift model NKG2005LU. Mapping Image Sci 3:4–12

Ågren J, Strykowski G, Bilker-Koivula M, Omang O, Märdla S, Forsberg R, Ellmann A, Oja T, Liepins I, Parseliunas E, Kaminskis J, Sjöberg L, Valsson G (2016) The NKG2015 gravimetric geoid model for the Nordic-Baltic region. https://doi.org/10.13140/RG.2.2.20765.20969

Bruinsma S, Foerste C, Abrikosov O, Marty J, Rio M-H, Mulet S, Sylvain B (2013) The new ESA satellite-only gravity field model via the direct approach. Geophys Res Lett 40:3607–3612

Dach R, Lutz S, Walser P, Fridez P (eds) (2015) Bernese GNSS Software Version 5.2. User manual. University of Bern, Bern Open Publishing, https://doi.org/10.7892/boris.72297

Drewes H, Kuglitsch F, Adám J, Rózsa S (2016) The geodesist's handbook 2016. J Geod 90(10):907–1205. https://doi.org/10.1007/s00190-016-0948-z

Ekman M (1996) A consistent map of the postglacial uplift of Fennoscandia. Terra Nova 8:158–165. https://doi.org/10.1111/j.1365-3121.1996.tb00739.x

Fan H (1997) Theory of errors and least squares adjustment. Royal Institute of Technology (KTH), Division of Geodesy and Geoinformatics

Häkli P, Lidberg M, Jivall L, Nørbech T, Tangen O, Weber M, Pihlak P, Aleksejenko I, Paršeliunas E (2016) The NKG2008 GPS campaign - final transformation results and a new common Nordic reference frame. J Geodetic Sci 6(1):1–33. https://doi.org/10.1515/jogs-2016-0001

Ihde J, Sánchez L, Barzaghi R, Drewes H, Foerste C, Gruber T, Liebsch G, Marti U, Pail R, Sideris M (2017) Definition and proposed realization of the international height reference system (IHRS). Surv Geophys 38(3):549–570. https://doi.org/10.1007/s10712-017-9409-3

Jivall L, Nilfouroushan F, Al Munaizel N (2022) Analysis of 20 years of GPS data from SWEREF consolidation points – using BERNESE and GAMIT-GLOBK software. Lantmäterirapport 2022:1. https://doi.org/10.13140/RG.2.2.25918.97609

Koch K-R (1999) Parameter estimation and hypothesis testing in linear models. Springer, Berlin. https://doi.org/10.1007/978-3-662-03976-2

Lantmäteriet (2021) Transformation between ITRF 2014/WGS 84 and SWEREF 99. https://www.lantmateriet.se/contentassets/bbc47979dfef4f338e3c4f8b139da2fb/transformation_itrf2014-sweref99.pdf. Accessed 26 Feb 2023

Mäkinen J (2021) The permanent tide and the international height reference frame IHRF. J Geod 95(9). https://doi.org/10.1007/s00190-021-01541-5

Mäkinen J, Lilje M, Ågren J, Engsager K, Eriksson P-O, Jepsen C, Olsson P-A, Saaranen V, Schmidt K, Svensson R, Takalo M, Vestøl

O (2006) The Baltic Levelling Ring. The Working Group for Height Determination of the Nordic Geodetic Commission. https://doi.org/10.13140/RG.2.2.33298.96961

Sánchez L, Cunderlík R, Dayoub N, Mikula K, Minarechová Z, Šíma Z, Vatrt V, Vojtíšková M (2016) A conventional value for the geoid reference potential W0. J Geod 90(9):815–835. https://doi.org/10.1007/s00190-016-0913-x

Sánchez L, Huang J, Ågren J, Barzaghi R, Vergos GS (2021a) Recovering potential values from regional (quasi-)geoid models. Unpublished guidelines. IAG joint working group 0.1.3

Sánchez L, Ågren J, Huang J, Wang YM, Mäkinen J, Pail R, Barzaghi R, Vergos GS, Ahlgren K, Liu Q (2021b) Strategy for the realisation of the international height reference system (IHRS). J Geod 95(3). https://doi.org/10.1007/s00190-021-01481-0

Sjöberg LE (1991) Refined least squares modification of Stokes' formula. Manuscripta Geodaetica 16:367–375

Sjöberg LE (2003) A computational scheme to model the geoid by the modified Stokes formula without gravity reductions. J Geod 77(7–8):423–432. https://doi.org/10.1007/s00190-003-0338-1

Tocho CN, Antokoletz ED, Gómez AR, Guagni H, Piñon DA (2022) Analysis of high-resolution global gravity field models for the estimation of International Height Reference System (IHRS) coordinates in Argentina. J Geodetic Sci 12(1):131–140. https://doi.org/10.1515/jogs-2022-0139

Vestøl O, Ågren J, Steffen H, Kierulf H, Tarasov L (2019) NKG2016LU: a new land uplift model for Fennoscandia and the Baltic Region. J Geod 93(9):1759–1779. https://doi.org/10.1007/s00190-019-01280-8

New Tidal Analysis of Superconducting Gravimeter Records at Metsähovi, Finland

Arttu Raja-Halli, Maaria Nordman, Hannu Ruotsalainen, and Heikki Virtanen

Abstract

Superconducting gravimeters are the most sensitive instruments for monitoring gravitational changes. At the Metsähovi Geodetic Research Station in southern Finland, a superconducting gravimeter has been operating since 1994. It can be used to monitor crustal loading effects affecting the other geodetic measurements made at the station. Gravimeters iGrav-013 and iOSG-022 replaced the old gravimeter SG-T020 at Metsähovi in 2016. The first step was to do a new local tidal gravity modelling for Metsähovi Geodetic Research Station based on the first 5.5 years of iGrav-013 and iOSG-022 superconducting gravimeter data. Here we present the first analysis of the gravity data and the results of tidal analysis of Earth body tides and ocean tidal loading.

Keywords

Gravity · Ocean tidal loading · Superconducting gravimeter · Tides

1 Introduction

The Finnish Geospatial Research Institute is operating the Metsähovi Geodetic Research Station (MGRS) which is a core site of the Global Geodetic Observing System (GGOS 2023). A superconducting gravimeter (SG) has operated at the station continuously since 1994. The first SG, GWR-T020, operated at the site from 1994 until 2016. A new dual sphere SG OSG-073 with two sensors, was installed in early 2014 to the same laboratory on a pier three meters apart from the SG-T020. Unfortunately, the OSG-073 operated only until May 2015 when it had to be sent back to the manufacturer for a total redesign. The solution was to

separate the two sensors of the iOSG-073 into two separate gravimeters: iGrav-013, replacing the SG-T020 on the original pier, and iOSG-022 installed on the second pier. SGs have proven to be very good instruments to study a variety of geophysical phenomena and offer a great tool to observe small periodical effects like free oscillations of the Earth and solid Earth and ocean tides (for a review see e.g., Hinderer et al. 2015).

Tidal signal is the largest periodic signal in the gravity time series and needs to be removed from the gravity data to be able to study other geophysical phenomena, e.g., crustal loading effects affecting other geodetic measurements like satellite laser ranging (SLR) and Global Navigation Satellite Systems (GNSS). To achieve best results, a local tidal model is necessary. In previous analysis of the SG gravity data, we have used a local observation based tidal gravity model referred as ME18, produced from the tidal analysis of the SG-T020 gravimeter data (see most recent results in Virtanen and Raja-Halli 2018). The old model ME18 included 45 tidal wave groups between annual Sa and quarter diurnal M4 tides. In the model ME18, the ocean tide loading was not separately analysed, hence the ocean tides were intervened with the body tides. To establish a new local tidal model

A. Raja-Halli (✉) · H. Ruotsalainen · H. Virtanen
Finnish Geospatial Research Institute, National Land Survey, Espoo, Finland
e-mail: arttu.raja-halli@nls.fi

M. Nordman
Finnish Geospatial Research Institute, National Land Survey, Espoo, Finland

School of Engineering, Aalto University, Aalto, Finland

© The Author(s) 2023

J. T. Freymueller, L. Sánchez (eds.), *Gravity, Positioning and Reference Frames*,
International Association of Geodesy Symposia 156, https://doi.org/10.1007/1345_2023_231

and study the contribution of ocean tides, we use in this study the 5.5 years of gravity data collected by the new SGs and the ETERNA-X-ET34-v80 (Wenzel 1996; Schüller 2015; Schüller 2020) Earth tide software to simultaneously compute the contributions of the body tides and five different ocean tide loading models. After removing the tidal signal, the largest remaining signal is due to environmental mass changes. In Metsähovi the residual environmental signal is mostly dominated by the non-tidal effects of the atmosphere and the Baltic Sea, and mass changes in the local hydrology. The effect of environmental mass changes on the gravity at Metsähovi have been previously studied in e.g., Virtanen (2001), Virtanen and Mäkinen (2003), Mäkinen et al. (2014) and Olsson et al. (2009). In this analysis we adopt a simpler approach and use local groundwater level and the Baltic Sea level height at the Helsinki tide gauge only as regression parameters in the tidal analysis. Further analysis of the hydrological gravity effects is out of scope of this study. Several tidal analyses have been made with using SG data from different gravimeters. However, this is the first tidal analysis from the data of the new SG's at Metsähovi also providing information on the drift and overall performance of the instruments.

An earlier very extensive analysis of ocean tidal loading at Metsähovi was done with the data from SG-T020 together with several SG stations around the globe by Boy et al. (2003). It was discussed that poorly modelled Baltic Sea and Arctic Sea might be the cause to discrepancies between the ocean tidal loading models and observations. Metsähovi is 15 km from the coast of the Baltic Sea which is a shallow estuary where tidal amplitudes are negligible compared to non-tidal sea level changes.

More recently, tidal analysis of SG data has been studied in Meurers et al. (2016) in which the temporal variation of the tidal parameters was analysed by using the data of several central European SGs. Recent local tidal gravity studies have been also carried out by Crossley et al. (2023) for the SG-046 at the Apollo Lunar Laser Ranging facility in USA, Luan et al. (2022) in Kunming, China, and Hinderer et al. (2020) in Djougou, Benin. In Hinderer et al. (2022) a comprehensive analysis is presented for eight SGs operated at the J9 gravity observatory in Strasbourg, France.

We adopt a similar approach as authors mentioned above and present the first tidal analysis of the new data from the iGrav-013 and iOSG-022, with a separate analysis where local groundwater and Baltic Sea level at Helsinki tide gauge were used as regression parameters. We compute the tidal amplitude factors and phases for the body tide wave groups from Sa to M4, and the ocean tidal loading of the 11 main tidal waves (Ssa, Mm, Mf, Q1, O1, P1, K1, N2, M2, S2, K2) for the different ocean tide models.

2   Data Processing

2.1   Gravity Data

The two new SGs were installed to the MGRS in 2016 three meters apart, inside the same laboratory, and have been operating and producing data with 1 Hz sampling rate since. For the tidal analysis we use the SG data from first of January 2017 until 22nd of August 2022 in total of 2059 days. The year 2016 was omitted from the processing due to the large initial drift of the instruments and disturbances caused by the repeated adjustments and installation procedures done on the instruments.

The raw 1 Hz voltage signal recorded by the gravity sensors is first converted to gravity variations in nm/s^2 with a scale factor acquired from simultaneous absolute gravity (AG) observations. In this study we use the calibration values of -945.27 ± 1.49 nm/s^2/V and -887.40 ± 1.40 nm/s^2/V for iGrav and iOSG, respectively, where V is the voltage recorded by the gravimeter. These values are the weighted means of a least squares adjustment of the SG gravity signal to 1825 absolute gravity set values measured during six AG campaigns made in 2018 between February and November. Absolute gravity measurements were made with the FG5X-221 and each measurement campaign lasted between 4–9 days, see Virtanen et al. (2014) for details of the calibration process. In addition to the scale factor, the instrumental time delays were determined with the help of the gravimeter manufacturer GWR (Richard Warburton 2017, personal communications). We followed the step procedure described in Van Camp et al. (2000), by injecting 6 step signals of known voltage to the gravimeter feedback coil during fifth of December 2017. From the resulting observations of these steps, we got time delays at zero frequency for iOSG-022 $\tau = 6.75s \pm 0.19s$, and $\tau = 6.83s \pm 0.09s$ for iGrav-013.

In the gravity pre-processing we follow the remove and restore workflow described in detail e.g., Virtanen (2006), Hinderer et al. (2015) and Virtanen and Raja-Halli (2018) to achieve a continuous and un-disturbed time series best suitable for tidal analysis. We have used the Tsoft-software package (Van Camp and Vauterin 2005) for the pre-processing of the data.

First, the empirical tidal gravity model ME18 is removed from the gravity signal and the air pressure related effects are reduced by subtracting the local barometric pressure changes with an admittance factor of -3.1 nm/s^2/hPa. Next step is to correct the time series for distinct steps and spikes, and occasional gaps by linear interpolation from the residual (Fig. 1). However, as pointed out in Hinderer et al. (2002), these pre-processing steps may cause significant differences in the

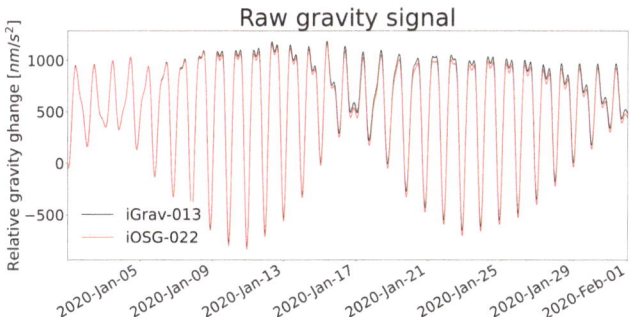
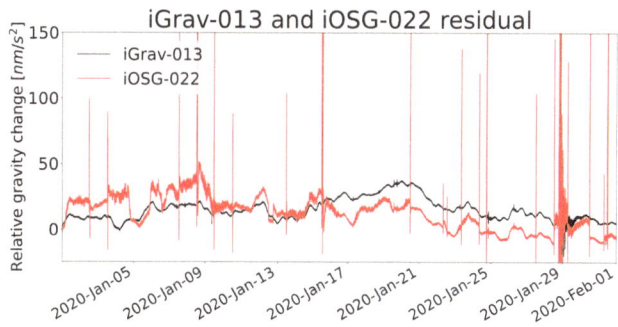

Fig. 1 A 1 month sample of the gravity data of iGrav-013 and iOSG-022. On the left, the raw gravity signal with full tidal signal, and on the right the residual after removing the tidal gravity model ME18 and air pressure effects. Adjustments of the cold head in the iOSG has produced the clear spikes and steps visible in the iOSG data which were removed in the data pre-processing

resulting time series depending on the chosen correction strategy, for instance, careless correction of a step in the data may cause significant change in the overall trend of the time series. Here, the benefit of two close-by SG's is evident, as we can compare the time series to cross-validate the changes in the gravity time series and distinguish even very small signals caused by instrumental disturbances. We have done the removal of outliers as an iterative process to minimize errors in the corrections and to achieve as clear time series as possible for tidal analysis: first we have corrected for spikes larger than 10 nm/s^2/min and offsets larger than 10 nm/s^2/min. These spikes and offsets are caused mainly by instrument maintenance like removing accumulated ice from the SG dewar or due to large earthquakes. Second, to clean the signal even further, we subtracted one gravimeter time series from the other to reveal additional smaller instrumental disturbances. This method allowed us to distinguish and correct the data for spikes larger than 3 nm/s^2/min and steps of 1 nm/s^2/min. However, the iOSG-022 has time periods lasting several days with overall noise level above 3 nm/s^2/min caused mainly by mechanical vibrations of the cold head, in these cases more conservative corrections were made. Also, in some cases even after comparing the time series it was hard to judge whether the difference between the two gravimeters is due to instrumental effect or due to a real physical phenomenon, e.g., snow accumulation on the roof of the laboratory can cause 20 nm/s^2 gravity effect which can be unevenly distributed (Virtanen 2001).

After the above-mentioned corrections, the removed tidal model and air pressure signal are restored to the gravity signal to produce a continuous gravity time series for the tidal analysis. The sampling rate is further reduced to 1 h with a least squares (LSQ) lowpass filter. First the 1 Hz data is decimated to 1 min sampling rate with a LSQ lowpass filter with 0.00833 Hz cutoff and 504 s window, and secondly to 1 h data with 12 cycles per day cutoff and 480 min window to avoid aliasing effects.

2.2 Environmental Data

To monitor the environmental mass changes in the vicinity of the gravity laboratory we have installed 11 boreholes for measuring the water table level in the sediments and in the crystalline bedrock (for details see Virtanen and Raja-Halli 2018). For this study we use only water table measurements from a borehole BH2 in the bedrock within few tens of meters of the gravimeter to give a proxy of the local hydrological changes. The measurements made at 0.1 Hz are resampled to 1 h after correcting for the atmospheric pressure and outliers. To account for the non-tidal gravity effects caused by the close-by Baltic Sea, we use hourly sea level data from the Helsinki tide gauge 30 km's from the station (Finnish Meteorological Institute Open data). The groundwater level and tide gauge data are used as regression parameters in the ETERNA-X analysis. For a more comprehensive description of the loading effects caused by the Baltic Sea we refer to Virtanen and Mäkinen (2003) and Virtanen (2004).

3 Ocean Tide Loading Models

In our analysis we compare five different ocean tide models, three recent models, EOT20 (Hart-Davis et al. 2021), FES2014b (Lyard et al. 2021) and TPX09v5a (Egbert and Erofeeva 2002), and two older models DTU10 (Cheng and Andersen 2010) and FES2004 (Lyard et al. 2006) which is still routinely used in the absolute gravity analysis in Finland. We used the ocean tide loading constituents calculated by the Onsala Ocean Tide Loading Provider (Scherneck 2022; Bos and Scherneck 2013) which determines the 11 main tidal load vectors for Ssa, Mm, Mf, Q1, O1, P1, K1, N2, M2, S2, K2 wave groups by using a visco-elastic Earth model (Kustowski et al. 2008) with Green's functions (Bos and Scherneck 2013). The ocean tide load vectors are directly implemented on the ETERNA-X analysis to get ocean tide contribution to the gravity signal.

4 Tidal Analysis

The tidal analysis was done with the comprehensive earth tide software package ETERNA-X-ET34-v80 (Schüller 2015; Schüller 2020) which allows the simultaneous analysis of body and ocean tides. In the analysis we have used the DEHANT-DEFRAIGNE-WAHR non-hydrostatic inelastic Earth model (DDW-NHi) (Dehant et al. 1999) and the HW95 tidal potential catalog (Hartmann and Wenzel 1995) with 12,361 constituents. In the analysis the gravity effect of the polar motion, i.e., pole tide, was removed by using the daily Earth Orientation Parameters (EOP) from the International Earth Rotation and Reference System Service (IERS 2022). The local air pressure was treated as a regression parameter in the analysis. For comparison we performed separate analysis with the local groundwater level and Helsinki tide gauge data included as regression parameters.

The ETERNA-X allows a wide range of possibilities on choosing the tidal wave groups to be analyzed depending on the length and quality of the time series. We have followed a wave grouping scheme proposed in the Ducarme and Schüller (2018) where a conservative wave grouping "Y04-R04-safe1" is proposed for time series with lengths from 4 to 9 years. This wave grouping includes 91 Earth body tide wave groups from Sa to MK4 wave groups. To assess the quality of the analysis of an individual wave group we have used the Correlation Root Mean Square Error Amplifier (CRA) values. ETERNA-X calculates CRAs for all analyzed wave groups. We inspected, that the CRA values for all wave groups were below 2 and hence acceptable under the criterions laid out in the detailed description in Ducarme and Schüller (2018) and ETERNA-X documentation (Schüller 2020).

5 Results

From the tidal analysis we get amplitude factors and phases for the body wave groups from the annual Sa-wave group, up to wave group M4 and a fit of the amplitudes and phases for the ocean tidal loading wave groups. The resulting tidal parameters for the main tidal constituents are shown in the Appendix 1, Tables 2 and 3. For the gravity time series processing the average of the amplitudes and phases of the five ocean tide models (shown as the mean in Fig. 3 and in Appendix 2) is used to overcome the possible shortcomings of individual models.

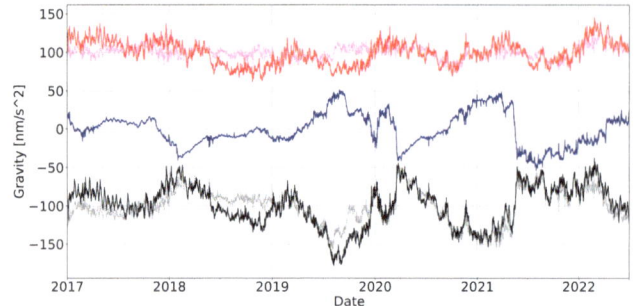

Fig. 2 The final residual gravity signals in nm/s^2 after removing the body and ocean tides, pole tide, pressure effects and instrumental drift. *Red*: iGrav+100 nm/s^2, *black*: iOSG-100 nm/s^2. Lighter coloured and dotted lines are the residuals reduced with groundwater and Helsinki sea level. *Blue*: The difference between iGrav and iOSG

The residual gravity series are presented in Fig. 2. The instrumental drift of the gravimeters was linear and stable as the first year of operation was omitted from the analysis. The drift was determined with a second order Chebychev polynomial in the ETERNA-X analysis. The drift rate includes both the instrumental drift but also the linear gravity change of approximately -7 nm/s^2/year caused by the post-glacial rebound (Bilker-Koivula et al. 2021). The resulting residual RMS, drift and regression parameters are presented in the Table 1.

A clear decrease in the RMS was achieved through reduction of groundwater and sea level, especially for the iGrav. However, there is a clear shortcoming in our analysis as we are using only the local groundwater as a proxy for all hydrological effects and omitting larger scale loading effects as well as local effects of snow and water in the soil layers above the bedrock. There also remains large differences between the gravity signal of the two instruments visible in the Fig. 2. Further investigation is required to understand whether these are due to e.g., local hydrology or some instrumental effects.

Metsähovi is 1,000 km from the ocean and the Baltic Sea is a shallow estuary with a very low tidal amplitudes, hence the amplitudes of the ocean tides at Metsähovi are mainly below 1 nm/s^2 with the maximum for the M2 being 4 nm/s^2. In the Fig. 3 we present the results for the five ocean tide models following the presentation used in Hinderer et al. (2020) and Luan et al. (2022): we plot the percentage rate of the vector **X** which is the excess in the gravity residual compared to the amplitude of the ocean tide model in the period range of the given wave group after removing the modelled body and ocean tide. The large remaining residuals, i.e., large **X**-vector, and differences between the models in the P1 and K1 ocean wave groups are believed to be due to poorly modelled Baltic Sea, which has been also discussed

Table 1 Gravity residual RMS, drift, regression coefficients and correlation between air pressure and tide gauge, for the two different analysis, where GW and TG represent the analysis where groundwater and Helsinki tide gauge were removed through regression

	Residual RMS [nm/s^2]	Drift [nm/s^2]	Atmospheric pressure (AP) [nm/s^2/hPa]	GW regression [nm/s^2/m]	TG regression [nm/s^2/m]	Correlation AP-TG
iGrav-013	15.38	59.52	-2.88 ± 0.05	–	–	–
iGrav-013 – GW-TG	7.91	45.25	-2.59 ± 0.03	17.77 ± 0.41	35.62 ± 1.46	0.44
iOSG-022	26.30	55.29	-3.14 ± 0.08	–	–	–
iOSG-022 – GW-TG	21.72	40.27	-2.87 ± 0.08	21.63 ± 1.16	30.75 ± 4.11	0.45

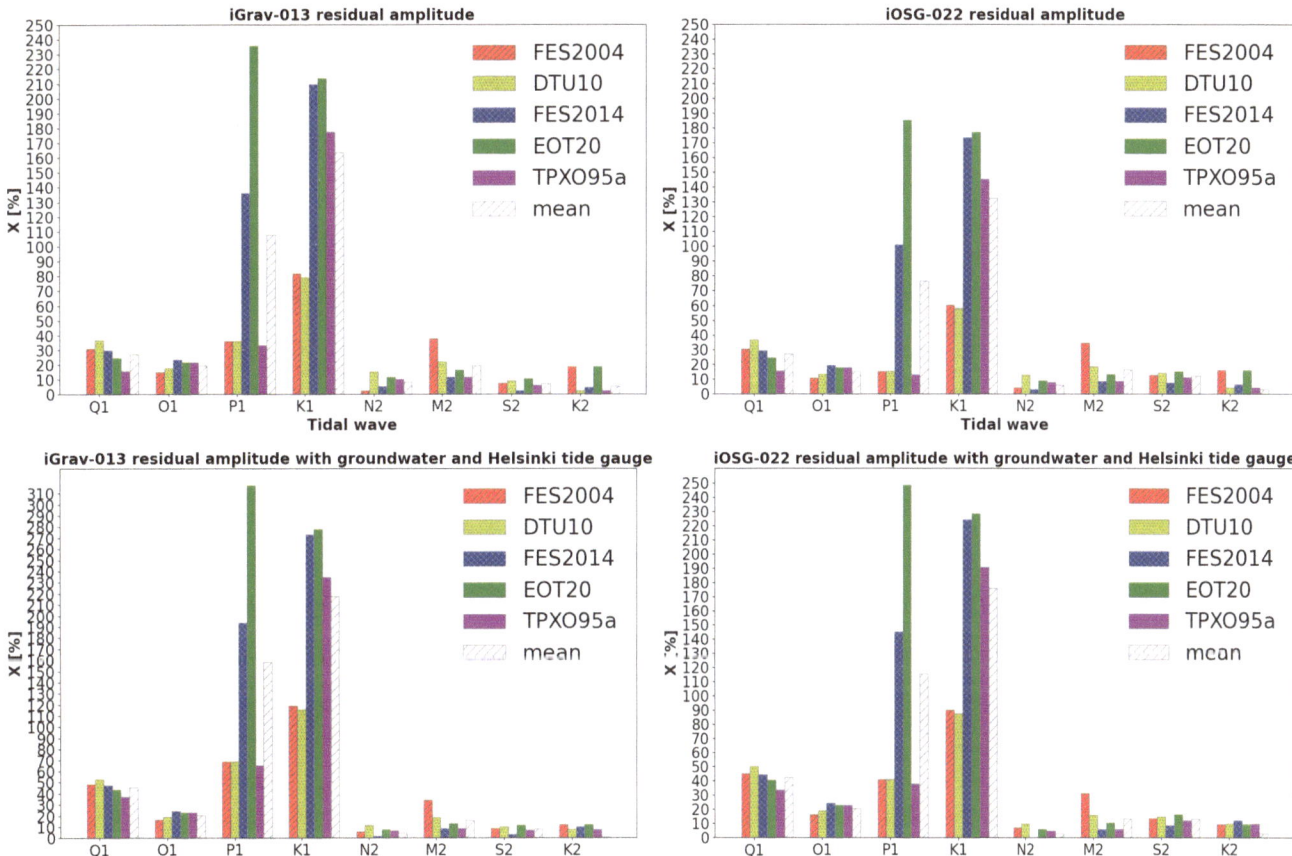

Fig. 3 The amplitude of the residual vector X for the iGrav-013 (left) and iOSG-022 (right) for the five ocean tide models. TOP: without GW and TG, BOTTOM: GW and TG as regression parameters. The X is the excess in the residual gravity signal as a percentage rate of the model amplitude after removing the modelled body and ocean tide factors

in Boy et al. (2003) and Lyard et al. (2021). The results agree well with the earlier results of Boy et al. (2003). In Appendix 2, Tables 4 and 5 show the resulting mean of the ocean tidal load vectors from the analysis and the residual vectors together with the **X**-vector. Results show clear differences between iGrav and iOSG which might be due to the disturbances and instrumental noise in the iOSG data, but also a small error in the calibration of the instruments might be the cause.

6 Discussion and Conclusions

In this study we have carried out the first tidal analysis of the new SG's iGrav-013 and iOSG-022 at the MGRS and a comparison with five ocean tide models with a regression with local groundwater and Helsinki tide gauge. The high noise level caused by the instrumental noise in the data of the iOSG disturbed the tidal analysis. The residual of the iOSG exhibited large deviations compared to the iGrav. A more careful analysis is required to distinguish whether these differences are due to data processing, instrumental or environmental origin. It is clear from the analysis that instrumental disturbances and mass changes in the close vicinity of the gravimeter can produce large discrepancies between the two gravimeters.

Differences between the gravimeters might also arise from errors in the scale factors which are now based on set values of AG measurements done in 2018. Further investigation on the accuracy and stability of the scale factors is required. Next step is to combine all absolute gravity measurements made during 2017–2022 for a more comprehensive scale factor and SG drift determination by using the individual drops in the AG measurement rather than sets.

We found large deviations between the ocean tide loading models and the observed signals especially in diurnal P1 and K1 wave groups, where remaining signal was in some cases more than double of the model tidal amplitude, Fig. 3 and Appendix 2. Boy et al. (2003) explain these discrepancies to be possibly due to poor modelling of the Baltic Sea and Arctic Ocean in the ocean tide models. In addition, the non-tidal effects in the Baltic Sea can reach up to 30 nm/s^2 (Olsson et al. 2009) and can have periodical properties through seiche waves which have periods close to diurnal and semi-diurnal bands (Metzner et al. 2000), hence interfering with the tidal analysis.

Seasonal signal was clearly decreased especially in the iGrav residual after removing groundwater and tide gauge signals through regression. However, this is not physically realistic to reduce hydrological effect by using only these data but served more as a proxy for this analysis. For further improvements it is necessary to include in the analysis the global hydrological and atmospheric contribution through e.g., by using EOST loading service (EOST 2023), a more detailed local hydrological modelling and a more detailed analysis of the loading effects caused by the close-by Baltic Sea.

Acknowledgements We are grateful for Richard Warburton for all his help on the installation and initial setup of the gravimeters. Special thanks for Mirjam Bilker-Koivula and Jyri Näränen for the absolute gravity measurements.

Appendix 1

Table 2 The main tidal constituents for iOSG-022, on the left without GW and TG regression, amplitudes in nm/s²

iOSG-022

Symbol	Theoretical	Analyzed	Amplitude factors	RMSE	Phase leads	RMSE	iOSG-022 – GW-TG Analyzed	Amplitude factors	RMSE	Phase leads	RMSE
Sa	33..90622	69.44684	2.0482	0.21548	23.288	4.622	41.7544	1.23147	0.16589	58.399	7.755
Ssa	37.61691	43.23295	1.1493	0.03558	−11.956	1.762	45.74213	1.216	0.02907	−6.792	1.389
Mm	42.71967	48.23862	1.12919	0.04003	0.996	2.027	48.18733	1.12799	0.03379	0.058	1.716
Mf	80.85344	95.20864	1.17755	0.01989	1.42	0.971	93.97344	1.16227	0.01689	0.414	0.834
Mtm	15.48088	18.68396	1.20691	0.0924	4.584	4.387	18.08124	1.16797	0.07716	1.715	3.791
Q1	51.33056	58.83475	1.14619	0.00056	0.095	0.028	58.91784	1.14781	0.00071	0.041	0.035
O1	268.09322	308.93084	1.15233	0.00011	0.259	0.005	309.45514	1.15428	0.00014	0.26	0.007
NO1	21.07402	24.35012	1.15546	0.00165	0.35	0.082	24.39881	1.15777	0.00213	0.262	0.105
P1	124.72331	143.83547	1.15324	0.00022	0.053	0.011	143.95165	1.15417	0.00028	0.064	0.014
K1	376.8968	429.46163	1.13947	0.00008	0.092	0.004	429.92842	1.14071	0.00009	0.063	0.005
N2	35.53436	41.82619	1.17706	0.00028	1.064	0.014	41.8168	1.1768	0.0003	1.027	0.015
M2	185.59068	218.92819	1.17963	0.00005	0.719	0.002	218.85298	1.17922	0.00005	0.7	0.003
S2	86.33887	101.44351	1.17495	0.00011	0.077	0.005	101.42796	1.17477	0.00012	0.09	0.006
K2	23.45576	27.63278	1.17808	0.00041	0.159	0.02	27.60938	1.17708	0.00044	0.165	0.021

Table 3 The main tidal constituents for iGrav-013, on the left without GW and TG regression, amplitudes in nm/s2

iGrav-013							iGrav-013 – GW-TG				
Symbol	Theoretical	Analyzed	Amplitude factors	RMSE	Phase leads	RMSE	Analyzed	Amplitude factors	RMSE	Phase leads	RMSE
Sa	33.90622	90.00709	2.65459	0.13471	−2.252	2.346	57.36783	1.69196	0.06469	−8.389	1.81
Ssa	37.61691	44.83348	1.19184	0.0227	−4.123	1.082	47.66.681	1.26716	0.01039	−2.173	0.475
Mm	42.71967	4.54933	2.06976	0.38518	8.374	10.624	48.50973	1.13554	0.01206	0.242	0.608
Mf	80.85344	48.71036	1.14023	0.02551	1.339	1.28	93.53857	1.15689	0.00603	0.413	0.299
Mtm	15.48088	94.78768	1.17234	0.01267	1.225	0.621	17.75856	1.14713	0.02753	1.14	1.377
Q1	51.33056	18.34416	1.18496	0.05889	3.678	2.848	58.93597	1.14817	0.00071	0.047	0.035
O1	268.09322	58.83728	1.14624	0.00047	0.103	0.023	309.7008	1.1552	0.00014	0.257	0.007
NO1	21.07402	309.08929	1.15292	0.00009	0.255	0.004	24.42656	1.15908	0.00213	0.163	0.105
P1	124.72331	24.37091	1.15644	0.00137	0.291	0.068	144.0819	1.15521	0.00028	0.071	0.014
K1	376.8968	143.93484	1.15403	0.00019	0.052	0.009	430.3137	1.14173	0.00009	0.062	0.005
N2	35.53436	429.74833	1.14023	0.00006	0.095	0.003	41.8366	1.17736	0.00023	1.034	0.011
M2	185.59068	41.84641	1.17763	0.00019	1.079	0.009	218.9991	1.18001	0.00004	0.691	0.002
S2	86.33887	219.08605	1.18048	0.00003	0.714	0.002	101.497	1.17557	0.00009	0.072	0.004
K2	23.45576	101.51098	1.17573	0.00007	0.064	0.004	27.61481	1.17731	0.00033	0.193	0.016

Appendix 2

Table 4 The mean of modelled ocean tide load vectors compared with the observed residual vector in nm/s², and the **X-vector** shown also in Fig. 3 for the iOSG-022

iOSG-022							iOSG-022-GW-TG			
	Ocean tidal loading (mean) [nm/s²]	Phase	Residual [nm/s²]	Phase	X	Phase difference	Residual [nm/s²]	Phase	X	Phasedifference
Q1	0.59	166.4	0.431	166.9	26.9	0.5	0.339	172.9	42.5	6.5
O1	1.767	128.3	1.498	111.4	15.2	16.9	1.404	90.9	20.5	37.3
P1	0.307	18.2	0.542	14.1	76.7	4.1	0.662	14.1	115.6	4.1
K1	0.917	11.8	2.133	18.8	132.7	7	2.531	10.7	176	1.1
N2	0.886	65.3	0.938	56	5.8	9.4	0.91	55.5	2.6	9.9
M2	3.654	41.9	4.247	40.3	16.2	1.6	4.144	40.2	13.4	1.7
S2	1.273	8.4	1.118	7	12.1	1.4	1.106	8.3	13.1	0.1
K2	0.371	11.1	0.383	11.5	3	0.5	0.361	12.8	3	1.7

Table 5 The mean of modelled ocean tide load vectors compared with the observed residual vector in nm/s², and the **X-vector** shown also in Fig. 3 for the iGrav-013

iGrav-013							iGrav-013-GW-TG			
	Ocean tidal loading (mean) [nm/s²]	Phase	Residual [nm/s²]	Phase	X	Phase difference	Residual [nm/s²]	Phase	X	Phase difference
Q1	0.59	166.4	0.43	165.8	27	0.6	0.322	171.3	45.4	4.9
O1	1.767	128.3	1.428	105.8	19.2	22.5	1.407	80.9	20.4	47.4
P1	0.307	18.2	0.638	11.8	108.1	6.4	0.792	13	158.2	5.2
K1	0.917	11.8	2.415	17.2	163.4	5.4	2.909	9.2	217.3	2.6
N2	0.886	65.3	0.958	55.3	8.1	10	0.926	54.7	4.5	10.7
M2	3.654	41.9	4.358	38.8	19.3	3.1	4.236	38.6	15.9	3.3
S2	1.273	8.4	1.183	5.5	7.1	2.9	1.171	6.3	8	2.1
K2	0.371	11.1	0.391	12.9	5.3	1.9	0.369	14.6	0.7	3.5

References

Bilker-Koivula M, Mäkinen J, Ruotsalainen H, Näränen J, Saari T (2021) Forty-three years of absolute gravity observations of the Fennoscandian postglacial rebound in Finland. J Geod 95:24. https://doi.org/10.1007/s00190-020-01470-9

Bos MS, Scherneck H-G (2013) Computation of Green's functions for ocean tide loading. In: Xu G (ed) Sciences of Geodesy - ii. Springer, Berlin, pp 1–52

Boy J-P, Llubes M, Hinderer J, Florsch N (2003) A comparison of tidal ocean loading models using superconducting gravimeter data. J Geophys Res 108:2193. https://doi.org/10.1029/2002JB002050

Cheng Y, Andersen O B (2010) Improvement in global ocean tide model in shallow water regions. Poster, SV.1–68 45, OSTST, Lisbon, Oct. 18–22

Crossley DJ, Murphy JT, Liang J (2023) Comprehensive analysis of superconducting gravimeter data, GPS, and hydrology modelling in support of lunar laser ranging at Apache Point Observatory, New Mexico. Geophys J Int 232(2):1031–1065. https://doi.org/10.1093/gji/ggac357

Dehant V, Defraigne P, Wahr J (1999) Tides for a convective Earth. J Geophys Res 104(B1):1035–1058

Ducarme B, Schüller K (2018) Canonical wave grouping as the key to optimal tidal analysis. Bull Inf Marees Terrestres 150:12131–12244

Egbert GD, Erofeeva SY (2002) Efficient inverse modeling of barotropic ocean tides. J Atmos Ocean Technol 19(2):183–204

GGOS (2023) Global geodetic observing system. https://ggos.org/item/ggos-core-sites/. Accessed 4 June 2023

Hart-Davis MG, Piccioni G, Dettmering D, Schwatke C, Passaro M, Seitz F (2021) EOT20: a global ocean tide model from multi-mission satellite altimetry. Earth Syst Sci Data 13(8):3869–3884. https://doi.org/10.5194/essd-13-3869-2021

Hartmann T and Wenzel H (1995) The HW95 tidal potential catalogue. Geophys Res Lett 22. https://doi.org/10.1029/95GL03324 ISSN: 0094-8276

Hinderer J, Rosat S, Crossley D, Amalvict M, Boy J-P, Gegout P (2002) Influence of different processing methods on the retrieval of gravity signals from GGP data. Bull Inf Marees Terrestres 123:9278–9301

Hinderer J, Crossley D, Warburton RJ (2015) Superconducting gravimetry. In: Herring T, Schubert G (eds) Treatise on geophysics, vol 3, 2nd edn. Elsevier, pp 59–115. https://doi.org/10.1016/B978-0-444-53802-4.00062-2

Hinderer J, Riccardi U, Rosat S, Boy J-P, Hector B, Calvo M, Littel F, Bernard J-D (2020) A study of the solid Earth tides, ocean and atmospheric loadings using a 8 year record (2010-2018) from superconducting gravimeter OSG-060 at Djougou (Benin, West Africa). J Geodyn 134:101692. https://doi.org/10.1016/j.j.g.o2019.101692

Hinderer J, Warburton RJ, Rosat S, Riccardi U, Boy JP, Forster F, Jousset P, Güntner A, Erbas K, Littel F, Bernard JD (2022) Intercomparing superconducting gravimeter records in a dense meter-scale network at the J9 Gravimetric Observatory of Strasbourg, France. Pure Appl Geophys 179(5):1701–1727. https://doi.org/10.1007/s00024-022-03000-4

International Earth Rotation and Reference System Service (IERS) (2022). http://hpiers.obspm.fr/eop-pc/index.php?index=C04&lang=en. Accessed 27 Feb 2023

Kustowski B, Ekström G, Dziewonski AM (2008) Anisotropic shear-wave velocity structure of the Earth's mantle: a global model. J Geophys Res 113:B6

Luan W, Shen W, Jia J (2022) Analysis of iGrav superconducting gravity measurements in Kunming, China, with emphasis on calibration, tides, and hydrology. Pure Appl Geophys 180(2):643–660. https://doi.org/10.1007/s00024-022-03036-6

Lyard F, Lefevre L, Letellier T, Francis O (2006) Modelling the global ocean tides: insights from FES2004. Ocean Dyn 56:394–415

Lyard F, Allain D, Cancet M, Carrère L, Picot N (2021) FES2014 global ocean tide atlas: design and performance. Ocean Sci 17(3):615–649

Mäkinen J, Hokkanen T, Virtanen H, Raja-Halli A, Mäkinen RP (2014) Local hydrological effects on gravity at Metsähovi, Finland: implications for comparing observations by the super-conducting gravimeter with global hydrological models and with GRACE. In: Proceedings of the International Symposium on Gravity, Geoid and Height Systems GGHS 2012, October 9–12, 2012, Venice, IAG Symposia 141

Metzner M, Gade M, Hennings I, Rabinovich A (2000) The observation of seiches in the Baltic Sea using a multi data set of water levels. J Mar Syst 24(1–2):67–84

Meurers B, Van Camp M, Francis O, Pálinkáš V (2016) Temporal variation of tidal parameters in superconducting gravimeter time-series. Geophys J Int 205(1):284–300. https://doi.org/10.1093/gji/ggw017

Olsson P-A, Scherneck H-G, Ågren J (2009) Effects on gravity from non-tidal sea level variations in the Baltic Sea. J Geodyn 48:151–156

Scherneck H-G (2022) Onsala ocean tide loading provider. http://holt.oso.chalmers.se/loading/index.html. Accessed 6 Sept 2022

Schüller K (2015) Theoretical basis for earth tide analysis with the new ETERNA34-ANA-V4.0 program. Bull Inf Marées Terrestres 149(1):12024–12061. http://maregraph-renater.upf.pf/bim/BIM/bim149.pdf

Schüller K (2020) Program System ETERNA-x et34-x-v80-* for Earth and Ocean Tides Analysis and Prediction, Documentation Manual 01: Theor. Technical report, Institution. http://ggp.bkg.bund.de/eterna?download=7283. Accessed Dec 2021

Van Camp M, Vauterin P (2005) Tsoft: graphical and interactive software for the analysis of time series and Earth tides. Comput Geosci 31(5):631–640. https://doi.org/10.1016/j.cageo.2004.11.015

Van Camp M, Wentzel H-G, Schott P, Vauterin P, Francis O (2000) Accurate transfer function determination for superconducting gravimeters. Geophys Res Lett 27(1):37–40

Virtanen H (2001) Hydrological studies at the gravity station Metsähovi, Finland. J Geod Soc Jpn 47(1):328–333

Virtanen H (2004) Loading effects in Metsähovi from the atmosphere and the Baltic Sea. J Geodyn 38:407–422

Virtanen H (2006) Studies of earth dynamics with the super-conducting gravimeter, academic dissertation in geophysics, Helsinki. Also published as No. 133 in the series of: publications of the Finnish Geodetic Institute. http://urn.fi/URN:ISBN:952-10-3057-7. Accessed 15 Feb 2023

Virtanen H, Mäkinen J (2003) The effect of the Baltic Sea level on gravity at the Metsähovi station. J Geodyn 35(4–5):553–565

Virtanen H, Raja-Halli A (2018) Parallel observations with three superconducting gravity sensors during 2014–2015 at Metsähovi Geodetic Research Station, Finland. Pure Appl Geophys 175:1669–1681. https://doi.org/10.1007/s00024-017-1719-3

Virtanen H, Bilker-Koivula M, Mäkinen J, Näränen J, Ruotsalainen H (2014) Comparison between measurements with the superconducting gravimeter T020 and the absolute gravimeter FG5-221 at Metsähovi, Finland in 2003–2012. Bull Inf Marees Terrestres 148:11923–11928

Wenzel HG (1996) The nanogal software: earth tide data preprocessing package. Bull Inf Marées Terrestres 124:9425–9439

Development of the National Gravimetric Geoid Model for the Kingdom of Saudi Arabia

Georgios S. Vergos, Rossen S. Grebenitcharsky, Abdullah Al-Qahtani, Sultan Al-Shahrani, Dimitrios A. Natsiopoulos, Suliman Al-Jubreen, Ilias N. Tziavos, and Juri Golubinka

Abstract

The development of a high-resolution and high accuracy geoid model is becoming nowadays a fundamental component of any modern geodetic infrastructure. The Kingdom of Saudi Arabia (KSA) has devoted the last decade a significant number of resources and manpower to collect high-quality land and airborne gravity data as well as GNSS/Levelling observations to create a state-of-the-art geoid model as a fundamental part of the Saudi Arabia National Spatial Reference System (SANSRS). In that frame, this work focuses on the collected gravity, terrain, and GNSS/Levelling data for the area under study, and their pre-processing in terms of horizontal, vertical and gravity reference system homogenization, blunder detection and removal. Given the availability of these data the latest gravimetric geoid model for the KSA is developed.

The gravity data pre-processing relied on the available metadata to collect information about the horizontal, vertical and gravity reference system. Hence, all this information has been homogenized to KSA-GRF17, tied to ITRF2014 at epoch 2017.0, and KSA-VRF14 which is tied to the geopotential number above the MSL of the Jeddah TGBM-B. Given that several data holdings of land gravity where either in the form of Bouguer anomalies or referred to some unknown horizontal datum, several tests have been carried out to identify the proper choices. Then, a least-squares collocation-based blunder detection and removal procedure has been conducted to identify blunders in the land data and possible biases between the various campaigns and the high-quality airborne gravity observations. The geoid prediction was carried out by the well-known remove-compute-restore technique evaluating Stokes' integral in the frequency domain via a 2D spherical Fast Fourier Transform and the Wang-Gore modification. After several tests with the latest GOCE/GRACE-based and combined Global Geopotential Models, XGM2019e has been used as a reference, while the residual terrain model correction was employed for the treatment of the topography. The validation of such a developed gravimetric geoid model has been performed for a set of 4,500 GNSS/Leveling benchmarks reaching external absolute accuracies at the 10–11 cm level and relative accuracies at the 1–5 ppm over distances ranging from 10 to 2,000 km.

G. S. Vergos (✉) · R. S. Grebenitcharsky · A. Al-Qahtani ·
S. Al-Shahrani · D. A. Natsiopoulos · S. Al-Jubreen · I. N. Tziavos ·
J. Golubinka
General Authority for Survey & Geospatial Information, Riyadh,
Kingdom of Saudi Arabia
e-mail: vergos@topo.auth.gr

© The Author(s) 2023
J. T. Freymueller, L. Sánchez (eds.), *Gravity, Positioning and Reference Frames*,
International Association of Geodesy Symposia 156, https://doi.org/10.1007/1345_2023_214

Keywords

FFT · GNSS/levelling validation · Gravimetric geoid · Kingdom of Saudi Arabia · Wang-Gore

1 Introduction

The availability of a high-accuracy and resolution gravimetric geoid model has gained increased importance as it is crucial for height and depth determination in a variety of applications such as construction, surveying, and geosciences while it is also essential for geophysical studies as it provides important information about the Earth's gravity field. In the Kingdom of Saudi Arabia (KSA), extensive related research has been carried out during the last 15 years in an effort to compute a Kingdom-wide geoid modeling using mainly gravity data from the General Authority for Geospatial information (GEOSA) and Arabian American Oil Company (ARAMCO). KSA-Geoid2009 was a geoid model based on EGM08 (Pavlis et al. 2012) and fitted to GNSS/Levelling geoid heights of 5,405 (5,028 from ARAMCO and 377 from GDMS) BMs. The collocation was the computation technique. The accuracy (STD of residuals to BMs) is higher than 10 cm around the GPS/Levelling BMs and increases to 1.3 m, may be 2.0 m within areas having sparse distribution of BMs. KSA-Geoid2015 computed by CC Tscherning and R Forsberg, is a gravimetric geoid based on land, ship-borne, satellite altimetry gravity data and EGM08 and DIR-R5 (Bruinsma et al. 2013) GGMs (up to degree/order 720). The SRTM30 PLUS (30$'$ × 30$'$) digital terrain mode has been used to compute terrain effects. The final KSA2015 geoid was obtained after fitting the gravimetric geoid to 4,157 GNSS/Levelling BMs. KSA-Geoid 2015 refers to the (old) SV71, tied to the MSL in Jeddah 1969, datum. KSA-Geoid2017, computed by R Forsberg and M Ayan, was based on the same principles of the KSA-Geoid2015 encompassing two new major data sets in the south-west Red Sea coastal region along with some advancements in terms of the RTM computations and estimation of geoid-quasi geoid differences. For KSA-Geoid2017 the RCR was used, EIGEN6C4 (Foerste et al. 2014) as a reference field, new DTU15 satellite altimetry data and more than a half million gravity data points from both new (GEOSA) and older (ARAMCO) data sources. The geoid was fitted to the new Jeddah2014 VRF system through a set of 280 GPS/levelling points along the new GEOSA first order levelling network.

Over the last years KSA has invested significant resources and manpower towards collecting various types of gravity data, as well as GNSS/Levelling observations aiming at the development of a new high-accuracy gravimetric geoid. However, when estimating a gravimetric geoid, it is important to ensure that the input data are homogeneous in terms of the horizontal, vertical, and gravity reference systems. As it is customary in all related work, when a geoid model is to be determined, it relies on all available gravity data, i.e., data from both historical and modern campaigns. This work is divided into two sections. In the first one, the creation of an accurate, consistent and homogeneous gravity database for both land and marine areas over KSA, by selecting and merging all the available gravity data sets is described, while the second one refers to the determination of a gravimetric only geoid model for KSA using the aforementioned new gravity database and the remove-compute-restore (RCR) technique evaluating Stokes' integral in the frequency domain with a 2D spherical Fast Fourier Transform (FFT) and the Wang-Gore modification.

2 Homogenization of Land and Marine Gravity Data

Initially, a pre-processing of the land and marine gravity data was carried out, following appropriate methodological and theoretical tools to achieve both the quality check and homogenization of the data regarding the geodetic reference system (GRS), the vertical reference datum and the tide conventions. Apart from the system homogenization, residuals of each dataset to GGMs have been evaluated to detect blunders, while a least-squares collocation (LSC) based blunder detection and removal procedure (Vergos et al. 2005) was carried out. The overall aim of this pre-processing analysis is the construction of a homogeneous and consistent gravity database, where the geodetic system (GRS) of all data will be GRS80 while the gravity reference system (GrGS) will be IGSN71.

In the frame of the geoid computation, all existing land and marine gravity data over KSA were collected along with the necessary information (format, reference system, defining standards, etc.) for each dataset. The land gravity data came from two major datasets, i.e., one from ARAMCO and another from the General Directorate for Military Survey (GDMS), with each one consisting of several independent gravity campaigns. First, the gravity data were cross-checked for the geodetic reference system (GRS), given that they can refer to GRS30, GRS67 or GRS80. As already mentioned, the entire geoid processing will be carried out in GRS80 and thus all the appropriate transformations were done. The second homogenization process referred to the transformation from the Potsdam to the IGSN71 gravity reference system. Since old gravity observations refer to the Potsdam system, rather than IGSN71, for common points in both systems relative transformation techniques were developed and applied. Finally, as some gravity observations where in the form of simple Bouguer anomalies, Bouguer anomaly (BA) corrections were computed and restored to form free-air gravity anomalies. The simple Bouguer plate correction as $-0.0419 \, \rho H$ was used with ρ being the average density and H the orthometric height of the station. As in most campaigns the used density value was not known, several tests with $\rho_1 = 2.2 \, g \cdot cm^{-3}$, $\rho_2 = 2.4 \, g \cdot cm^{-3}$ and $\rho_3 = 2.67 \, g \cdot cm^{-3}$ have been performed. The so-formed free-air gravity anomalies were then tested against free-air gravity anomalies from XGM2019e and other local data to conclude on the density value that provided the closest, in terms of the std. of the differences, results. The same pre-processing strategy was followed for the marine gravity data that were divided into two datasets depending on the wider areas that they are located, i.e., the Arabian Gulf and the Red Sea. After the evaluation and validation of the data for the GRS, the vertical datum and the tide conventions used, a final homogeneous gravity database referring to GRS80 and IGSN71 was created. At the end, all gravity data (old and new) referred to the KSA Gravity Reference Frame (KSA-GrRF), defined by absolute gravity values at absolute gravity stations of the KSA Gravity Base Network (KSA-GBN) observed over the entire KSA territory. An additional transformation from IGSN71 to the KSA-GrRF has been conducted for all old gravity data and all gravity values utilized for gravimetric geoid computations are in one unified KSA-GrRF. Table 1 tabulates the data holdings for the land and marine gravity datasets available before and after the data homogenization, clean-up and removal of double entries.

The method of spectral evaluation (Gruber et al. 2012; Vergos et al. 2014) of GGMs using GPS/Levelling data is a standard tool during the last decade to achieve a fast evaluation of the spectral contribution of GGMs w.r.t. in-situ data. In the frame of the KSA-Geoid2021, the latest GOCE-only, GRACE/GOCE and combined GGMs have been evaluated using EGM2008 as ground truth in order to evaluate which GOCE-based GGM and to what d/o provides the overall best improvement relative to EGM2008. Among the GGMs evaluated, XGM2019e (Zingerle et al. 2019) provided the overall best results and was used as the reference field for the determination of the KSA-Geoid21 gravimetric geoid. The evaluation of the land gravity data consisted of two steps. First, LSC was employed in a blunder detection and removal step, during which the land gravity data were splitted in two halves, using the one as ground truth and the other as observations. Then the test was repeated in the opposite way (see Tscherning 1991; Vergos et al. 2005). During that test, 70 points in the ARAMCO database have been identified as blunders and 18 points in the GDMS one. Finally, for the areas where the newly acquired airborne gravity data (SGL 2021) overlapped with historic gravity campaigns, a similar LSC-based scenario for blunder detection and removal has been followed, using the airborne gravity data as ground truth to evaluate the land observations. In both LSC-based tests, the comparisons are performed with residual free-air gravity anomalies, using XGM2019e (Zingerle et al. 2019) as the reference field and modelling the topographic effects with a residual terrain model (RTM) reduction based on high-order (up to d/o 90,000) effects (Rexer et al. 2016). After all these pre-processing steps, the common and homogeneous database contained a total number of 2,010,766 land, airborne and shipborne points to be used for the determination of the gravimetric geoid model for the Kingdom as depicted in Fig. 1. At sea, the available marine gravity data have been complemented by altimetry-derived gravity anomalies from DTU2018 (Andersen and Knudsen 2019), up to 20 km from the coastline, and SIO29.1 (Sandwell et al. 2014) for the rest of the marine areas. In the neighbouring countries were no gravity data were available, EGM2008 to its full d/o has been used as fill-in. Table 2 tabulates the statistics for the original, reduced and residual gravity data in the final database.

3 Geoid Determination with the Remove-Compute-Restore Procedure

The practical determination of the gravimetric geoid was performed using the remove-compute-restore approach (RCR) in the frequency domain employing the FFT evaluation of Stokes' kernel function and a Wong-Gore modification (Wong and Gore 1969; Sideris 2013). The latter was mandatory especially for the geoid modelling over the Kingdom, given the large extent of the study area in both latitude and longitude. The modification by Wong-Gore accounts for long-wavelength errors in the residual gravity anomalies, after the reduction of the original gravity data to a GGM

Table 1 Land and marine gravity datasets available and data holdings

Dataset	Original	Final
Land	774,682	753,270
Marine	245,813	245,167

Fig. 1 Distribution of complete land (magenta), airborne (grey), altimetry (blue from SIO and red from DTU), shipborne marine (black) and fill-in (brown) gravity data for the KSA-Geoid21 project

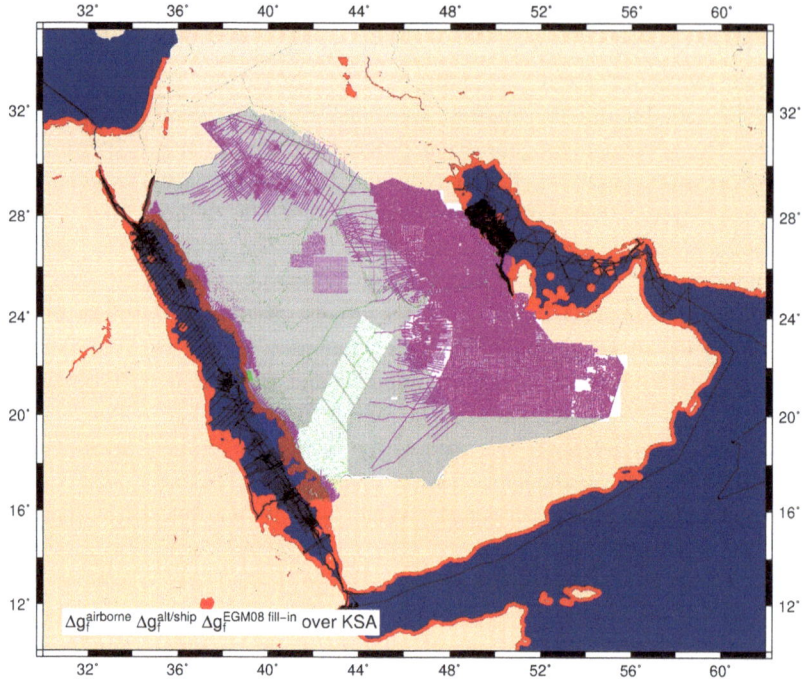

Table 2 Statistics of the original, reduced and residual gravity data in the Aramco database (565,752 point values), GDMS database (5,492 point values) in the shipborne marine database (245,813 point values) and in the altimetry database (771,500 point values). Unit: [mGal]

	Max	Min	Mean	RMS	Std
Δg_f^{Aramco}	139.113	−66.929	−2.476	29.003	28.897
$\Delta g_{f\,red}^{Aramco}$	136.006	−156.058	7.989	12.172	9.183
$\Delta g_{f\,res}^{Aramco}$	136.222	−163.764	7.899	12.161	9.071
Δg_f^{GDMS}	237.812	−97.686	16.628	37.143	33.214
$\Delta g_{f\,red}^{GDMS}$	81.696	−85.606	−0.168	13.331	13.329
$\Delta g_{f\,res}^{GDMS}$	53.245	−72.562	0.208	10.362	10.360
Δg_f^{ship}	204.018	−202.193	−24.560	38.099	29.126
$\Delta g_{f\,red}^{ship}$	81.736	−81.577	−4.980	13.951	13.032
$\Delta g_{f\,res}^{ship}$	83.511	−81.577	−4.844	13.923	13.010
$\Delta g_f^{altimetry}$	312.800	−190.850	−10.599	37.106	35.660
$\Delta g_{f\,red}^{altimetry}$	62.228	−68.356	0.108	4.508	4.507
$\Delta g_{f\,res}^{altimetry}$	66.950	−67.692	0.051	4.309	4.309

Table 3 Statistics of the final gravimetric geoid, quasi-geoid, and their validation [m]

	Max	Min	Mean	RMS	Std
N^{grav}	29.763	−65.355	−10.330	22.140	19.582
ζ^{grav}	29.544	−65.355	−10.277	22.132	19.601
ζ to N	0.247	−0.977	−0.053	0.109	0.096
$\zeta^{grav} - N^{grav}$	0.181	−1.451	−0.050	0.113	0.101
$N^{grav\,(from\,\zeta)} - N^{grav}$	0.111	−0.082	0.000	0.001	0.001
$N^{grav} - N^{GPS/Lev}$	0.465	−0.563	−0.102	0.170	0.136

and the removal of the topographic effects. Then we can determine residual geoid heights (N_{res}) and restore the contribution of the GGM (N_{GGM}) and the topography (N_{topo}), to derive the final gravimetric geoid with RCR as:

$$N_{grav} = N_{res} + N_{GGM} + N_{topo}. \qquad (1)$$

The FFT evaluation was carried out with GravSoft's spfour program, during which the number of reference parallels can be selected, as well as the modification of the Stokes kernel. For the number of the of reference parallels used four options were tested (1, 3, 6 and 9). For the Wong-Gore

modification, which is performed for a specific d/o and then linearly tapered to another higher d/o, all pairs formed from d/o 60 to d/o 300 have been tested. Since FFT needs gridded residual gravity anomalies, the grid was generated based on the irregular residual gravity anomalies over the Kingdom and prediction on a grid with LSC. To evaluate the different gravimetric geoid models resulting from the combination of number of parallels and modification degrees, evaluation with a set of available, high-accuracy, GNSS/Leveling dataset by GEOSA was performed. The best results were achieved with a Wong-Gore modification between d/o 80 and 100 and a multiband solution with 3 bands. Figure 3 depicts the final geoid height differences between the final gravimetric geoid and the GEOSA GNSS/Leveling geoid heights. In the same processing line, the quasi-geoid over the Kingdom has been determined and from that the geoid was once again estimated using the analytical evaluation of the quasi-geoid to geoid separation by Flury and Rummel (2009). Table 3 tabulates the final gravimetric geoid (see Fig. 2), the quasi-geoid, the difference between the gravimetric geoid and that

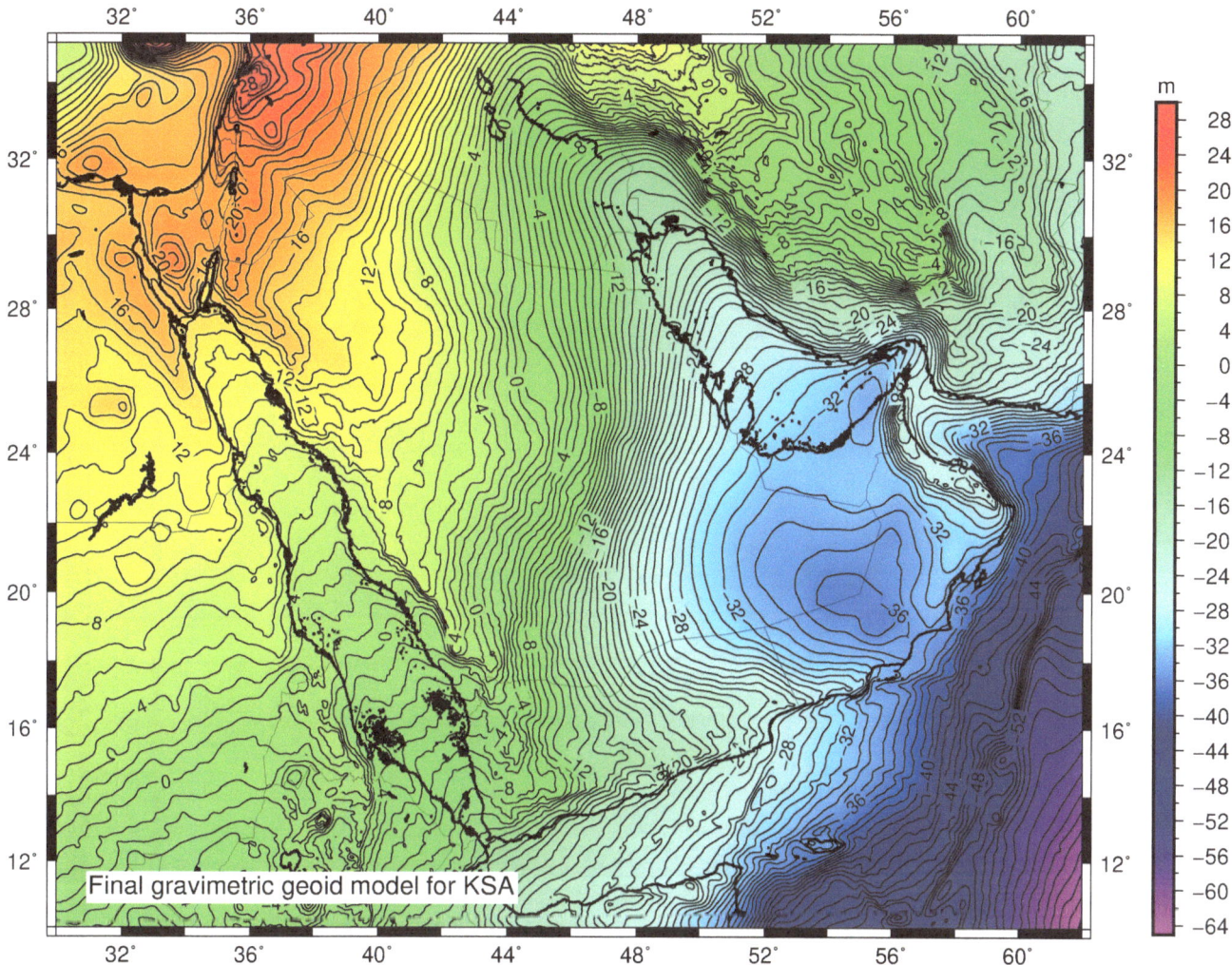

Fig. 2 The final gravimetric geoid model KSA-Geoid21GRA for KSA

determined from the quasi-geoid model. The latter has a std. of 1 mm only, showing the consistency of the processing steps followed. KSA-Geoid21GRA shows an absolute difference to the 3,522 GEOSA GNSS/Leveling at the 13.6 cm level (see Fig. 3), while the relative difference is at the 6 ppm for distances up to 10 km and 1–5 ppm over distances ranging from 10 to 2,000 km. These results are achieved before any deterministic and/or stochastic fit. Compared to

the previous KSAGeoid2017 gravimetric geoid model, the refined KSAGeoid21 model shows an improvement by 7 cm in terms of the std. to the GNSS/Levelling data, despite the fact that this is not directly comparable as KSAGeoid2017 was validated against a much smaller number of 287 BMs. The deterministic and stochastic treatment of the residuals to the GNSS/Leveling BMs are discussed in the development of the KSA-Geoid21 Hybrid model (KSA-GEOID21GEOSA).

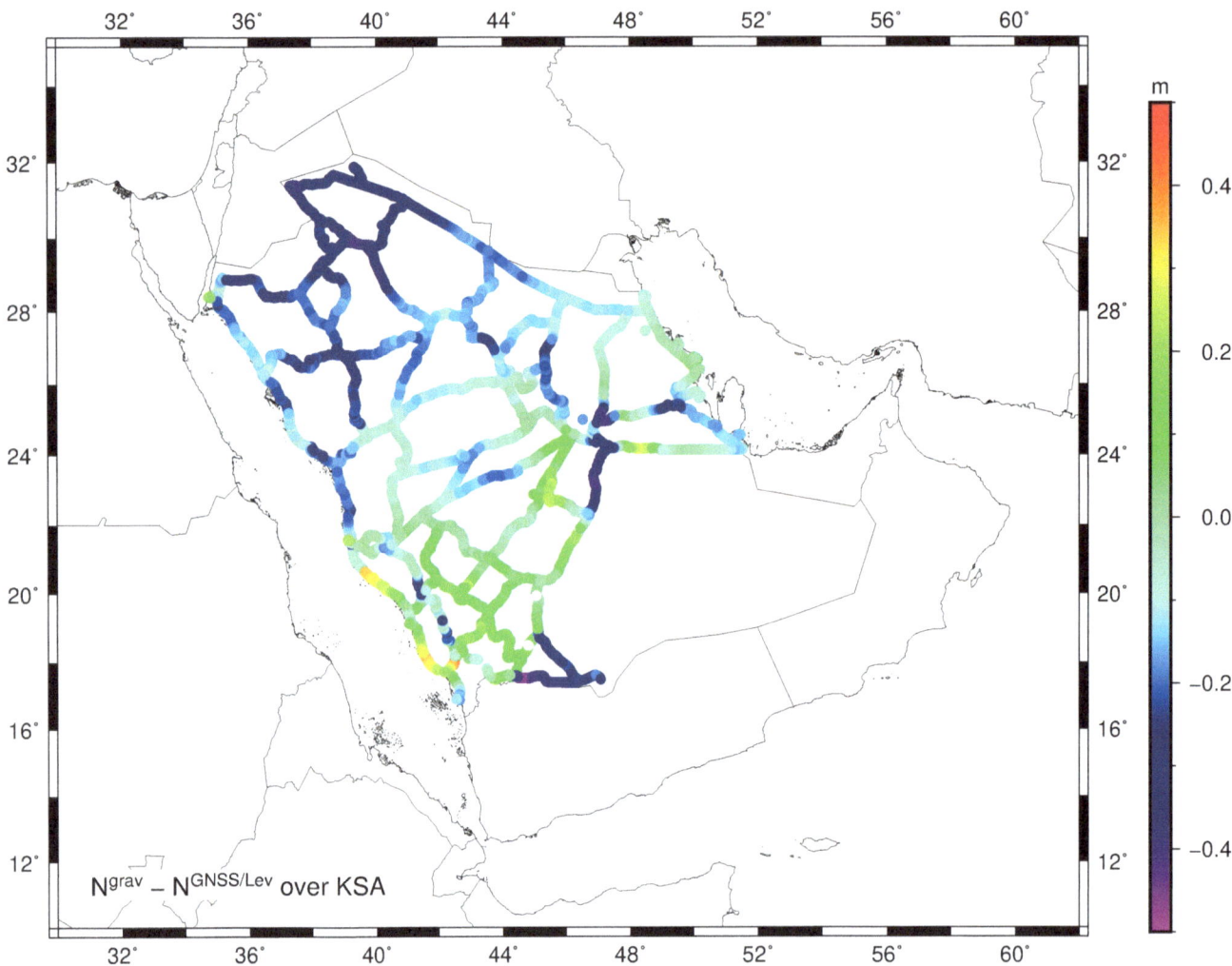

Fig. 3 Geoid height differences between the final gravimetric geoid and the GEOSA GNSS/Leveling geoid heights

4 Conclusions

The creation of an accurate, consistent, and homogeneous gravity database for both land and marine areas over the Kingdom of Saudi Arabia (KSA) has been outlined, followed by the determination of a gravimetric-only geoid model for KSA using the new gravity database. To construct the final gravity database, many pre-processing steps have been conducted to quality control the data and homogenize them in terms of the geodetic reference system, the vertical reference datum, and the tide conventions. The geoid prediction was carried out with an RCR approach and was based on an FFT evaluation of Stokes' integral with a Wang-Gore modification. The validation of the developed gravimetric geoid over 3,522 GNSS/Leveling benchmarks resulted in external absolute accuracies at the 13.6 cm level and relative accuracies at the 1–5 ppm over distances ranging from 10 to 2,000 km.

References

Andersen OB, Knudsen P (2019) The DTU17 global marine gravity field: first validation results. In: Mertikas S, Pail R (eds) Fiducial reference measurements for altimetry, International Association of Geodesy Symposia, vol 150. Springer, Cham. https://doi.org/10.1007/1345_2019_65

Bruinsma S, Foerste C, Abrikosov O, Marty J-C, Rio M-H, Mulet S, Bonvalot S (2013) The new ESA satellite-only gravity field model via the direct approach. Geophys Res Let 40(14):3607–3612. https://doi.org/10.1002/grl.50716

Foerste C, Bruinsma SL, Abrykosov O, Lemoine J-M, Marty JC, Flechtner F, Balmino G, Barthelmes F, Biancale R (2014) EIGEN-6C4 The latest combined global gravity field model including GOCE data up to degree and order 2190 of GFZ Potsdam and GRGS Toulouse. GFZ Data Services. https://doi.org/10.5880/icgem.2015.1

Flury J and Rummel R (2009) On the geoid-quasigeoid separation in mountain areas. J Geod 83: 829–847. https://doi.org/10.1007/s00190-009-0302-9

Gruber T, Gerlach C, Haagmans R (2012) Intercontinental height datum connection with GOCE and GPS-levelling data. J Geod Sci 2(4):270–280. https://doi.org/10.2478/v10156-012-0001-y

Pavlis NK, Holmes SA, Kenyon SC, Factor JK (2012) The development and evaluation of the Earth Gravitational Model 2008 (EGM2008). J Geophys Res 117:B04406. https://doi.org/10.1029/2011JB008916

Rexer M, Hirt C, Claessens S, Tenzer R (2016) Layer-based modelling of the Earth's gravitational potential up to 10km scale in spherical harmonics and ellipsoidal approximation. Surv Geophys 37(6):1035–1074

Sandwell DT, Müller RD, Smith WHF, Garcia E, Francis R (2014) New global marine gravity model from CryoSat-2 and Jason-1 reveals buried tectonic structure. Science 346(6205):65–67. https://doi.org/10.1126/science.1258213

SGL (2021) Final project report – airborne gravity measurements within the Kingdom of Saudi Arabia 2018–2021 for GASGI, Project Number 4/141/38

Sideris MG (2013) Geoid determination by FFT techniques. In: Sansò F, Sideris M (eds) Geoid determination, Lecture notes in earth system sciences, vol 110. Springer, Heidelberg. https://doi.org/10.1007/978-3-540-74700-0_10

Tscherning CC (1991) The use of optimal estimation for gross-error detection in databases of spatially correlated data. BGI, Bull d' Inf 68:79–89

Vergos GS, Tziavos IN, Andritsanos VD (2005) Gravity database generation and geoid model estimation using heterogeneous data. In: Sansò F (ed) A window on the future of Geodsy, International Association of Geodesy Symposia, vol 128. Springer, Heidelberg, pp 155–160. https://doi.org/10.1007/3-540-26932-0_27

Vergos GS, Grigoriadis VN, Tziavos IN, Kotsakis C (2014) Evaluation of GOCE/GRACE global geopotential models over Greece with collocated GPS/levelling observations and local gravity data. IAG Symp 141:85–92. https://doi.org/10.1007/978-3-319-10837-7_11

Wong L, Gore R (1969) Accuracy of geoid heights from modified Stokes kernels. Geophys J R Astron Soc 18:81–91. https://doi.org/10.1111/j.1365-246X.1969.tb00264.x

Zingerle P, Pail P, Gruber T, Oikonomidou X (2019) The experimental gravity field model XGM2019e. GFZ Data Serv. https://doi.org/10.5880/ICGEM.2019.007

Comparisons of Absolute Gravimeters as a Key Component of the International Terrestrial Gravity Reference Frame (ITGRF) Shown on the Example of the WET-CAG2021 at Wettzell, Germany

Hartmut Wziontek, Reinhard Falk, Vojtech Pálinkáš, Andreas Engfeldt, Julian Glässel, Andreas Hellerschmied, Domenico Iacovone, Jakub Kostelecky, Marvin Reich, Ludger Timmen, Christian Ullrich, Alessandro Valluzzi, and Barbara Zehetmaier

Abstract

Comparisons of absolute gravimeters are essential to guarantee their traceability to the International System of Units (SI) and their compatibility and will be a key component of the upcoming International Terrestrial Gravity Reference Frame (ITGRF) of IAG.

The results of the regional comparison of absolute gravimeters WET-CAG2021 hosted at the Geodetic Observatory Wettzell, Germany, in autumn 2021 are presented. Seven FG5/X absolute gravimeters and—for the first time—two commercial AQG absolute quantum gravimeters took part. Temporal gravity variations during the comparison period of 12 weeks were monitored with the superconducting gravimeter GWR OSG-030. The equivalence of each absolute gravimeter is evaluated against a common reference level derived from the measurements during this comparison period. Although the comparison is outside the scope of CIPM MRA it is linked to the EURAMET.M.G-K3 2018 at the same site and the CM.G-K2.2017 in Beijing, China, which ensures the traceability.

Keywords

Absolute gravimeter · Comparison · Gravity reference system

H. Wziontek (✉) · R. Falk · J. Glässel
Federal Agency of Cartography and Geodesy (BKG), Leipzig and Frankfurt/Main, Germany
e-mail: hartmut.wziontek@bkg.bund.de

V. Pálinkáš · J. Kostelecky
Research Institute of Geodesy, Topography and Cartography (VÚGTK/RIGTC), Geodetic Observatory Pecný, Ondřejov, Czech Republic

A. Engfeldt
Lantmäteriet, Gävle, Sweden

A. Hellerschmied · C. Ullrich · B. Zehetmaier
Federal Office of Metrology and Surveying (BEV), Wien, Austria

D. Iacovone · A. Valluzzi
Agenzia Spaziale Italiana (ASI) / e-geos, Matera Space Center/Operations, Contrada Terlecchie, Matera, Italy

M. Reich
GFZ German Research Centre for Geosciences, Wissenschaftspark Albert Einstein, Potsdam, Germany

L. Timmen
Leibniz University Hannover, Institute of Geodesy, Hannover, Germany

1 Introduction

The International Terrestrial Gravity Reference System (ITGRS) of IAG will be defined based on the instantaneous acceleration of free fall expressed in the International System of Units (SI) (Wziontek et al. 2021). The conventional quantity "acceleration of gravity" is then derived by a set of corrections. The International Terrestrial Gravity Reference

© The Author(s) 2023
J. T. Freymueller, L. Sánchez (eds.), *Gravity, Positioning and Reference Frames*,
International Association of Geodesy Symposia 156, https://doi.org/10.1007/1345_2023_226

Frame (ITGRF) should be realized by observations using absolute gravimeters (AG) and a set of conventional models for the correction of temporal changes (tides, atmosphere, polar motion). Accessibility to the users is ensured by a compatible infrastructure with reference stations as the main components. Reference stations provide a long-term stable absolute gravity reference function by monitoring the seasonal gravity variations using a superconducting gravimeter in combination with repeated AG observations or in future by continuously operated AQG, respectively.

Comparisons of absolute gravimeters are essential for the ITGRF to guarantee the traceability and compatibility of observations with AGs. The gravity reference is realized based on a set of precise absolute measurements during comparisons (e.g. Jiang et al. 2012).

The Regional Comparison of Absolute Gravimeters WET-CAG2021, organized as an Additional Comparison beyond the scope of the CIPM MRA (CCM 2015), was held in Germany at the Geodetic Observatory Wettzell (GOW) of the German Federal Agency for Cartography and Geodesy (BKG) in autumn of 2021. All measurements in the New Gravity Laboratory of 2010 were collected between September 7 and December 2, 2021 (Falk et al. 2022).

Due to the pandemic, the number of participants and the measurement schedule at GOW could not be fixed and optimized in advance. Finally, 9 absolute gravimeters of 4 different types were compared. Overall, two teams from National Metrology Institutes (NMIs) or Designated Institutes (DIs) and 7 teams from geodetic institutions participated in WET-CAG2021. The DI from Czech Republic (VÚGTK/RIGTC) participated here with FG5X-251H, but in past key comparisons with FG5-215H. A statistically significant and stable difference of $(1.8 \pm 0.3)\,\mu$Gal between these gravimeters has been estimated from almost 70 repeated absolute measurements at the station Pecný (Czech Republic) between 2017–2021. Therefore, the difference has to be taken into account for linking the WET-CAG2021 to key comparisons.

The comparison results were calculated independently by BKG and VÚGTK/RIGTC. Two processing strategies were followed as described in Pálinkáš et al. (2021). In one solution, labelled as ICN, all compatible gravimeters contribute to the reference values according to their harmonized uncertainties. In the other solution, labelled as KCN, the absolute level of the comparison reference values is determined only by gravimeters belonging to NMI/DIs which also provide the link to a key comparison while all other instruments stabilize the solution only with gravity differences. Although WET-CAG2021 is not a metrological key comparison and therefore a separation into groups of AGs is not mandatory, both approaches were followed to demonstrate the impact on the mean comparison level.

Six instruments have participated in both, the EURAMET.M.G-K3 (Falk et al. 2020) and WET-CAG2021. The comparison WET-CAG2021 was conducted to ensure the compatibility of the AGs, to check the long-term stability of the FG5/X gravimeters and for the first time to evaluate two commercial quantum gravimeters AQG (Ménoret et al. 2018). In contrast to the AQG-A laboratory device, the AQG-B has also been designed for field use. The characterization of the uncertainty budget for these instruments is still under investigation.

The deviation of a particular gravimeter from the reference values (RV) is usually expressed by the Degree of Equivalence (DoE). Since this additional comparison is not within the scope of CIPM MRA, the standard uncertainty, denoted by u, is given with a coverage factor of $k = 1$ as is usual in geodesy. In metrology $k = 2$ is common, which was applied in recent key comparisons of AGs. Sigma (σ) denotes the standard deviation or the error estimates, respectively obtained from error propagation for the parameters from the least-squares adjustment.

All measurement data and processing results can be found in the comparison report (Falk et al. 2022). Results are given in microGal (μGal) as unit of acceleration of gravity, 1 μGal is equal to 1×10^{-8} m/s^2.

2　Absolute Gravity Measurements

Each gravimeter measured at a minimum of two or generally, three different sites. Some gravimeters occupied the same site at multiple times to estimate the setup error. The absolute gravity measurement g_{raw} is the mean free-fall acceleration at the specific measurement height and was corrected for the Earth and ocean tides (zero-tide system), the effect of atmospheric mass variations using the local measured air pressure record and an admittance factor of $-0.3\,\mu$Gal/hPa, and the reference air pressure, polar motion, vertical gravity gradients above the measurement site, in accordance with the IGRS Conventions 2020 (Wziontek et al. 2021). The tidal parameters are the same as in previous comparisons (e.g. Falk et al. 2020) and were estimated from 10 years of continuous measurements of the superconducting gravimeter GWR SG-029 (2000–2010). This device is operated at the Old Gravity Laboratory, about 200 m away from the comparison site. Further, systematic instrumental corrections, e.g. for FG5/X the speed-of light correction, the self-attraction correction and the laser beam diffraction correction, were applied individually by each participant of the comparison together with processing of their observations. Corrections to the AQG measurements were applied following the latest instructions of the manufacturer.

In total, 43 final gravity values with associated uncertainties at the measurement height have been submitted. The transfer to the common comparison height of 1.250 m was based on vertical gravity gradients (VGGs) determined

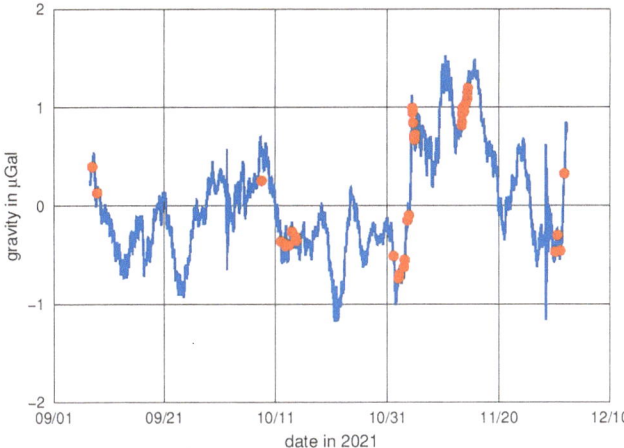

Fig. 1 The residual gravity variations observed during the comparison with the superconducting gravimeter GWR SG-030 referred to October 20, 2021 12:00 UT as the comparison reference time. Red dots mark the mean values at the reference time of absolute gravity observations

before and during EURAMET.M.G-K3 comparison in 2018 (Falk et al. 2020). The used comparison height of 1.25 m above the bench mark was defined as approximate mean of the effective instrumental heights of FG5 and FG5X gravimeters (Wziontek et al. 2021). As long as the AQG have distinct larger uncertainties as FG5/X gravimeters the comparison reference height should be close to 1.25 m to minimize the contribution of the transfer errors to the comparison reference values.

Residual temporal gravity variations not covered by models are accounted for by the continuous record of the super-conducting gravimeter SG-030 as shown in Fig. 1. A correction for each AG observation epoch has been applied according to Fig. 1, together with including a contribution of 0.3 μGal to the uncertainty budget. This correction resolved the effect of temporal gravity variations reaching up to 2 μGal and allowed for a duration of the comparison over 12 weeks.

3   Data Elaboration

A least-squares adjustment was performed with the gravity values at the reference comparison height (g) and their associated uncertainties (u) as input. The direct observation equation for each gravimeter i with a bias δ_i at the site j is

$$g_{ij} = g_j + \delta_i + \varepsilon_{ij} \qquad (1)$$

The weights w_{ij} for the stochastic model are derived from the respective uncertainties $w_{ij} = u_o^2/u_{ij}^2$ where u_o is the unit weight. As the set of observation equations has no unique solution for δ, a constraint, which can be interpreted as

definition of the CRV is required (Pálinkáš et al. 2021). Here, the weighted constraint

$$\sum_{i=1}^{n} w_i\, \delta_i = d \qquad (2)$$

was used, where the weights w_i are normalized by the condition $\Sigma\ w_i = 1$, and d is the linking converter (Jiang et al. 2012).

Similar to Pálinkáš et al. (2021), two different solutions (ICN/KCN) were processed. All gravimeters contribute to the definition of reference values of the ICN solution. This solution is independent on other comparisons, thus $d = 0$. For this solution, in case of Czech DI (VÚGTK/RIGTC) the reported data of FG5X-251H have been used without considering the bias to FG5-215H. The corresponding bias of (1.8 ± 0.3) μGal needs to be applied only when the link to previous key comparisons is accounted for, where FG5-215H and not FG5X-251H took part. The KCN solution is considering only the group of gravimeters belonging to NMI/DIs for the definition of comparison reference values in accordance with CCM (2015). The other gravimeters, practically, are only contributing with gravity differences and are thus neither included into the constraint nor in the determination of the linking converter d. Consequently, weights for non-NMI/DI gravimeters are all set to zero in Eq. (2). As proposed by Pálinkáš et al. (2021) declared uncertainties of those FG5/X gravimeters lower than 2.4 μGal were changed to this limit in the ICN solution. In case of the KCN solution, only the uncertainties of non-NMI/Dis were harmonized.

Following Pálinkáš et al. (2021) we used a correlation coefficient of 0.75 to account for correlations of repeated observations of the same instrument, reflecting the typical ratio between repeatability and uncertainty for all gravimeters included in this comparison. This includes the AQGs, as a similar ratio between uncertainty and reproducibility as for the FG5/X was found (Table 3). Note, that the parameter named repeatability in Pálinkáš et al. (2021), in this paper will be changed to a more correct term short-term reproducibility, because it also describes the variability of results due to setup error. The respective covariances for a particular AG i are then obtained from the harmonized u_{ij} as $cov = 0.75\ u_{ij,min}^2$, where $u_{ij,min}$ is the minimum of all u_{ij} of that AG. This approach can be understood as if the measurements of a particular gravimeter carried out within a few days are affected by the same systematic errors. So multiple measuring results of one instrument at one site after a new independent setup can be included with negligible influence on CRV, DoE and associated uncertainties, but providing more precise information about the instrument's short-term reproducibility.

Table 1 Linking converters as weighted mean of DoEs determined at the EURAMET.M.G-K3 (Falk et al. 2020) and CCM.G-K2.2017 (Wu et al. 2020) of joint NMI/DI participants of the three comparisons. Due to the bias of 1.8 ± 0.3 μGal applied to FG5X-251H original measurements, these results are in the level of FG5-215H

	EURAMET.M.G-K3		CCM.G-K2.2017	
	DoE	u	DoE	u
Gravimeter	/μGal	/μGal	/μGal	/μGal
FG5-242	−1.1	2.75	−0.1	3.45
FG5X-251H*/(FG5-215H)	−1.2	2.50	−1.0	2.25
Linking converter d	−1.15	1.85	−0.73	1.88

The linking converter d in Eq. (2) is conventionally taken to be zero in CCM key comparisons. However, regional comparisons have to be linked to a CCM comparison by at least two AGs (CCM 2015), therefore d was computed as weighted average of the biases of the respective gravimeters. Links were established to EURAMET.M.G-K3 held 2018 also at Wettzell (Falk et al. 2020) and to CCM.G-K2.2017 at Beijing, China (Wu et al. 2020) by two gravimeters (Table 1). Here, the link of FG5X-251H to FG5-215H was ensured by applying a bias of (1.8 ± 0.3) μGal, determined based on four-years of repeated measurements by both gravimeters at the station Pecný, Czech Republic.

4 The Geodetic Approach (ICN-Solution)

For the ICN solution (Table 2), all gravimeters were included in the weighted constraint (Eq. 2) with weights related to harmonized uncertainties. The weighting matrix of the observation equation (Eq. 1) introduced the correlation coefficient of 0.75 for all observations of a particular AG.

The consistency of measurements was checked based on the reported uncertainties using the compatibility index En which is defined as the ratio between the difference of the measured gravity value (g_{ij}) and the reference value RV (g_j) at a site and its uncertainty

$$E_{ij} = \frac{\left(g_{ij} - g_j\right)}{u\left(d_{ij}\right)} \qquad (3)$$

Here, the uncertainty of deviations $u(d_{ij})$ was achieved from error propagation accounting for the correlations between observations (Pálinkáš et al. 2021).

As in previous comparisons, the expanded uncertainty was used here and, therefore, an absolute value of En larger than 2 indicates that a measurement is incompatible at a 95% confidence level, as the difference is not covered by the (expanded) uncertainties.

Table 2 ICN-solution comparison reference values (ICN-RV). The constant value 980,836,900.0 μGal was subtracted from the ICN-RV. u is the uncertainty at 68.3% confidence level computed as root mean square of standard deviations σ (from the least-squares adjustment). The reference height is 1.250 m. The ICN-RVs refer to October 20, 2021 12:00 UT as the mean time of the comparison

ICN-solution results			
WET-CAG2021 comparison			
	ICN-RV	σ	u ($k = 1$)
Site	/μGal	/μGal	/μGal
CA	49.88	0.93	0.93
DA	39.33	0.94	0.94
EA	47.71	0.95	0.95
FA	58.18	0.97	0.97

With the exception of one measurement the harmonized compatibility indexes are all below 2, indicating consistency with the harmonized uncertainties. Only one observation of FG5-218 (DA) was found to be incompatible and consequently, the uncertainties of all measurements with this instrument were increased further by 50% to reduce the impact on the mean reference level. No measurements have been removed, as all showed a good short-term reproducibility (Eq. 5), and therefore not affecting the gravity differences. The Degree of Equivalence (DoE, Jiang et al. 2012)

$$D_i = \left[\sum w_{ij} \left(g_{ij} - g_j\right)\right] / \sum w_{ij} \qquad (4)$$

is computed as the weighted average difference between the measurements of a gravimeter i and the RV at site j. Its uncertainty is computed by error propagation, again accounting for correlations (Pálinkáš et al. 2021). The DoEs are identical with the biases δ estimated from least squares adjustment when no measurement is excluded from the adjustment.

The short-term repeatability (Jiang et al. 2012) was computed for each AG from the differences between each observation with the respective CRV as the standard deviation

$$R_i = \sqrt{\sum \left(g_{ij} - g_j\right)^2 / (n - 1)} \qquad (5)$$

It allows to assess whether an incompatibility is caused by systematic deviation of an instrument. The measurements of FG5-218 show with reproducibility of 0.7 μGal a high precision, but with DoE of −4.6 μGal a lower accuracy. Since its DoE exceeds significantly the (harmonized) uncertainty, the FG5-218 was excluded from the constraint (Eq. 2). The final DoEs with an uncertainty at the 68.3% confidence level are presented in Table 3 and Fig. 2.

Table 3 Degrees of Equivalence (DoE) for ICN-solution of the gravimeters participating in the WET-CAG2021. The standard uncertainty U_{DoE} is obtained from error propagation considering correlation of 0.75 between measurements of a particular gravimeters. The short-term reproducibility of each gravimeter during the comparison is also presented

ICN-solution results WET-CAG2021 comparison			
	Degree of Equivalence		
	DoE	u_{DoE} ($k = 1$)	Repeatability
Gravimeter	/μGal	/μGal	/μGal
FG5-242	1.42	2.03	0.7
FG5X-251H	−1.89	1.93	0.8
AQG-A02	6.77	6.67	2.7
AQG-B02	−5.79	6.99	3.9
FG5-101	3.06	1.93	0.6
FG5-218	−4.55	1.89	0.7
FG5-301	−1.52	1.98	0.3
FG5X-220	2.74	1.78	1.2
FG5X-233	−1.96	1.98	1.3

5   The Metrological Approach (KCN-Solution)

The KCN solution is obtained similar to the ICN solution. Nevertheless, only two gravimeters were included in the constraint Eq. (2) with the weights 0.54 for FG5X-251H

and 0.46 for FG5-242. By setting the weights of the other gravimeters to zero in the constraint, they actually only contribute as relative gravimeters. As for the ICN solution underestimated uncertainties were harmonized. Declared uncertainties below 2.40 μGal were set to this value, but only for non-NMI/DI gravimeters. The final DoE with uncertainty at the 68.3% confidence level are presented in Fig. 2 and Table 5, applying linking converters to both, EURAMET.M.G-K3 and CCM.G-K2.2017.

6   Results and Conclusions

The primary objective of the Additional comparison WET-CAG2021 was to validate the long term stability of the FG5/X gravimeters relative to EURAMET.M.G-K3 and CCM.G-K2.2017. Also, for the first time, two commercial quantum gravimeters were included in such a comparison, which allows for an assessment of the compatibility of these two fundamentally different principles of measurement. However, as the characterization of systematic effects for both AQGs by the manufacturer represents a work-in-progress, we consider this a preliminary result.

Three solutions including nine gravimeters are presented which mainly differ by the definition of the absolute level of the comparison reference values. The ICN solution is independent of previous comparisons and documents DoE

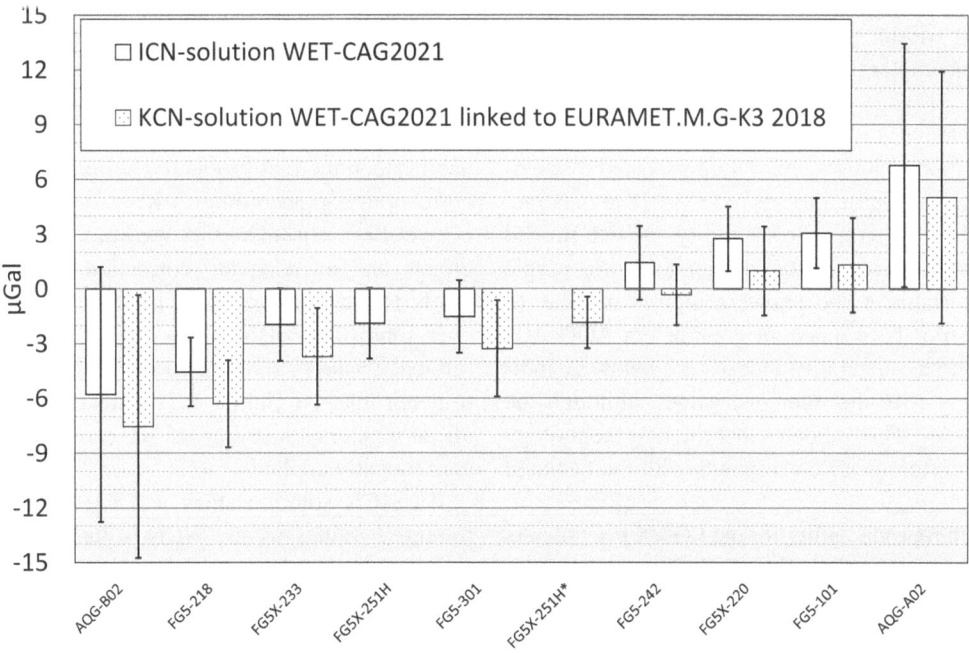

Fig. 2 Joint presentation of Degrees of Equivalence (DoE) in μGal of the gravimeters participating in the WET-CA2021 comparison calculated from the difference between the gravimeter measurements and the CRV or KCN-RV for the corresponding pillar for both solutions. The error bars represent the standard uncertainties (U_{DoE}) at 68.3%

confidence. In the ICN-solution the incompatibility of FG5-218 with its DoE has been solved by down weighting it in the constraint. These KCN results are linked to EURAMET.M.G-K3. FG5X-251H* is given in the level of FG5-215H (FG5X-251H + 1.8 μGal)

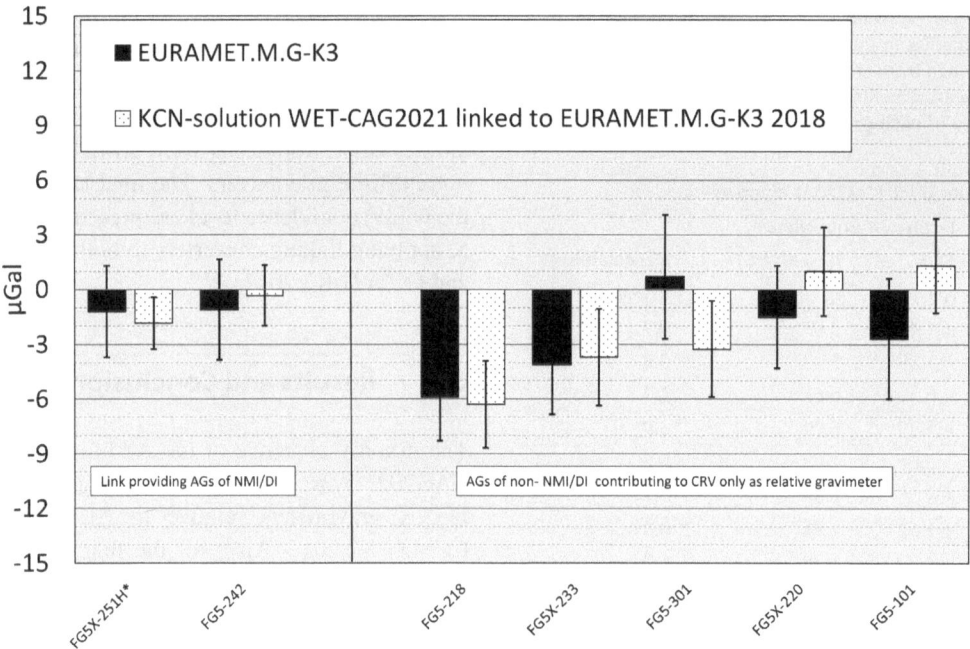

Fig. 3 Joint presentation of the Degrees of Equivalence (DoE) of EURAMET.M.G-K3 and of the KCN-solution (linked to EURAMET.M.G-K3) of WET-CAG2021. FG5X-251H* is given in the level of FG5-215H (FG5X-251H + 1.8 μGal), which participated at EURAMET.M.G-K3

of the seven FG5X gravimeters in the range of −4.6 μGal and +3.1 μGal, while for the quantum gravimeters the DoE ranges between −5.8 μGal and +6.8 μGal.

Both KCN-solutions differ by only 0.4 μGal, where the link to EURAMET.M.G-K3 results have higher RVs and lower DoEs. Using that link, the DoE of FG5/X gravimeters vary within −6.3 μGal and +1.3 μGal while the quantum gravimeters range between −7.5 μGal and +5.0 μGal.

Eight participating gravimeters are equivalent in all presented solutions taking into account their associated uncertainties. Although the measurements of FG5-218 are characterized by a high short-term reproducibility, the determined DoE is neither within twice of the declared nor the simple harmonized uncertainties and shows a significant bias of about −5 μGal. For both quantum gravimeters AQG-A02 and AQG-B02 the compatibility indexes are below 2, indicating consistency with the reference values, although the deviations are larger. This validates that the new technology based in atom interferometry corresponds with the declared uncertainties.

The short-term reproducibility for the FG5/X gravimeters varies between 0.3 to 1.3 μGal. The AQGs short-term reproducibility is with 2.7/3.9 μGal, resp. slightly lower than half of the declared uncertainties. These results confirming the assumption of a correlation between measurements of the same instrument of 75%.

For most of the FG5/X gravimeters a stable DoE (within 1 μGal) between both Wettzell comparisons (2018 and 2021)

is demonstrated as shown in Fig. 3. This suggests that biases different for each gravimeter are reproducible and a stability within a few years could be achieved. This is important for the realization of the ITGRF and for studies, where the gravity rate of changes plays a crucial role (Van Camp et al. 2011; Olsson et al. 2019). However, biases may change when key components of an AG are replaced or maintained. For instance, the obvious changes of the DoE for FG5-101, FG5-301 and FG5X-220 between comparisons may be related to an exchange of the original collimator by another commercial product. The impact on the diffraction correction, e.g. according to Kren and Pálinkáš (2022), has not yet been considered. A rigorous quantification of such effects and an adequate correction of the measurements needs to be established to improve the stability of AGs over time. This also affects the accuracy of comparison reference values (Tables 2 and 4) that can only be enhanced to better than 1 μGal ($k = 1$) if instrumental corrections of all participating instruments are carefully investigated and consequently applied.

The ICN solution show a 1.8 μGal lower comparison reference values as the KCN solution, which seems not to be negligible. Nevertheless, this difference is within the uncertainties of CRVs. The reference values of the ICN solution are lower mainly due to the fact that the participating gravimeters are providing lower gravity values than those used for the link to key comparisons. Generally, the reference values always depend on the set of gravimeters participating at comparisons (Table 5).

Table 4 KCN-solution comparison reference values linked to EURAMET.M.G-K3 using linking converter of (-1.15 ± 1.85) μGal and also to CCM.G-K2.2017 using linking converter of (-0.73 ± 1.88) μGal related to 2 NMI/DI gravimeters. The constant value 980,836,900.0 μGal was subtracted from the KCN-RV. The uncertainty u is at 68.3% confidence level computed as root mean square of the squared standard deviations σ (from the least-squares adjustment) and the squared uncertainty of the linking converter. The reference height is 1.250 m. The KCN-RVs refer to October 20, 2021 12:00 UT as the mean time of the comparison

KCN-solution results						
WET-CAG2021 comparison						
	Linked to EURAMET.M.G-K3 in 2018			*Linked to* CCM.G-K2.2017		
	KCN-RV	σ	$u\,(k=1)$	KCN-RV	σ	$u\,(k=1)$
Site	/μGal	/μGal	/μGal	/μGal	/μGal	/μGal
CA	51.65	1.58	2.43	51.23	1.58	2.45
DA	41.07	1.56	2.42	40.65	1.56	2.44
EA	49.46	1.60	2.45	49.04	1.60	2.47
FA	59.94	1.61	2.45	59.52	1.61	2.48

Table 5 Degrees of Equivalence (DoE) for solution KCN of the gravimeters participating in WET-CAG2021. The standard uncertainty U_{DoE} is obtained from error propagation considering a correlation of 0.75 between measurements of a particular gravimeter. FG5X-251H* is given in the level of FG5-215H. Results using different links (Table 1) are presented

KCN-solution results				
WET-CAG2021 comparison				
	Degree of Equivalence			
	Linked to EURAMET.M.G-K3		Linked to CCM.G-K2.2017	
	DoE	$u_{DoE}\,(k=1)$	*DoE*	$u_{DoE}\,(k=1)$
Gravimeter	/μGal	/μGal	/μGal	/μGal
FG5-242	**−0.33**	**1.66**	**0.09**	**1.66**
FG5X-251H*	**−1.84**	**1.41**	**−1.42**	**1.41**
AQG-A02	5.02	6.90	5.44	6.90
AQG-B02	−7.54	7.20	−7.12	7.20
FG5-101	1.31	2.59	1.73	2.59
FG5-218	−6.30	2.39	−5.88	2.39
FG5-301	−3.27	2.64	−2.85	2.64
FG5X-220	0.99	2.44	1.41	2.44
FG5X-233	−3.71	2.65	−3.29	2.65

The WET-CAG2021 documents a high reproducibility for all FG5/X (0.8 μGal in average), even if a bias exceeds the uncertainty and that the biases of most FG5/X are stable over more than 3 years, compared to EURAMET.M.G-K3 (2018). For the first time it has been demonstrated that quantum gravimeters AQG are in equivalence with their declared uncertainties and reach short-term reproducibility of about 3 μGal (AQG-A02: 2.7 μGal, AQG-B02: 3.9 μGal).

WET-CAG2021 demonstrates the importance of additional (or non-metrological) comparisons to confirm the long-term stability of AGs and their traceability (by the KCN solution). Therefore, additional comparisons and comparison stations should become a key component of ITGRF in order to monitor AGs used for its realization. Finally, it is demonstrated again that the duration of a comparison can extend over several weeks by including gravity variations recorded by a SG.

7   Outlook

The research on AQG systematic effects is ongoing and the manufacturer is actively improving its procedure for the characterization of systematic biases and uncertainties. In spring of 2022, an improved characterization of both the AQG-A02 and AQG-B02 was preformed, resulting in an increased systematic uncertainty of 11 μGal for both instruments. Furthermore, with the new systematic bias corrections, the results of AQG-A02 and AQG-B02 are reduced by 5.6 μGal and 11.9 μGal, respectively. As these characterizations took place after the comparison data submission deadline, they were not considered in the analysis of the comparison. However, due to the larger measurement uncertainty of the AQGs compared to the FG5/X gravimeters, the impact on the RVs, and thereby DoE results for the other instruments, is negligible.

Acknowledgements We thank Axel Rülke and Thomas Klügel of BKG for the valuable support of WET-CAG2021 and the staff of the Geodetic Observatory Wettzell for their hospitality. Vojtech Pálinkáš and Jakub Kostelecký acknowledge the financial support from the project "Research related with International Gravity Reference System" funded by the Ministry of Education, Youth and Sports of the Czech Republic (M˘SMT) under grant no. LTT19008.

References

CCM (2015) CCM - IAG Strategy for Metrology in Absolute Gravimetry. https://www.bipm.org/wg/CCM/CCM-WGG/Allowed/2015-meeting/CCM_IAG_Strategy.pdf

Falk R, Pálinkáš V, Wziontek H, Rülke A, Val'ko M, Ullrich C, Butta H et al. (2020) Final report of EURAMET.M.G-K3 regional comparison of absolute gravimeters. Metrologia 57:07019. https://doi.org/10.1088/0026-1394/57/1A/07019

Falk R, Wziontek H, Pálinkáš V, Rülke A (2022) Wettzell-comparison of absolute gravimeters 2021 WET-CAG2021. Zenodo. https://doi.org/10.5281/zenodo.7437360

Jiang Z, Pálinkáš V, Arias FE, Liard J, Merlet S, Wilmes H, Vitushkin L et al (2012) The 8th International Comparison of Absolute Gravimeters 2009: the first Key Comparison (CCM.G-K1) in the field of absolute gravimetry. Metrologia 49:666. https://doi.org/10.1088/0026-1394/49/6/666

Kren P, Pálinkáš V (2022) Estimation of the effective wavenumber for a collimated beam in an interferometer, case study for FG5/X absolute gravimeters. Appl Opt 61:1811–1817. https://doi.org/10.1364/AO.451498

Ménoret V, Vermeulen P, Le Moigne N, Bonvalot S, Bouyer P, Landragin A, Desruelle B (2018) Gravity measurements below 10^{-9} g with a transportable absolute quantum gravimeter. Sci Rep 8:12300. https://doi.org/10.1038/s41598-018-30608-1

Olsson P-A, Breili K, Ophaug V, Steffen H, Bilker-Koivula M, Nielsen E, Oja T, Timmen L (2019) Postglacial gravity change in Fennoscandia—three decades of repeated absolute gravity observations. Geophys J Int 217(2):1141–1156. https://doi.org/10.1093/gji/ggz054

Pálinkáš V, Wziontek H, Val'ko M, Křen P, Falk R (2021) Evaluation of comparisons of absolute gravimeters using correlated quantities: reprocessing and analyses of recent comparisons. J Geod 95:21. https://doi.org/10.1007/s00190-020-01435-y

Van Camp M, de Viron O, Scherneck H-G, Hinzen K-G, Williams SDP, Lecocq T, Quinif Y, Camelbeeck T (2011) Repeated absolute gravity measurements for monitoring slow intraplate vertical deformation in western Europe. J Geophys Res 116:B08402. https://doi.org/10.1029/2010JB008174

Wu S et al (2020) The results of CCM.G-K2.2017 key comparison. Metrologia 57:07002. https://doi.org/10.1088/0026-1394/57/1A/07002

Wziontek H, Bonvalot S, Falk R, Gabalda G, Mäkinen J, Pálinkáš V, Rülke A, Vitushkin L (2021) Status of the international gravity reference system and frame. J Geod 95:7. https://doi.org/10.1007/s00190-020-01438-9

Newly Acquired Gravity Data in Support of the GeoNetGNSS CORS Network in Northern Greece

D. A. Natsiopoulos, E. G. Mamagiannou, A. Triantafyllou, E. A. Tzanou, G. S. Vergos, I. N. Tziavos, D. Ramnalis, and V. Polychronos

Abstract

The main purpose of the GeoNetGNSS project, funded by the European Union and National Funds through the Region of Central Macedonia (RCM), is to establish a dense network of Continuously Operating Reference Stations (CORS) in northern Greece to support geodetic, surveying, engineering, and mapping applications. A regional, high-accuracy and high-resolution gravimetric geoid model is essential for the accurate determination of physical heights from CORS so as to transform the geometric heights into orthometric ones. In that frame and given the geological complexity and topographic peculiarities of the region, gravity campaigns have been designed and carried out around the newly established CORS stations to densify the already available land gravity database. The observations have been carried out employing the GravLab CG5 relative gravity meter and have been referred to GRS80/IGSN71, relative to the absolute gravity stations established by GravLab at the AUTH premises using the A10 (#027) absolute gravity meter. Moreover, dual-frequency GNSS receivers in network real time kinematic (NRTK) mode were used for orthometric height determination. This work also leverages a database of previous gravity measurements to ensure the data coverage for the region. The XGM2019e Global Geopotential Model (GGM) has been used to model the low frequencies. Moreover, as the development of the geoid model is based on the Remove-Compute-Restore (RCR) technique and the Least Squares Collocation (LSC), the topographic corrections were calculated by the spectral Residual Terrain Model (RTM) method. In this work, the gravity anomalies derived from terrestrial gravity observations over the wider region of Central Macedonia are analyzed and compared with gravity anomalies derived from the XGM2016e GGM. The evaluation of the terrestrial gravity data was performed over six separate traverses, at various heights, in order to investigate the effect of height on the measurements. This technique allows for the comparison of the magnitude of gravity anomalies and the correlation with height, providing a more comprehensive understanding of the region's gravitational field and possible improvement with the newly acquired data.

Keywords

CG5 · Gravity field · Relative gravity · Terrestrial gravity · XGM2016e

D. A. Natsiopoulos · E. G. Mamagiannou · A. Triantafyllou · G. S. Vergos (✉) · I. N. Tziavos
Laboratory for Gravity Field Research and Applications (GravLab), Department of Geodesy and Surveying, Aristotle University of Thessaloniki, Thessaloniki, Greece
e-mail: vergos@topo.auth.gr

E. A. Tzanou
Department of Surveying and Geoinformatics Engineering, School of Engineering, International Hellenic University, Serres, Greece

D. Ramnalis · V. Polychronos
GeoSense PCo, Thessaloniki, Greece

J. T. Freymueller, L. Sánchez (eds.), *Gravity, Positioning and Reference Frames*, International Association of Geodesy Symposia 156, https://doi.org/10.1007/1345_2023_213

1   Introduction

The main goal of the GeoNetGNSS project (GeonetGNSS n.d.) is to establish a dense network of Continuously Operating Reference Stations (CORS) in Northern Greece to support high-accuracy horizontal and vertical position determination for engineering and geodetic applications. The construction of an efficient and cost-effective system to determine physical heights with GPS requires a highly accurate gravimetric geoid model, which is derived from a multitude of gravimetric observations obtained from various sensors and platforms. To achieve this, it is crucial to understand the characteristics of each type of gravity measurement, stemming from historical to recent campaigns, different instrumentation used, etc. This also requires a very good knowledge of the topography in order to determine accurate topographic effects. In this frame, and given both the geological complexity and topographic peculiarities of the area as well as gaps in the exiting gravity database, dedicated gravity campaigns have been designed and carried out around the newly established CORS. The aim was to acquire new gravity observations, both around the new CORS and to fill-in areas where the existing free-air gravity anomaly database has voids and gaps (Grigoriadis 2009). The latter have been utilized in the latest calculated geoid models for Greece (Tziavos et al. 2010, 2013).In the frame of the present work, and depending on the gravity data used, two free-air gravity anomaly models are determined. The first one, called original, where only existing data are used, and a second one, called merged, where the original point gravity anomalies are merged with the newly acquired ones. In this paper, we first summarize the collection and post-processing of gravity and GNSS/Leveling data to densify the available land gravity database in Northern Greece and investigate the effect of height on the gravity data, in order to quantify whether the newly acquired data over areas with gaps improve the gravity field representation. Then, as the estimation of the geoid in a next step will be based on the RCR concept, and in order to compute residual free-air gravity anomalies, the XGM2019e GGM (Zingerle et al. 2019) was used as a reference for modeling the low-frequency part of the spectrum. The topographic effects were calculated through a spherical harmonics representation of the Earth's potential and high-resolution residual terrain corrections from a global model (Hirt et al. 2014; Rexer et al. 2018).The such derived residual gravity anomalies are used to predict gravity anomalies using the Least-Squares Collocation (LSC) technique (Sansò and Sideris 2013; Tscherning 2013), based on the analytical covariance functions of the collocation. The LSC was employed to estimate gravity anomalies over six test traverses spanning the entire region, in an effort to evaluate the improvement brought by including the newly acquired data to the available gravity database.

2   Study Area and Gravity Measurements

The Region of Central Macedonia (RCM) is located in northern Greece and bounded by $39.4° \leq \varphi \leq 42°$ and $21.2° \leq \lambda \leq 24.6°$ (see Fig. 1). It is mainly a low-land area, but on the other hand some of the highest mountains in Greece like Mount Olympus (to the south), Mount Voras (to the north) and Mount Athos (in the third leg of Chalkidiki peninsula) are situated within its bounds. RCM has an extensive coastline with the characteristic peninsula of Chalkidiki in its central part (see Fig. 1) showing varying topographic and morphological characteristics. These pose significant challenges in the accurate determination of the gravity field and the geoid, as quality data are needed over areas with steep terrain which are succeeded by coastal areas with totally different topographic characteristics. Several measurement campaigns were initiated to collect gravity data at various selected sites along the RCM region, aiming to gather high precision gravimetric data and fill-in the gaps in the existing database of the Laboratory of Gravity Field Research and Applications (GravLab) of the Aristotle University of Thessaloniki (AUTh). In this study, we use GravLab's Scrintex CG-5 relative gravity meter, which is one of the standard instrumentations used in relative campaigns with a standard deviation (std) of measurement less than 5 μGal and a low drift of the order of 0.02 mGal/day (Lederer 2009; Yushkin 2011). Before the start of each measurement campaign, survey parameters, corrections, and filters including Tide Correction, Continuous Tilt Correction, Auto Rejection Filter and Seismic Filter have been set. The terrain correction estimation provided by the software of CG-5 has not been used, as we will model the topographic effects through a spherical harmonics expansion of the topographic potential to degree and order (d/o) 2,190 (Hirt et al. 2014) and estimate high-order residual terrain model effects as it is suggested by (Rexer et al. 2018). The field campaigns were carried out in most cases along the road network of the area under study, collecting gravity measurements at a spatial resolution of 1 km, in order to observe the local variations of the gravity field. The reading time for each occupation was set at 60 s with a 5 s start delay for each observation set. At the absolute gravity stations which have been used at the beginning of each daily campaign, five sets of observations have been taken in order to increase the accuracy of the mean observation. In all relative campaigns we have used as reference point the AUTh1 absolute gravity benchmark (BM)

Fig. 1 Distribution of original (red dots) and new (blue dots) gravity data and the locations of the new CORS stations (black stars) and test traverses (green lines)

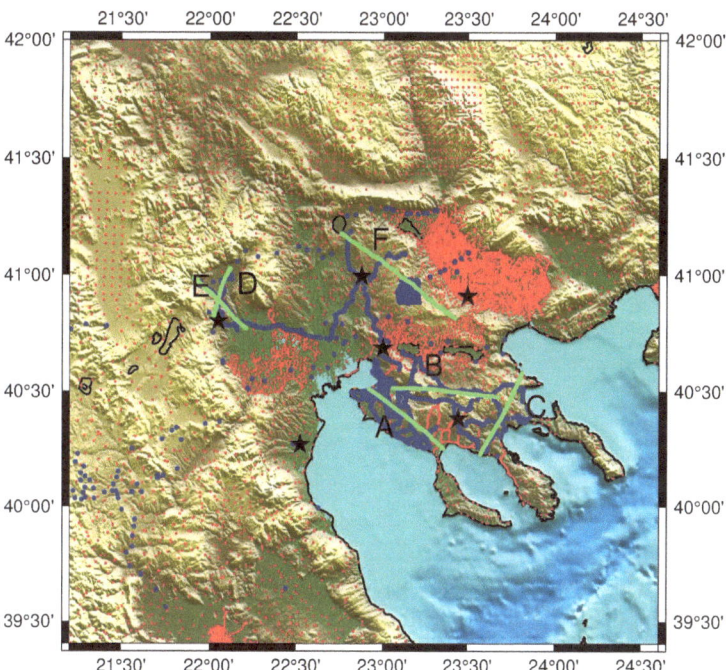

Fig. 1 Distribution of original (red dots) and new (blue dots) gravity data and the locations of the new CORS stations (black stars) and test traverses (green lines)

located at the premises of the University campus. This BM has been established by GravLab using the Microg-Lacoste A10 (#027) absolute gravity meter with a gravity value of 980,276,178.42 ± 10.05 μGal.

In the frame of the gravity campaigns, 2,156 new gravity densification points have been established with their position determined through geodetic-grade GNSS receivers measuring in Network Real Time Kinematic (NRTK) mode. For each new station, 10 epochs of 1 Hz GNSS observations have been collected employing the Virtual Reference Station (VRS) mode and acquiring differential corrections from the Hellenic Positioning Service (HEPOS). These data were merged with the already available 13,961 land free-air gravity anomalies (Grigoriadis 2009) resulting in a total number of 16,117 irregularly distributed observations in the area under study. Figure 1 depicts the old (red) and newly acquired (blue) gravity observations along with the newly established CORS (black stars), while the six validation traverses are shown in green and are numbered from A to F. The six traverses are located on regions that exhibit diverse topographic and land/sea characteristics as well as over regions with sparse original gravity data. Three traverses are situated in the southern part of RCM close to Chalkidiki, one in the west, one in the center, and one in the east part of the test area (traverse A–C respectively), two in the northwest part of RCM (traverse D and E) and one in the north-central part (traverse F). It should be mentioned that the collection of gravity data is a work still in progress, as we need to fill-in observations for the two westernmost stations in Katerini (south-west star in Fig. 1) and Edessa (north-west star in Fig. 1). Especially, the station over Katerini is in close proximity

with Mt. Olympus and the Olympus mountain range, where very few observations exist in our historical database.

3 Relative Gravity Data Processing

In the frame of each measurement campaign, and during each occupation, the quality of the instrument gravity readings is monitored continuously in order to inspect the reading accuracy. In the event that measurements were degraded due to anthropogenic or weather-related factors (e.g. heavy traffic, wind, pedestrians walking by, etc.), then they were immediately discarded and the measurements were repeated. The observations quality of each 60 s measurement was ensured by setting a threshold of 20 μGal for the observation standard deviation, i.e., the instrument accuracy which is the mean of the variances of each 1 s individual observation. In that sense, if the precision of the measurement exceed the 20 μGal level, then the observation was repeated. All campaigns started from the aforementioned AUTh1 absolute gravity station and ended each day at the same station. Therefore, each daily misclosure at AUTh1, as the difference of the observations at the reference station at the beginning and the end of each daily campaign, has been treated as an additional drift correction. From the determined gravity value g for each station, the normal gravity in GRS80 (γ_0) is subtracted and finally the free-air gravity anomalies (Δg_F) are derived using the free-air reduction (δg_F):

$$\Delta g_F = g - \gamma_0 + \delta g_F. \tag{1}$$

Table 1 Statistics of the local free-air gravity anomaly field and their residuals. Units: [mGal]

	Max	Min	Mean	Std
Δg_f merged data	−137.985	258.979	17.022	±53.172
Δg_{XGM} merged data	−126.789	212.762	22.213	±49.116
Δg_{TOPO} merged data	−108.448	81.144	−5.454	±16.542
Δg_{fres} merged data	−63.302	116.492	0.263	±13.722
Δg_f orig. data	−137.985	258.979	11.109	±53.632
Δg_{XGM} orig. data	−126.789	212.762	17.156	±49.835
Δg_{TOPO} orig. data	−108.448	81.144	−5.418	±17.301
Δg_{fres} orig. data	−63.302	113.095	−0.629	±13.012

The main objective of the GeoNetGNSS project is to finally determine a high resolution and accuracy gravimetric and hybrid geoid model through the application of the RCR technique (Barzaghi et al. 2019; Tscherning and Forsberg 1987). This involves the removal of both the long and short wavelengths of the gravity field spectrum from the input data. XGM2019e complete to d/o 2,190 (Zingerle et al. 2019) has been selected as a reference to model the long-wavelength component of the spectrum (Forsberg and Tscherning 2008), while the topographic effects are calculated through a spherical harmonics expansion of the Earth's potential and the residual terrain correction (RTC) from a global model (Hirt et al. 2014; Rexer et al. 2018). Thus, residual free-air gravity anomalies can be calculated as:

$$\Delta g_{fres} = \Delta g_f - \Delta g_{GGM} - \Delta g_{topo}, \qquad (2)$$

where, Δg_{fres} denote the residual gravity anomalies, Δg_f the available free-air gravity anomalies, Δg_{GGM} the contribution of the GGM and Δg_{topo} the contribution of the topography. Table 1, tabulates the corresponding statistics for both the original and merged gravity datasets. The merged residual free-air gravity anomalies show a higher standard deviation by 0.7 mGal and a smaller mean by 0.4 mGal compared with the original free-air gravity anomalies.

In order to evaluate the influence of the recent gravity data in the representation of the Earth's gravity field over the study area, LSC has been used to predict free-air gravity anomalies from the original and merged datasets to the six aforementioned traverses. LSC is frequently used in physical geodesy to interpolate and estimate gravity-related quantities with the challenge being to model appropriately the analytical covariance function to be used. Two different empirical

covariance functions have been estimated (see Fig. 2), one for the merged gravity data set and one for the original gravity data set, while in both cases the analytical model was that of Tscherning and Rapp (1974). The empirical covariance functions have been estimated with an in-house developed software in Matlab © while the analytical model has been fitted using the covfit module of the Gravsoft gravity field modeling software (Forsberg and Tscherning 2008).

As it can be seen in Fig. 2, the merged residual gravity anomalies present higher power compared to the original ones, which can be attributed to the better representation of the local gravity variations in the area under study. The merged data has a correlation length (denoted by ξ in Fig. 2) that is nearly half that of the original datasets (6.2 km compared to 12.78 km), thus indicating that the merged dataset presents a smoother signal of the local gravity field after the removal of the XGM2019e and RTM contributions. The so-determined models of analytical covariance functions have been used to estimate, from the irregularly distributed original and merged gravity data, residual gravity anomalies at the test traverses, while then the effects of the topography and that of the GGM were restored. The prediction has been carried out both over regions that exhibit diverse topographical and land/sea attributes and regions with sparse original gravity data. For all six test traverses mentioned above, the differences among the original, merged, and XGM2019e gravity anomaly data were computed, with their statistics tabulated in Table 2 and their differences depicted in Fig. 3, along with their corresponding statistics presented in Table 3. It should be noted that the original, historic, gravity data have been incorporated in the development of EGM2008 (Pavlis et al. 2012), hence this should influence the statistics achieved. The analysis of their differences indicates that the merged dataset exhibits a stronger representation of the real gravity signal than the original dataset, especially in traverse A between points 25–35, in traverse B between points 30–35, and in traverse F between points 30–40. It is evident that the merged dataset (represented by the blue dashed line) provides a significantly better representation of local gravity variations in the study area for these specific points, introducing higher frequencies compared to the old dataset. In most cases, the original database provides a smaller standard deviation to XGM2019e, but this is something expected, as XGM2019e is based on the EGM2008 terrestrial dataset to a resolution of $0.25 \times 0.25°$ (corresponding to d/o 720). Hence, both the original gravity dataset and XGM2019e do

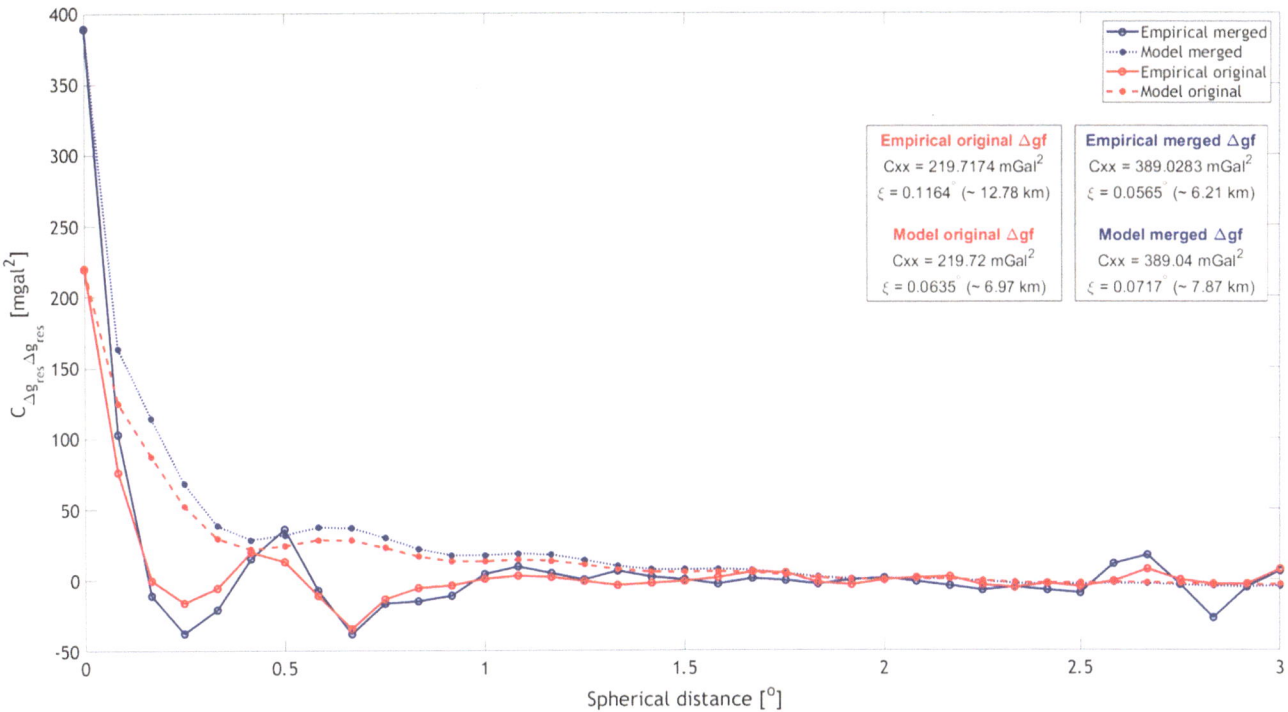

Fig. 2 Empirical covariance functions of the residual gravity anomalies and fitted analytical models for the original and merged datasets (where ξ in the figure text denotes the correlation length)

Table 2 Statistics of the differences between the original, merged and XGM2019e gravity anomalies for each of the traverses studied. Units: [mGal]

Traverse	Differences to XGM2019e	Max	Min	Mean	Std
A	Original	20.242	−24.614	−0.496	11.168
	Merged	19.788	−24.011	−2.250	13.285
B	Original	40.401	−24.114	0.710	15.248
	Merged	44.208	−22.841	−1.420	16.777
C	Original	41.223	−20.046	5.366	13.310
	Merged	53.037	−23.928	2.748	15.858
D	Original	9.747	−21.661	−5.123	10.598
	Merged	6.853	−26.521	−7.876	10.481
E	Original	14.260	−49.964	−6.309	15.658
	Merged	12.372	−55.989	−8.783	16.873
F	Original	36.748	−16.783	3.871	11.654
	Merged	36.503	−21.294	3.123	13.366

not manage to represent some fine details in the local gravity features of the area, given the undersampling of original dataset, especially over rugged terrain. Of course, this is not a problem over areas of lower terrain like traverse D (see after gravity point 8).

Additional insights can be derived by estimating the correlation coefficient between free-air gravity anomalies and height, as, primarily, the higher frequencies depicted by the merged dataset should lead to improved correlation coefficients. This has been done for both the two local gravity anomaly datasets and XGM2019e, the latter setting the threshold for the global models. Table 3 summarizes the correlation coefficients estimated, where it can be seen that in all cases the new merged dataset provides a, slightly or significantly, higher correlation. The results indicate that the incorporation of the new gravity data yielded a significantly enhanced correlation throughout all regions, with a notable increase of 0.2 (37.5%) observed in the western part of Chalkidiki (traverse A), where very few observations existed in the original database, while now extensive data have been collected. A slightly smaller improvement is found over traverse B in the central part of Chalkidiki (from 0.78 to 0.84 or 13.5%), which is expected as in that region the original database contains observations. Over traverse C the improvement reaches 26.4%, 3% over traverse D, 4% over traverse E and 3.6% over traverse F. XGM2019e presents in all cases slightly lower correlation with topography but very close to the original dataset, which is an indication of the high-quality of the GGM. This is something expected as most of the historical terrestrial gravity data in Greece have been used in the development of EGM2008 and hence in XGM2019e as the latter incorporates the terrestrial gravity data of EGM2008 to a resolution of 0.25 × 0.25°, which corresponds to spherical harmonics degree 720. On the other hand, especially over traverse A in the western part of Chalkidiki, XGM2019e shows a correlation with topography

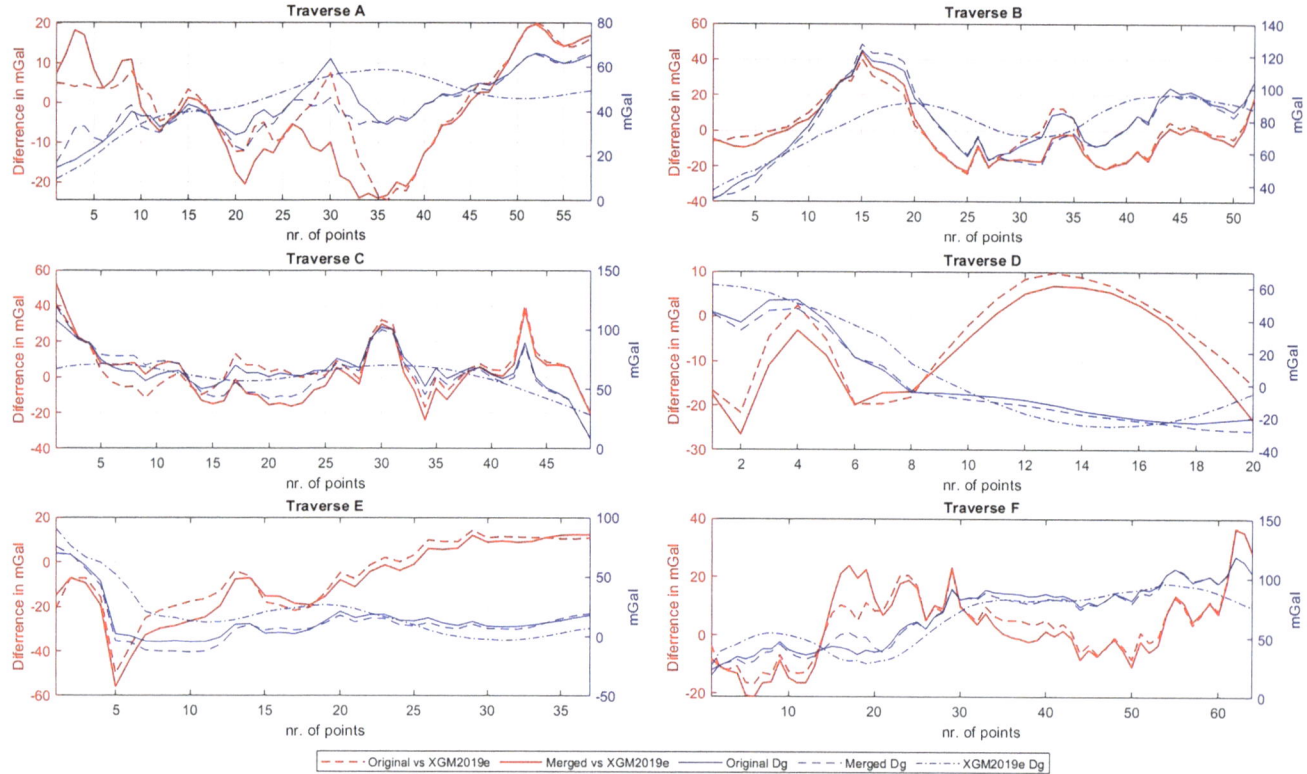

Fig. 3 Original, merged and XGM2019e gravity anomalies, and their differences, along the traverses

Table 3 Correlation between gravity anomalies and elevation

TRAVERSE	Original data	Merged data	XGM2019e
A	0.557	0.766	0.159
B	0.784	0.844	0.803
C	0.486	0.614	0.452
D	0.936	0.962	0.860
E	0.845	0.888	0.851
F	0.831	0.857	0.944

at the 0.159 level only, while the original dataset is at the 0.557. This is quite peculiar and might be attributed to some of the old data not being used in its development, but it is still an issue that requires further investigation. What is evident though is that especially over that part of the area under study, where dense new observations have been collected, the improvement is significant. Therefore, taking more gravity measurements and filling the gaps in existing databases over a region, can help to better capture and understand the variations of the underlying topography, model the underlying mass distribution and thus derive a more precise representation of the gravity field and the geoid.

4 Conclusions

The GeoNetGNSS project aims to establish a dense network of CORS in Northern Greece for the construction of an accurate gravimetric geoid model. To achieve this, gravity campaigns have been designed and carried out around the newly established CORS, and these data were combined with already available free-air gravity anomalies resulting in a total number of 16,117 irregularly distributed point values. Details on the study area, measurement equipment, and processing of the data, including the calculation of free-air gravity anomalies have been given. After removing both the long and short wavelengths of the gravity field spectrum least, the LSC method was used to predict over six separate traverses, towards the evaluation of the terrestrial gravity data. By incorporating the new denser gravity dataset, the correlation in all regions improved significantly, particularly in the western part of Chalkidiki, where few gravity observations were available in the previously available dataset. This area showed a noticeable increase of 0.2 in the correlation coefficient which is an improvement by 37.5%. The current campaigns will be enhanced with additional gravity acquisitions over the western part of the area under study where some dominant topographic features, like Mount Olympus are found. It is expected that the recently acquired datasets will substantially enhance the accuracy in the gravimetric and hybrid geoid models to be determined. The latter are to serve to the ultimate goal of the GeoNetGNSS project, which is to support accurate GNSS-based orthometric height determination over RCM.

Acknowledgements This research was carried out as part of the project "GeoNetGNSS" (Project code: KMP6-0071139) under the framework of the Action "Investment Plans of Innovation" of the Operational Program "Central Macedonia 2014-2020", that is co-funded by the European Regional Development Fund and Greece.

References

Barzaghi R, Carrion D, Vergos GS, Tziavos IN, Grigoriadis VN, Natsiopoulos DA, Bruinsma S, Reinquin F, Seoane L, Bonvalot S, Lequentrec-Lalancette MF, Salaün C, Andersen O, Knudsen P, Abulaitijiang A, Rio MH (2019) GEOMED2: high-resolution geoid of the Mediterranean. In: Freymueller J, Sánchez L (eds) International Association of Geodesy Symposia. pp 43–50. https://doi.org/10.1007/1345_2018_33

Forsberg R, Tscherning CC (2008) An overview manual for the GRAV-SOFT geodetic gravity field modelling programs. Contract report for JUPEM, 2nd edn. - References - Scientific Research Publishing

GeonetGNSS (n.d.) GeonetGNSS [WWW Document]

Grigoriadis V (2009) Geodetic and geophysical approximation of the Earth's gravity field and applications in the Hellenic area, PhD Thesis, Aristotle University of Thessaloniki (in Greek)

Hirt C, Kuhn M, Claessens S, Pail R, Seitz K, Gruber T (2014) Study of the earth/s short-scale gravity field using the ERTM2160 gravity model. Comput Geosci 73:71–80. https://doi.org/10.1016/j.cageo.2014.09.001

Lederer M (2009) Accuracy of the relative gravity measurement. Acta Geodyn Geomater 6:383–390

Pavlis NK, Holmes SA, Kenyon SC, Factor JK (2012) The development and evaluation of the Earth Gravitational Model 2008 (EGM2008). J Geophys Res Solid Earth 117. https://doi.org/10.1029/2011JB008916

Rexer M, Hirt C, Bucha B, Holmes S (2018) Solution to the spectral filter problem of residual terrain modelling (RTM). J Geod 92:675–690. https://doi.org/10.1007/S00190-017-1086-Y/FIGURES/12

Sansò F, Sideris MG (2013) The local modelling of the gravity field by collocation. Lecture notes in earth system sciences, vol 110, pp 203–258. https://doi.org/10.1007/978-3-540-74700-0_5/COVER

Tscherning CC (2013) Geoid determination by 3D least-squares collocation. Lecture notes in earth system sciences, vol 110, pp 311–336. https://doi.org/10.1007/978-3-540-74700-0_7/TABLES/6

Tscherning CC, Forsberg R (1987) Geoid determination in the Nordic countries from gravity and height data. Bollettino di geodesia e Scienze Affinni 46:21–43

Tscherning CC, Rapp R (1974) Closed covariance expressions for gravity anomalies, geoid undulations, and deflections of the vertical implied by anomaly degree variance models. Report No 208, pp 1–89

Tziavos IN, Vergos GS, Grigoriadis VN (2010) Investigation of topographic reductions and aliasing effects on gravity and the geoid over Greece based on various digital terrain models. Surv Geophys 31. https://doi.org/10.1007/s10712-009-9085-z

Tziavos IN, Vergos GS, Mertikas SP, Daskalakis A, Grigoriadis VN, Tripolitsiotis A (2013) The contribution of local gravimetric geoid models to the calibration of satellite altimetry data and an outlook of the latest GOCE GGM performance in Gavdos. Adv Space Res 51. https://doi.org/10.1016/j.asr.2012.06.013

Yushkin VD (2011) Operating experience with CG5 gravimeters. Meas Tech 54:486–489. https://doi.org/10.1007/S11018-011-9753-5/FIGURES/1

Zingerle P, Pail R, Gruber T, Oikonomidou X (2019) The experimental gravity field model XGM2019e. GFZ Data Serv. https://doi.org/10.5880/ICGEM.2019.007

Strapdown Airborne Gravimetry Based on Aircrafts and UAVs: Postprocessing Algorithms and New Results

Vadim S. Vyazmin and Andrey A. Golovan

Abstract

The paper describes a new methodology for postprocessing raw data from a strapdown airborne gravimeter based on a navigation-grade inertial measuring unit and global navigation satellite system receivers (one is on board the aircraft and the others are placed on the ground). The key aspects of the methodology's algorithms are outlined. We also present the numerical results (gravity estimates) from two airborne gravimetry surveys. The surveys were carried out using state-of-the-art strapdown airborne gravimeters on board a fixed-wing aircraft (An-3T) and helicopter-type unmanned aerial vehicle (UAV). In the first survey, the flights were flown in the draped mode with extreme vertical accelerations reaching 2.5 g, which appears to be the first case in airborne gravimetry. In the second survey, the UAV was flying at a constant altitude. The gravity estimation accuracy (RMS) varies from the sub-mGal up to 2-mGal level depending on campaign, with larger values corresponding to the draped flight survey.

Keywords

Airborne gravimetry · GNSS · Gravity vector · IMU · Postprocessing · Strapdown gravimeter

1 Introduction

In airborne gravimetry, measurements of the Earth's gravity are traditionally collected by means of a gravimeter based on a gyro-stabilized gimbal platform. The current trend is the use of strapdown airborne gravimeters based on a navigation-grade strapdown inertial navigation system or inertial measurement unit (IMU) (Stepanov and Peshekhonov 2022). The principle of measurements of such a system is based on measuring the specific force and angular velocity vector by the inertial sensors (accelerometer and gyroscope triads) of the gravimeter's IMU. The gravimeter is also supplemented by a thermal stabilization system and global navigation satellite system (GNSS) receivers (one is onboard the aircraft and the others are the ground-based reference stations).

The well-known advantages of strapdown airborne gravimeters are light weight, small size, and low power consumption (Jensen and Forsberg 2018), which allows to install them in a small aircraft or drone. In addition, there are almost no technical limitations for using strapdown systems, in contrast to the traditional airborne gravimeters, in harsh dynamic conditions (draped flights over terrain of any complexity).

However, postprocessing of raw gravimeter data becomes more challenging in the case of strapdown gravimetry because the aircraft accelerations (during motion and manoeuvring) affect directly on the IMU inertial sensors' measurements. Moreover, one needs to take into account the systematic errors of the IMU inertial sensors (bias, drifts, scale factor errors, etc.) in order to obtain reliable results (Becker 2016; Jensen and Forsberg 2018).

V. S. Vyazmin (✉) · A. A. Golovan
Department of Mathematics and Mechanics, Lomonosov Moscow State University, Moscow, Russia
e-mail: v.vyazmin@navlab.ru

© The Author(s) 2023
J. T. Freymueller, L. Sánchez (eds.), *Gravity, Positioning and Reference Frames*,
International Association of Geodesy Symposia 156, https://doi.org/10.1007/1345_2023_219

In 2020–2022, Lomonosov Moscow State University has developed the postprocessing methodology and algorithms for strapdown airborne gravimetry (Vyazmin and Golovan 2023) and at the moment completes developing the postprocessing software package. This work is on the base of our experience in developing (and supporting) postprocessing software for the GT-2A airborne gravimeter (manufactured by Gravimetric Technologies, LLC, Russia) (Parusnikov et al. 2008; Bolotin and Golovan 2013).

The developed strapdown airborne gravimetry methodology and algorithms were tested in a number of airborne gravimetry campaigns with using state-of-the-art strapdown gravimeters (manufactured by iMAR GmbH and a domestic company) and various carriers – fixed-wing aircrafts (An-30, An-3T, Cessna 208 B, and others) and helicopter-type unmanned aerial vehicles (UAVs) (Vyazmin and Golovan 2023).

In the paper, we briefly outline the key stages of the postprocessing methodology and present the numerical results from two airborne gravimetry campaigns carried out by a Russian geophysical company (Aerogeophysica JSC) using an aircraft (An-3T) flown in the draped mode and UAV (BAS-200) flown at a constant altitude.

2 Postprocessing Methodology and Algorithms

Postprocessing strapdown airborne gravimeter data is different from that developed for the traditional airborne gravimeters. First, in strapdown gravimetry, one has to process raw data (IMU inertial sensors' measurements) recorded at a high sampling rate (300–400 Hz). Second, installing a strapdown gravimeter in a small aircraft or drone leads to higher angular velocities and accelerations during the flight than in the case of large-size aircrafts traditionally used in airborne gravimetry. This means that the systematic errors in the IMU inertial sensors' readings must be determined as accurately as possible. For example, when flying in the draped mode, the aircraft's roll and pitch angles can reach 30° and, hence, the horizontal accelerometer biases will noticeably affect the gravity disturbance estimates. For instance, the 10 mGal bias in one of the horizontal accelerometers will produce a 5 mGal error in the gravity estimate in the case of 30° rolls.

Third, the IMU initial alignment procedure (determination of the IMU attitude at the aircraft's stop before the flight) is strongly affected by external disturbances (caused by wind at the aerodrome, turning on the aircraft's engines, work of the crew, etc.). Hence, the IMU initial alignment algorithm should be operable under such conditions (vibrations).

Below are the key stages of the developed methodology for postprocessing raw data from a strapdown airborne gravimeter (Vyazmin and Golovan 2023):

1. quality control of raw data collected by the gravimeter's IMU inertial sensors and GNSS receivers (check for possible data losses, analysis of system performance indicators, etc.);
2. raw GNSS data processing (determining velocities and position of the onboard GNSS receiver in the differential mode using pseudorange, Doppler pseudorange rate and carrier phase multi-frequency measurements) (Golovan and Vavilova 2007);
3. IMU initial and final alignment (determining the IMU attitude at the aircraft stops before and after the flight and the accelerometers biases and linear drifts) (Vyazmin et al. 2023);
4. IMU-GNSS integration (using the horizontal channels only) (Vavilova et al. 2020);
5. estimation of the gravity disturbance (or all three components of the gravity vector) on the aircraft's flight path.

2.1 Key Aspects of the Algorithms

The IMU initial alignment algorithm (stage 3 of the methodology) is based on approximating the specific force in the inertial frame and admits angular motion of the IMU.

IMU-GNSS integration is performed using the Kalman filtering and optimal smoothing technique (Kailath et al. 2000). At this stage, the estimates of the attitude errors and systematic errors of the IMU inertial sensors are obtained. The system's state vector also includes the GNSS antenna offsets with respect to the IMU and time-synchronization errors in the IMU and GNSS data.

The gravity estimation on the flight path (stage 5 of the methodology) is performed using the basic equation of airborne gravimetry (equation of the gravimeter's proof mass motion projected onto the vertical axis of the navigation geodetic (east, north, up) frame) (Bolotin and Golovan 2013). Other choices of the navigation frame are possible (Becker 2016).

By introducing the IMU accelerometer errors (residual biases $\mathbf{b}_f$ and random noise $\mathbf{q}_f$) and residual attitude errors k_e, k_n (misalignments of the vertical axis of the computed geodetic frame), we rewrite the airborne gravimetry basic equation in the form (Bolotin and Golovan 2013):

$$\Delta \dot{v}_{up} = -\delta g + k_e f_n - k_n f_e + \mathbf{n}^T \left(\mathbf{b}_f + \mathbf{q}_f \right), \quad (1)$$

where Δv_{up} is the IMU vertical velocity error, $\mathbf{n}$ is the unit vector of the vertical axis of the geodetic frame expressed in the IMU frame, and f_e, f_n are the accelerometers readings projected onto the east and north directions. Additional error parameters (time-synchronization errors, GNSS antenna offsets, and the scale factor error of the vertical accelerometer) are omitted in Eq. (1) for simplicity sake. The estimate of the gravity disturbance δg and IMU systematic errors on the flight path are provided by the Kalman filter and optimal smoothing. The algorithm for estimating all three components of the gravity vector is presented in (Vyazmin and Golovan 2023).

3 Numerical Results

Below are the numerical results from two strapdown airborne gravimetry surveys carried out in 2021–2022 by Aerogeophysica JSC (Russia) for geophysical applications. A state-of-the-art strapdown gravimeter (Fig. 1) was used in both surveys. Postprocessing of the gravimeter raw data was performed by Lomonosov Moscow State University.

3.1 Aircraft-Based Survey

The survey was based on the Antonov An-3T aircraft (Fig. 2) flown in the drape mode over the mountainous area. The flight altitude above the ellipsoid varied from 300 m up to 1,200 m during the flights. The aircraft's average speed at a flight line was 40 m/s. The flights were characterized by harsh dynamics conditions leading to roll and pitch angles of up to 30° at flight lines and extreme vertical accelerations of up to 2.5 g, which appears to be the first case in airborne

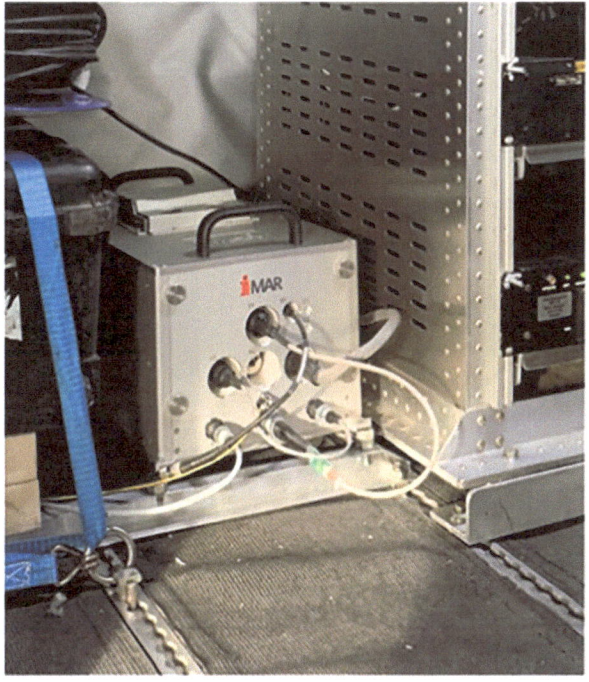

Fig. 1 A strapdown airborne gravimeter on board the aircraft

gravimetry (Fig. 3). Three GNSS receivers from JAVAD (one is onboard and two placed on the ground) were used with the baseline lengths reaching 150 and 200 km.

Fig. 2 The Antonov An-3T aircraft at the aerodrome

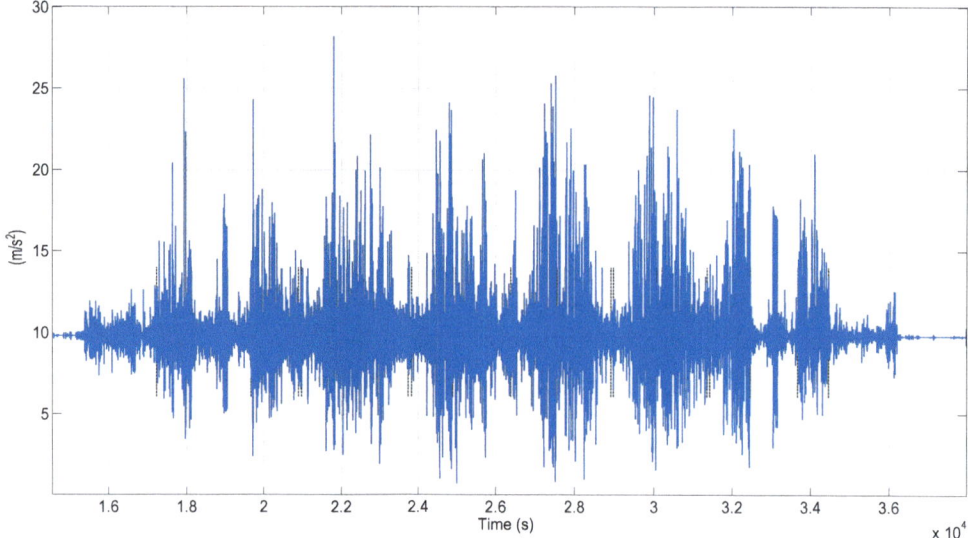

Fig. 3 The gravimeter's IMU vertical accelerometer readings, m/s^2

Postprocessing of raw GNSS and IMU data was performed using the developed algorithms and software. The gravity estimation results and flight altitudes are presented in Fig. 4. The estimation accuracy from the cross-over statistics (without levelling) is 0.5 (mean) and 2.3 mGal (RMS error). The half-wavelength spatial resolution of the gravity estimates is 2.4 km (on the average), which was determined from the average flight speed and the Kalman filter cutoff frequency (equal to 1/120 Hz). The RMS of the flight height discrepancies at the intersection points is about 7 m.

3.2 UAV-Based Survey

The survey flight was carried out on 19.08.2022 using a large-size helicopter-type UAV (BAS-200) shown in Fig. 5. Four repeat lines were flown at the altitudes of 340 m (the L1 and L2 lines) and 420 m (the L3 and L4 lines) above the WGS-84 ellipsoid. The line length was about 20 km. The flight speed was 25 m/s.

The flights were characterized by strong vibrations (the STD of the vertical velocity at a line is 0.2 m/s), which were damped by shock absorbers inside a box containing the gravimeter (the black box in Fig. 5).

The gravity estimates at four repeat lines are shown in Fig. 6. The STD is 1.12 mGal for the difference between the L1 and L2 lines and 0.66 mGal for the difference between the L3 and L4 lines. The STD for the difference between the gravity estimates at the L1–L4 lines and the averaged estimate is

0.61 mGal. The Kalman filter cutoff frequency is 1/100 Hz, which is equivalent to the half-wavelength spatial resolution of 1.25 km.

4 Conclusions

A new methodology and algorithms for strapdown airborne gravimeter data postprocessing are outlined. The key stages of the methodology are the IMU-GNSS integration (using the horizontal channels) and gravity estimation on the flight path (using the IMU vertical channel and IMU-GNSS results, namely, the estimates of the IMU attitude angles). The algorithms from the both stages are based on the Kalman filtering and optimal smoothing technique.

The methodology's algorithms were tested in a number of strapdown airborne gravimetry campaigns (80,000 line km in total) carried out in 2020–2022. State-of-the-art strapdown gravimeters (by iMAR and a domestic company) and various carriers (fixed-wing aircrafts and UAVs) were used in the campaigns. For the UAV-based survey (one of the first in Russia), the gravity estimation accuracy at the sub-mGal level was reached, which is promising for surveying with this type of a carrier. The 2-mGal accuracy was achieved for the draped flight survey (using the Antonov An-3T aircraft), which is a good result for the flights with extreme vertical accelerations up to 2.5 g. The latter appears to be the first case in airborne gravimetry.

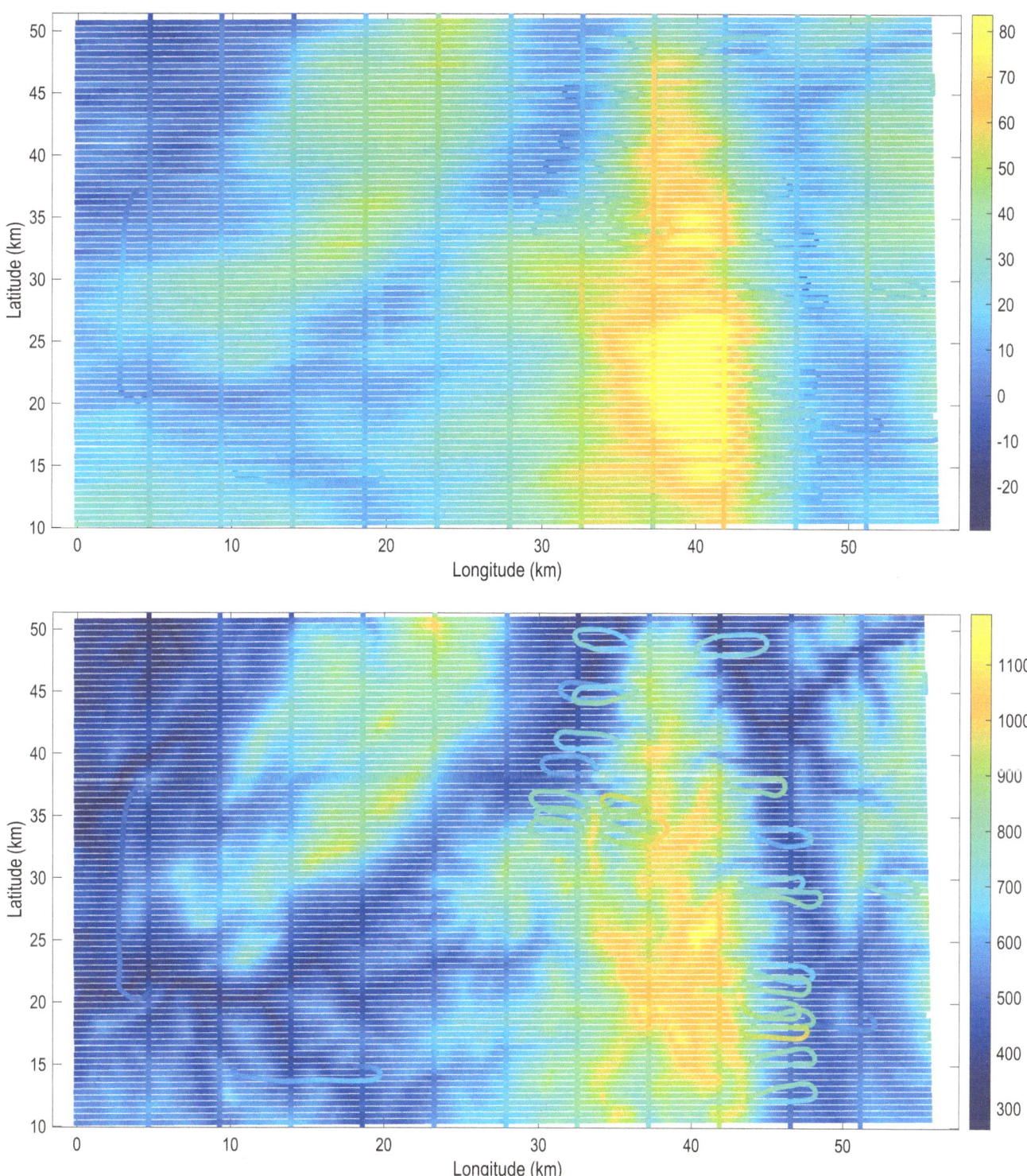

Fig. 4 The results of the aircraft-based survey: (**a**) the gravity disturbance estimates, mGal; (**b**) flight altitudes, m

Fig. 5 The BAS-200 UAV at the aerodrome

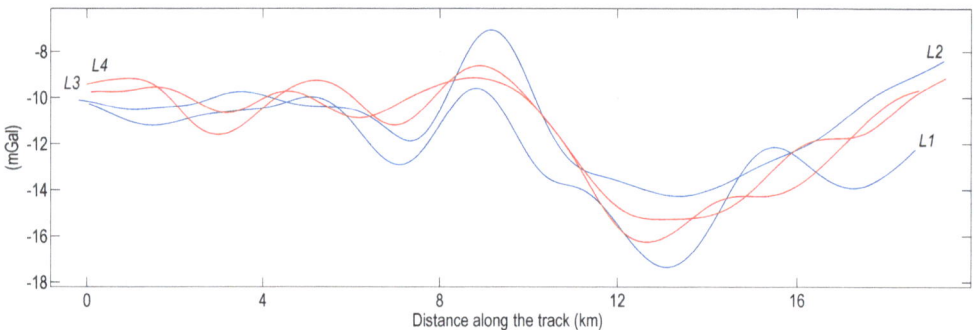

Fig. 6 The gravity disturbance estimates at repeat flight lines, mGal

Acknowledgements Aerogeophysica JSC is acknowledged for providing access to gravimeter measurements. We thank the reviewer for the valuable comments.

References

Becker D (2016) Advanced calibration methods for strapdown airborne gravimetry. Ph.D. Thesis, Technische Universität Darmstadt, Darmstadt, 188 pp

Bolotin YV, Golovan AA (2013) Methods of inertial gravimetry. Mosc Univ Mech Bull 68:117–125

Golovan AA, Vavilova NB (2007) Satellite navigation. Raw data processing for geophysical applications. J Math Sci 146:5920–5930

Jensen TE, Forsberg R (2018) Helicopter test of a strapdown airborne gravimetry system. Sensors 18:3121

Kailath T, Sayed AH, Hassibi B (2000) Linear estimation. Prentice Hall, Englewood Cliffs, p 440

Parusnikov NA, Bolotin YV, Golovan AA (2008) Experience in post processing software elaboration for Russian airborne gravimetry systems GT1A and Graviton-M. In: Proceedings of international symposium on terrestrial gravimetry: static and mobile measurements, Concern CSRI Elektropribor, St. Petersburg, pp 95–95

Stepanov OA, Peshekhonov VG (eds) (2022) Methods and technologies for measuring the earth's gravity field parameters. Springer, Cham, p 388

Vavilova NB, Golovan AA, Parusnikov NA (2020) Mathematical foundations of inertial navigation systems. Moscow University Press, 164 p (in Russian)

Vyazmin VS, Golovan AA (2023) Scalar and vector strapdown airborne gravimetry on aircraft and UAV: Methodology of surveying and data processing, In: 2023 30th Saint Petersburg international conference on integrated navigation systems (ICINS), Saint Petersburg, Russian Federation, 2023, pp. 1–6

Vyazmin VS, Golovan AA, Govorov AD (2023) Initial and final alignment of a strapdown airborne gravimeter and accelerometer bias determination. Gyroscopy Navig 14(1):48–55

Estimation of Temporal Variations in the Earth's Gravity Field Using Novel Optical Clocks Onboard of Low Earth Orbiters

Akbar Shabanloui, Hu Wu, and Jürgen Müller

Abstract

The current generation of optical atomic clocks has reached a fractional frequency uncertainty of 1×10^{-18} (and beyond) which corresponds to a geopotential difference of $0.1 \ m^2/s^2$. Those gravitational potential differences can be observed as gravitational redshift when comparing the frequencies of optical clocks. Even temporal potential variations might be determined with precise novel optical atomic clocks onboard of low-orbiting satellites such as SLR-like (e.g. LAGEOS-1/2) and GRACE-like missions.

In this simulation study, the potential of precise space-borne optical clocks for the determination of temporal variations of low-degree Earth's gravity field coefficients are investigated. Different configurations of satellite orbits, i.e. at different altitudes (between 400 and 6000 km) and inclinations, are selected as well as certain assumptions on the clock performance are made. A particular focus is put on how well degree-2 coefficients can be estimated from those optical clock measurements and how it compares to results from SLR.

Keywords

Optical clock measurements in space · Relativistic geodesy · Temporal long-wavelength Earth's gravity field variations

1 Introduction

In this article, the application of accurate novel optical clocks for the determination of temporal variations of the Earth's gravity field is described. In the past decade, the performance of optical atomic clocks have made spectacular progress in the laboratories of metrological institutes. Over many decades Cs atomic clocks provided microwave frequency standards with superior long-term stability and accuracy, which today are approaching relative uncertainties of 10^{-16}

(Guéna et al. 2017). Nowadays, optical clocks have become 100 times more precise than the best cesium clocks (Godun 2021). An optical atomic clock generates a frequency reference in the form of light stabilized to an atomic transition frequency in the optical frequency range (Sören et al. 2022). The analysis of the clock measurements has to be done in the framework of general relativity (Philipp 2018). Figure 1 depicts the progress of Cs microwave clocks and optical clocks over the last three decades. It should be mentioned the atomic and optical clocks have steadily improved since the emergence of laser-cooled fountain clocks in the early 1990s, but two distinct types of optical atomic clocks, i.e. optical lattice based and based on trapped ion based clocks, currently compete at a fractional frequency uncertainty of approximately 10^{-18} which corresponds to gravitational potential difference of $0.1 \ m^2/s^2$ (for more details, we refer to Alonso et al. (2022)). The optical lattice clocks at Physikalisch-Technische Bundesanstalt (PTB) in Germany are currently upgraded from a fractional frequency uncertainties of few

Authors Hu Wu and Jürgen Müller contributed equally to this work.

A. Shabanloui (✉) · H. Wu · J. Müller
Institute of Geodesy, Leibniz University of Hannover, Hannover, Germany
e-mail: shabanloui@ife.uni-hannover.de; wuhu@ife.uni-hannover.de; mueller@ife.uni-hannover.de

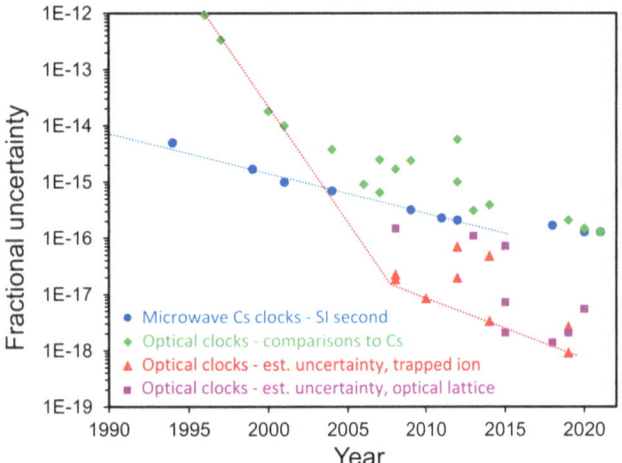

Fig. 1 Progress in the relative accuracy of microwave Cs atomic clocks as well as atomic optical clocks (Alonso et al. 2022)

10^{-17} (Schwarz et al. 2020) to the low 10^{-18} regime. As optical lattice clocks approach a fractional frequency uncertainty 10^{-20} in the near to far future, this will dramatically improve the accuracy for unifying height and the determination of temporal variations of the Earth's gravity field.

The concept to obtain the physical height value with the gravitational redshift measurement with clocks is called chronometric levelling which was originally proposed by Bjerhammar (1985). Furthermore, that concept is extended as chronometric geodesy (Delva et al. 2019). In addition, the high-performance optical clocks connected with dedicated frequency or time links are novel promising network in geodesy (Müller et al. 2018; Mehlstäubler et al. 2018). Also, the frequency transfer via optical fibers in optical clock network has been reached the level of 10^{-19} (Lisdat et al. 2016; Xu et al. 2018; Dix-Matthews et al. 2021) which fulfills the scientific requirement for the comparison of terrestrial optical clocks with 10^{-18} uncertainty. Based on optical clock networks, the physical height differences are estimated from optical clock measurements by observing the gravitational redshift effect through the ultra-precise comparison of their frequencies which is called relativistic geodesy (Müller et al. 2018; Mehlstäubler et al. 2018). Moreover, transportable clocks (Grotti et al. 2018; Takamoto et al. 2020) which are developed with high accuracy and stability can be exploited for clock network densification (Wu and Müller 2020). Those dense optical clock networks are suitable candidate for the realization of the International Height Reference System (IHRS) and detection of time-variable Earth's gravity field signals (Wu and Müller 2020, 2021). The development and progress trend of novel optical atomic clocks make it feasible in the near future to recover the temporal long-wavelength of the Earth's gravity field from space, to establish a frequency-

based physical height reference system as well as the unification of geodetic height systems. Several investigations on affecting the optical clock measurements by mass variations in the Earth system or height differences have been run at Institute of Geodesy, Hannover (Voigt et al. 2016; Denker et al. 2018).

The objective of this paper is to present a description of estimation of lower degree and order spherical harmonic coefficients of the Earth's gravity field using novel optical clock measurements onboard of low Earth orbiters such as GRACE-like and LAGEOS-like missions. Section 2 introduces the concept and methodology of determining temporal long-wavelength variations of Earth's gravity field by observing the gravitational redshift with optical atomic clocks. Section 3 gives details on the simulation scenarios with different configurations of satellite orbits at different altitudes for estimation of mass variations with optical atomic clocks. Section 4 presents the numerical results. We first show the results of the analysis of nearly 2 years of optical clock measurements onboard LAGEOS-1 and LAGEOS-2 for the determination of temporal variations of lower degree and order spherical harmonic coefficients. Then, the mass variations from optical clocks onboard a GRACE-like satellite mission is addressed. Finally, the combined solution is discussed.

2 Temporal Variations of the Earth's Gravity Field from Clock Measurements

2.1 Methodology

The optical clocks measure the gravitational redshift (GRS) within a gravitational potential field. According to general relativity theory (GRT), the optical clocks readings reflect the effect of a potential field on frequency. In GRT, it is essential to distinguish between proper time which is locally measurable and coordinate time which is based on convention. In fact, an ideal clock observes local time as proper time (Soffel and Langhans 2013; Müller et al. 2008).

The relation between proper (relative) time τ of an atomic clock within a potential field W such as Earth's gravity field and coordinate time t at point s can be written as (Mai 2013; Mai and Müller 2013):

$$\frac{d\tau_s}{dt} = \sqrt{1 - \frac{2W_s}{c^2} - \frac{v_s}{c^2}} = 1 - \frac{W_s}{c^2} - \frac{v_s}{2c^2} + \varepsilon\left(c^{-4}\right). \quad (1)$$

W_s represents the gravitational potential at point s which depends only on the positions within the potential field in an

Earth-fixed system, v_s is the clock velocity, c is the speed of light as fixed value and $\varepsilon\left(c^{-4}\right)$ stands for omitting higher order terms. The relativistic time dilation according to Eq. (1) is closely related to the relativistic red shift. Equation (1) can also be applied for a second clock position, replacing s by p. By assuming that the velocities of two stations (or rovers) were precisely determined via Global Navigation Satellite Systems (GNSS), the relativistic time dilation between two optical clocks is then obtained as:

$$\frac{d\tau_s}{d\tau_p} = \left(1 - \frac{W_s}{c^2}\right) \Big/ \left(1 - \frac{W_p}{c^2}\right) = \left(1 - \frac{W_s}{c^2}\right)\left(1 + \frac{W_p}{c^2}\right) + \varepsilon\left(c^{-4}\right).$$
(2)

For a clock located on the Earth surface, W also includes the effect due to Earth rotation, then called gravity potential, whereas in satellites just reflects the gravitational potential. Since the proper frequency is inversely proportional to the proper time, the following Eq. (3) can be used to derive the relativistic red shift observation equation for two optical clocks as:

$$1 - \frac{f_p}{f_s} = 1 - \frac{d\tau_s}{d\tau_p} = \frac{W_s - W_p}{c^2} + \varepsilon\left(c^{-4}\right)$$
(3)

or

$$\frac{f_s - f_p}{f_s} = \frac{\Delta f}{f_s} = \frac{W_s - W_p}{c^2} + \varepsilon\left(c^{-4}\right) = \frac{\Delta W}{c^2} + \varepsilon\left(c^{-4}\right)$$
(4)

where f_s and f_p are the proper frequencies of an electromagnetic signal as observed at two points s and p. By multiplying the relative frequency difference $\frac{\Delta f}{f_s}$ by c^2 and define it as

$\frac{\Delta f^*}{f_s}$, Eq. (4) is simplified as:

$$\frac{\Delta f^*}{f_s} = \Delta W + \varepsilon\left(c^{-4}\right).$$
(5)

Equation (5) is the backbone formula which relates the frequency differences and gravitational potential differences where a fractional frequency difference of one part in 10^{18} corresponds to about 0.1 m^2/s^2 in terms of gravitational potential differences.

2.2 Setup of Optical Clock Observation Equations for the Estimation of Temporal Gravity Field Variations

The gravitational red shift effect which is observed by an optical clock onboard a low earth orbiter is directly related to the gravitational potential difference. Based on this new measurement technique, for the first time in geodesy, it is possible to directly observe the gravitational potential differences. Based on Eq. (5), the optical clock observations as gravitational potential differences between two points s and p can be written as:

$$\frac{\Delta f^*}{f_s} = W_p\left(r, \lambda, \phi; t\right) - W_s\left(r, \lambda, \phi; t\right) + \varepsilon\left(c^{-4}\right)$$

where r, λ, ϕ represents the spherical coordinates i.e. radial distance, longitude and latitude of point along the satellite orbit at time t.

On the other hand, the disturbing potential at point s can be formulated as:

$$W\left(r, \lambda, \phi; t\right) = \frac{GM}{R}\sum_{n=0}^{n_{max}}\left(\frac{R}{r}\right)^{n+1}\sum_{m=0}^{n}\left(\bar{c}_{nm}\left(t\right)\cos\left(m\lambda\right) + \bar{s}_{nm}\left(t\right)\sin\left(m\lambda\right)\right)P_{nm}\left(\sin\phi\right)$$
(6)

where G, M and R are the gravitational constant, mass of the Earth and the reference radius of Earth. $P_{nm}\left(\sin\phi\right)$ is the associated Legendre polynomial of degree and order n and m at latitude ϕ and $\bar{c}_{nm}\left(t\right)$ and $\bar{s}_{nm}\left(t\right)$ are the normalized geopotential coefficients at time t.

The objective of this paper is to see the performance of optical clocks onboard low Earth orbiters such as LAGEOS-

and GRACE-like missions for the estimation of lower degree and order spherical harmonic coefficients of the Earth's gravity field. Figure 2 depicts a schematic diagram of clocks onboard of LAGEOS- and GRACE-like satellite missions with different altitudes and configuration for the estimation of lower degree/order harmonic coefficients. Therefore, here we simplify Eq. (7) up to degree and order 2:

$$W\left(r, \lambda, \phi; t\right) = \frac{GMR^2}{r^3}\begin{pmatrix}\bar{c}_{20}\left(t\right)P_{20}\left(\sin\phi\right) + \\ \bar{c}_{21}\left(t\right)\cos\lambda\, P_{21}\left(\sin\phi\right) + \bar{s}_{21}\left(t\right)\sin\lambda\, P_{21}\left(\sin\phi\right) + \\ \bar{c}_{22}\left(t\right)\cos\left(2\lambda\right)P_{22}\left(\sin\phi\right) + \bar{s}_{22}\left(t\right)\sin\left(2\lambda\right)P_{22}\left(\sin\phi\right)\end{pmatrix}.$$
(7)

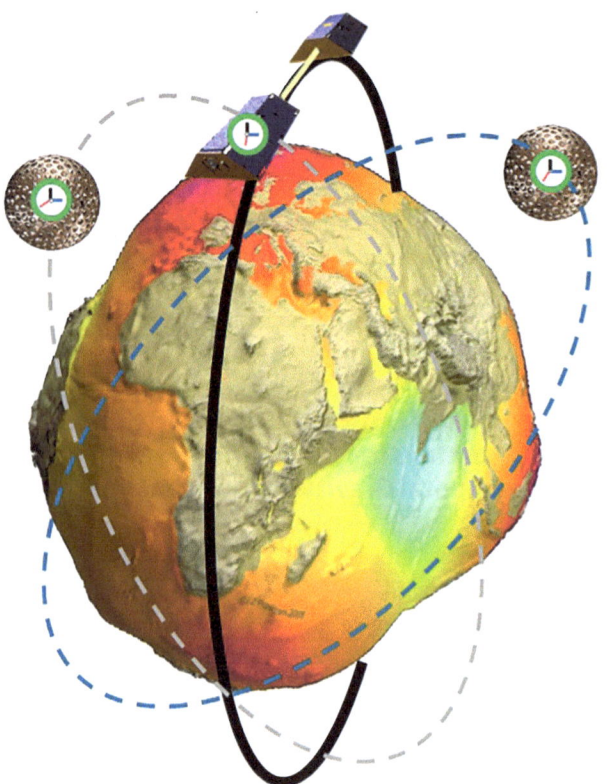

Fig. 2 Schematic diagram of clocks onboard of LAGEOS- and GRACE-like satellite missions with different altitudes and configuration for the estimation of lower degree/order harmonic coefficients

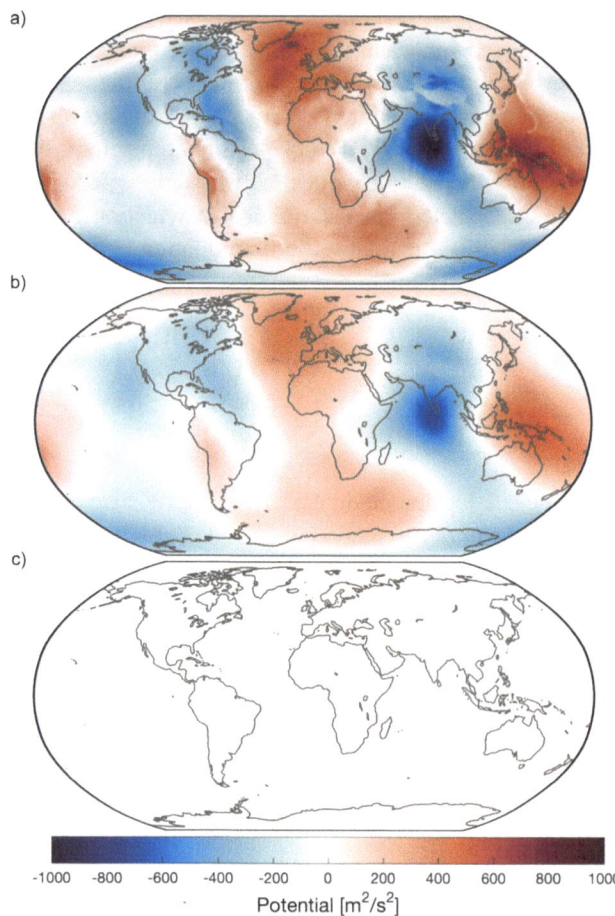

Fig. 3 The gravitational disturbing potential attenuation at different altitudes. (**a**): disturbing potential at altitude of zero with a mean value of 282.57 m^2/s^2, (**b**): disturbing potential at an altitude of a GRACE-like mission of 450 km with a mean value of 210.04 m^2/s^2, (**c**): disturbing potential at the altitude of 35,786 km for geo-stationary satellites with a mean value of 0.52 m^2/s^2

The gravitational disturbing potential attenuates with respect to altitude as shown in Fig. 3. The gravitational disturbing potential has a value of 282.57 m^2/s^2 at an altitude of zero, at an altitude of a GRACE-like mission (450 km), it is 210.04 m^2/s^2, and at an altitude of geo-stationary satellites (35,786 km), it is 0.52 m^2/s^2. With further improvement of optical clock uncertainties into the 10^{-18} to 10^{-19} regimes, the higher satellite altitudes, e.g., the geostationary orbit are good choices for the establishment of a reference optical atomic clocks in space.

The observation equations for optical clock observations i.e. the gravitational potential differences along the satellite orbits with the sampling rate of Δt can be written as:

$$
\begin{pmatrix} \Delta W\,(t_0) \\ \vdots \\ \Delta W\,(t_n) \end{pmatrix} = \begin{pmatrix} P_{20}\,(t_0)\ \cos\lambda\,P_{21}\,(t_0)\ \sin\lambda\,P_{21}\,(t_0)\ \cos2\lambda\,P_{22}\,(t_0)\ \sin2\lambda\,P_{22}\,(t_0) \\ \vdots \qquad \vdots \qquad \vdots \qquad \vdots \qquad \vdots \\ P_{20}\,(t_n)\ \cos\lambda\,P_{21}\,(t_n)\ \sin\lambda\,P_{21}\,(t_n)\ \cos2\lambda\,P_{22}\,(t_n)\ \cos2\lambda\,P_{22}\,(t_n) \end{pmatrix} \begin{pmatrix} \Delta\bar{c}_{20}\,(t_0) \\ \Delta\bar{c}_{21}\,(t_0) \\ \Delta\bar{s}_{21}\,(t_0) \\ \Delta\bar{c}_{22}\,(t_0) \\ \Delta\bar{s}_{22}\,(t_0) \\ \Delta\bar{c}_{20}\,(t_1) \\ \vdots \end{pmatrix} \quad (8)
$$

or

$$\Delta \mathbf{W} = \mathbf{A} \Delta \mathbf{x} + \varepsilon \quad (9)$$

where $\Delta \mathbf{W}$ is the vector of the observables as potential difference of dimension (n), $\mathbf{A}$ is the design matrix of dimension $(n \times m)$ and $\Delta \mathbf{x}$ represents the monthly spherical harmonic coefficients of dimension (m).

The overall set of spherical harmonic coefficients as unknowns in Eq. (9) is estimated by least-squares adjustment.

2.3 Optical Atomic Clock Noise

For this study, the stochastic model for optical clock observations along the satellite orbit is considered as white noise with zero mean and known variance σ_c^2 as:

$$\mathbf{E}(\varepsilon(t)) = \mathbf{0}; \quad \mathbf{C}_c = \sigma_c^2 \mathbf{I}. \quad (10)$$

The operator $\mathbf{E}(\varepsilon(t))$ is the expectation of optical clock noise and $\mathbf{C}_c$ is the diagonal matrix with known variance σ_c^2 and unit matrix $\mathbf{I}$.

It should be mentioned that different averaging periods of 15 min or 30 min along the satellite orbit are assumed to achieve clock accuracies of 10^{-18} or 10^{-19}.

3 Simulation Scenarios for Estimating Mass Variations with Optical Atomic Clocks

Figure 4 depicts the simulation chain of optical clock observations along the satellite orbits as potential differences and the recovery of the Earth gravity field based on those observations. Based on constant degree-2 Stokes coefficients and a-priori secular and annual variations from SLR monthly gravity solutions, the coefficients are synthesized. The optical clock measurement are computed as gravitational potential differences along the satellite orbits. In the second step, white noise for the optical clock measurements is added. Monthly gravity field solutions are determined from the gravitational potential differences. In the final step, a-posteriori secular and annual variations of the degree- and order-2 coefficients are estimated by least-squares adjustment. The zonal coefficient c_{20} represents the dynamic flattening of the Earth. The temporal variations of that coefficient reflect the hydrostatic balance between gravitational and centrifugal force variations as global scale mass redistribution. The temporal variations of the tesseral harmonic coefficients c_{21}, s_{21} represent the Earth's principal figure axis variations related to polar motion or rotational deformation. The sectorial c_{22}, s_{22} coefficients describe the flattening of the equator.

3.1 Data

To simulate the optical clock measurements along the satellite orbits, the geodetic satellite missions LAGEOS-1, LAGEOS-2 and GRACE-FO are utilized. Table 1 summarizes the orbital parameters such as altitude, inclination and revolution of satellites. Figure 5 depicts the periodic altitude variations of satellite orbit LAGEOS-1 for three days. For this study, two years of real satellite orbits from Sep. 2018 to Aug. 2020 are used. Due to GRACE and GRACE-FO orbit designs, specific configuration and polar gaps, the lower degrees of the Earth's gravity field can not be estimated with good accuracy. Therefore, the low degree monthly gravity field coefficients are taken from satellite laser ranging (SLR) observations such as LAGEOS-1 and LAGEOS-2 (Cheng et al. 2013) to be used as a-priori values for this study.

The gravitational potential differences observed by optical clocks along the satellite orbits LAGEOS-1, LAGEOS-2 and GRACE-FO are computed based on Eqs. (9) and (10). Table 2 shows the different white noise cases for the simulation scenarios of the optical clocks measurements.

4 Numerical Results

Temporal variations i.e. seasonal variations and secular trend of spherical harmonic coefficients up to degree and order 2 from 24 months noise-free SLR observations is shown in Fig. 6. Figure 7 depicts the temporal variations of spherical harmonic coefficients up to degree and order 2, i.e. $\Delta \bar{c}_{20}$, $\Delta \bar{c}_{21}$, $\Delta \bar{s}_{21}$, $\Delta \bar{c}_{22}$, $\Delta \bar{s}_{22}$, for 24 months which are estimated based on two years of optical clock observations along LAGEOS-1 and LAGEOS-2 orbits considering different clock uncertainties and different averaging times of 10, 2 and 60 min.

An averaging time of 60 min is needed to achieve frequency uncertainties of 1×10^{-19}. With these measurements, the spherical harmonic coefficients $\Delta \bar{c}_{20}$, $\Delta \bar{s}_{22}$ can accurately be estimated and are comparable to SLR-derived monthly gravity field solutions. However, for the averaging times of 2, 10 and 60 min with fractional frequency uncertainties of 4.52×10^{-18}, 4.08×10^{-18} and 1×10^{-18} which correspond to gravitational potential differences of 0.452, 0.408 and 0.100 m^2/s^2, the temporal variations of $\Delta \bar{c}_{21}$, $\Delta \bar{s}_{21}$, $\Delta \bar{c}_{22}$ can not as precisely be estimated as with SLR.

The temporal variations of spherical harmonic coefficients up to degree and order 2 from 24 months optical clock observations onboard GRACE-FO is shown in Fig. 8. Again, the monthly solutions are estimated with least-squares adjustment for different frequency uncertainties and different averaging times of 10, 2 and 60 min. The averaging times of

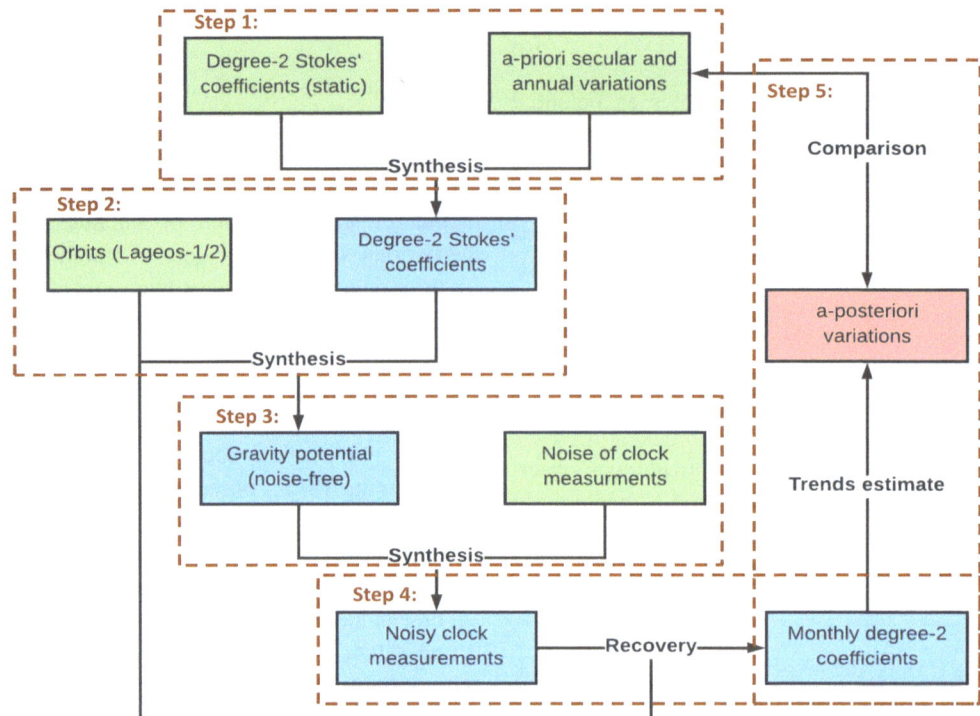

Fig. 4 Flowchart of the simulation of optical clock observations as potential differences and recovery of the Earth gravity field

Table 1 Orbital information of geodetic satellite missions as used for the simulation of optical clock measurements

Satellite	Simulation period	Alt. [km]	Inc. [deg.]	Rev. [min.]
LAGEOS-1	2 years [2018–2020]	5860	109.84	225
LAGEOS-2	2 years [2018–2020]	5620	52.64	223
GRACE-FO	2 years [2018–2020]	490	89.0	94.5

2 and 10 min are selected to demonstrate the performance of optical clocks onboard low earth orbiters for precise determination of temporal long-wavelength variations of the Earth's gravity field. With an averaging time of 60 min to achieve the frequency uncertainties of 1×10^{-19} along the satellite orbit GRACE-FO, the spherical harmonic coefficients $\Delta \bar{c}_{20}$, $\Delta \bar{s}_{21}$, $\Delta \bar{s}_{22}$ can be accurately estimated and are comparable to SLR-derived monthly gravity field solutions. But the temporal variations of $\Delta \bar{c}_{21}$, $\Delta \bar{c}_{22}$ are not obtained accurate enough. The same holds for the other GRACE cases, where a poorer clock performance has been assumed.

Figure 9 shows the temporal variations of spherical harmonic coefficients up to degree and order 2 from 24 months of optical clock observations along the orbits of LAGEOS-1, LAGEOS-2 and GRACE-FO. The monthly solutions are estimated for the same cases as before. For the averaging time of 60 min enabling frequency uncertainties of 1×10^{-19}, the spherical harmonic coefficients $\Delta \bar{c}_{20}$, $\Delta \bar{s}_{21}$, $\Delta \bar{s}_{22}$ are accurately obtained and comparable

to SLR-derived monthly gravity field solutions. But the temporal variations of $\Delta \bar{c}_{21}$, $\Delta \bar{c}_{22}$ are less accurately obtained than the SLR monthly gravity field solutions. For taveraging times of 2, 10 and 60 min with fractional frequency uncertainties of 4.52×10^{-18}, 4.08×10^{-18} and 1×10^{-18}, the temporal variations of $\Delta \bar{c}_{22}$ is improved relative to the GRACE-FO case, but $\Delta \bar{c}_{21}$ is still worse.

5   Conclusions

Changes of the low-degree spherical harmonic coefficients, such as the zonal term $\bar{c}_{20}$, reflect significant mass variations in the Earth system. Nowadays, SLR observations, e.g., from LAGEOS-1 and LAGEOS-2 are routinely used for the estimation of temporal variations of lower degree/order spherical harmonic coefficients. Moreover, as the low-degree zonal coefficients of the Earth's gravity field are poorly recovered with GRACE and GRACE-FO satellite missions, their temporal variations are taken from SLR observations to supplement the GRACE and GRACE-FO estimates. In future, also optical lattice clocks onboard of low earth orbiters have the potential to determine temporal variations of those low-degree gravity field coefficients with good accuracy. Different configurations of satellite orbits such as GRACE-FO, LAGEOS-1 and LAGEOS-2 between 400 and 6000 km

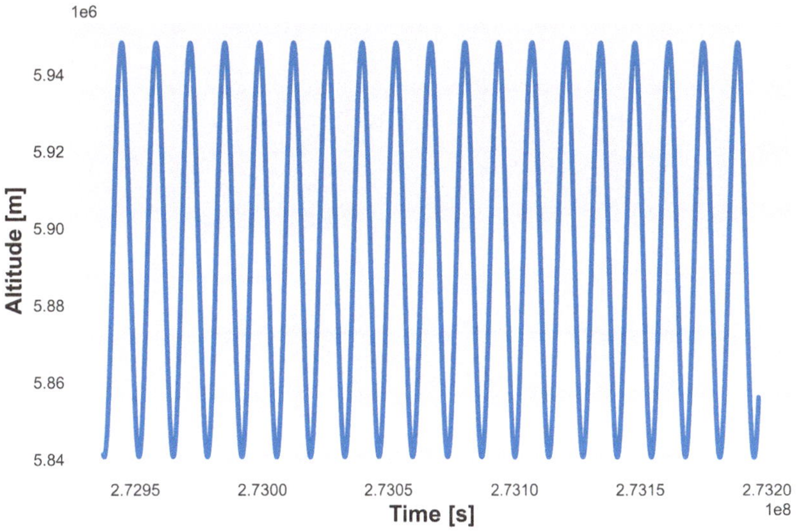

Fig. 5 Altitude variations of satellite orbit LAGEOS-1 for three days

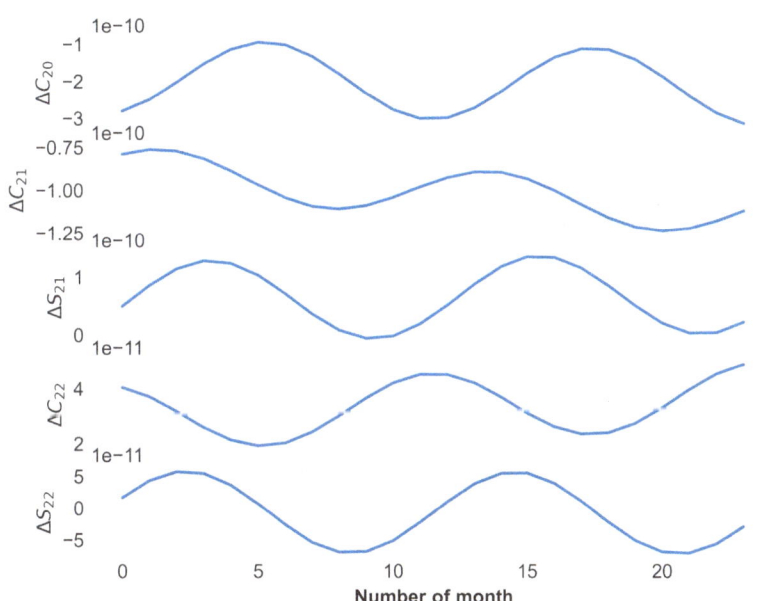

Fig. 6 Temporal variations of spherical harmonic coefficients up to degree and order 2 from 24 months noise-free SLR observations

Table 2 White noise cases with different average time, different frequency uncertainties and corresponding potential differences used in this simulation study

Case	Frequency uncertainties [—]	Potential differences [m²/s²]	Average time [min]
1	4.08×10^{-18}	0.408	10
2	4.52×10^{-18}	0.452	2
3	1.00×10^{-18}	0.100	60
4	1.00×10^{-19}	0.01	60

with certain assumptions on the optical clock errors have been studied to quantify this application. Optical clocks with instabilities of 1.0×10^{-19} in 60 min can reach the SLR accuracy in the future.

Assuming some progress in the development of optical atomic clocks in the future, the precise determination of temporal long-wavelength variations of the Earth's gravity field from space is possible.

Acknowledgements The authors would like to acknowledge the Deutsche Forschungsgemeinschaft (DFG, German Research Foundation) under Germany's Excellence Strategy EXC 2123 QuantumFrontiers, Project-ID 390837967 and and the Collaborative Research Centre CRC-1464 "TerraQ - Relativistic and Quantum-based Geodesy", Project-ID 434617780.

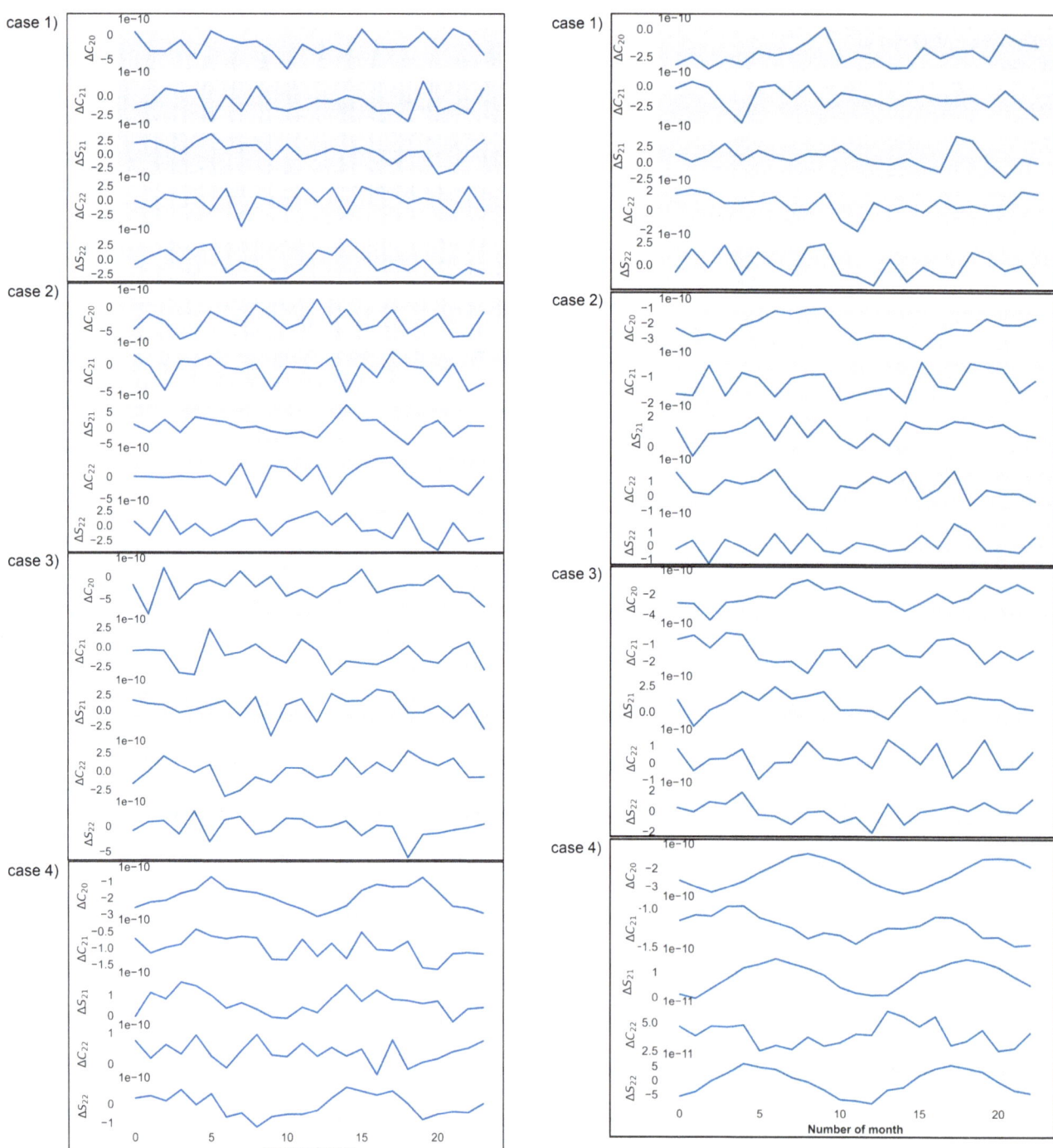

Fig. 7 Estimated spherical harmonic coefficients of degree/order 2 from optical clock measurements along LAGEOS-1 and LAGEOS-2 orbits, from top to bottom corresponding to cases 1–4

Fig. 8 Estimated spherical harmonic coefficients of degree/order 2 from optical clock measurements along GRACE-FO satellite orbits, from top to bottom corresponding to cases 1–4

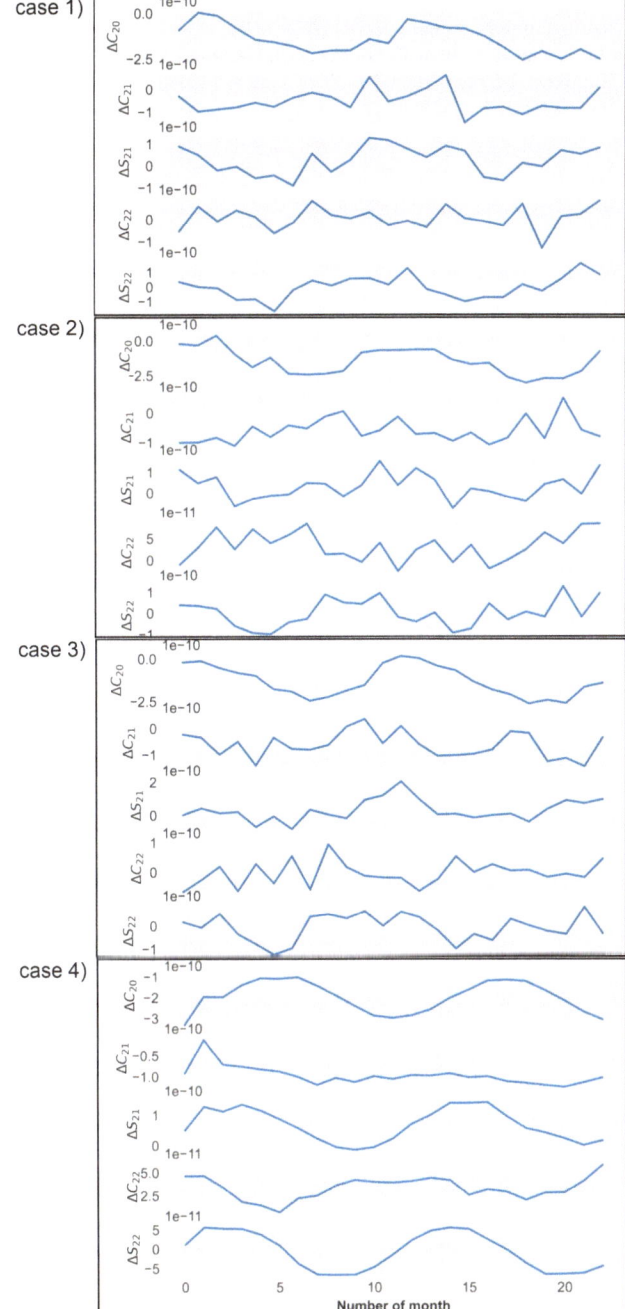

Fig. 9 Estimated spherical harmonic coefficients of degree/order 2 from optical clock measurements along LAGEOS-1, LAGEOS-2 and GRACE-FO orbits, from top to bottom corresponding to cases 1–4

References

Alonso I, Alpigiani C, Altschul B, Araújo H, Arduini G, Arlt J, Badurina L, Balaz A, Bandarupally S, Barish B, Barone M, Barsanti M, Bass S, Bassi A, Battelier B, Baynham C, Beaufils Q, Belic A, Bergé J, Zupaniè E (2022) Cold atoms in space: Community workshop summary and proposed road-map. EPJ Quant Technol 9(10):40507–022

Bjerhammar A (1985) On a relativistic geodesy, bulletin. Géodésique 59:207–220

Cheng MK, Tapley BD, Ries JC (2013) Deceleration in the Earth's oblateness. J Geophys Res 118:1–8. https://doi.org/10.1002/jgrb.50058

Delva P, Denker H, Lion G (2019) Chronometric geodesy: methods and applications. Springer International Publishing, Cham, pp 25–85

Denker H, Timmen L, Voigt C, Weyers S, Peik E, Margolis HS, Delva P, Wolf P, Petit G (2018) Geodetic methods to determine the relativistic redshift at the level of 10^{-18}. J Geodesy 92(5):487–516

Dix-Matthews, BP, Schediwy SW, Gozzard DR, Savalle E, Esnault F-X, Lévèque T, Gravestock C, D'Mello D, Karpathakis S, Tobar M, et al (2021) Point-to-point stabilized optical frequency transfer with active optics. Nat Commun 12(1):1–8

Godun R (2021) Atomic clocks compared with astounding accuracy. Nature 591(534–535):738–0

Grotti J, Koller S, Vogt S, Häfner S, Sterr U, Lisdat C, Denker H, Voigt C, Timmen L, Rolland A, Baynes FN, Margolis HS, Zampaolo M, Thoumany P, Pizzocaro M, Rauf B, Bregolin F, Tampellini A, Barbieri P, Zucco M, Costanzo GA, Clivati C, Levi F, Calonico D (2018) Geodesy and metrology with a transportable optical clock. Nat Phys 14:437–441

Guéna J, Weyers S, Abgrall M, Grebing C, Gerginov V, Rosenbusch P, Bize S, Lipphardt B, Denker H, Quintin N, Raupach SMF, Nicolodi D, Stefani F, Chiodo N, Koke S, Kuhl A, Wiotte F, Meynadier F, Camisard E, Chardonnet C, Le Coq Y, Lours M, Santarelli G, Amy-Klein A, Le Targat R, Lopez O, Pottie PE, Grosche G (2017) First international comparison of fountain primary frequency standards via a long distance optical fiber link. Metrologia 54:348. https://doi.org/10.1088/1681-7575/aa65fe

Lisdat C, Grosche G, Quintin N, et al (2016) A clock network for geodesy and fundamental science. Nat Commun 7:12443. https://doi.org/10.1038/ncomms12443

Mai E (2013) Time, atomic clocks, and relativistic geodesy. München, Deutsche Geodätische Kommis-sion, Reihe A: Theoretische Geodäsie, Munich. ISBN: 978-3-7696-8204-5. https://publikationen.badw.de/en/023386397

Mai E, Müller J (2013) General remarks on the potential use of atomic clocks in relativistic geodesy. ZFV 138(4):257–266

Mehlstäubler T, Grosche G, Lisdat C, Schmidt P, Denker H (2018) Atomic clocks for geodesy. Rep Prog Phys 064401:81. https://doi.org/10.1088/1361-6633/aab409

Müller J, Soffel M, Klioner SA (2008) Geodesy and relativity. J Geodesy 82:133–145

Müller J, Dirkx D, Kopeikin SM, Lion G, Panet I, Petit G, Visser PNAM (2018) High performance clocks and gravity field determination. ISSI Book on High Performance Clocks, Space Science Reviews 214:5. https://doi.org/10.1007/s11214-017-0431-z

Philipp D (2018) Theoretical aspects of relativistic geodesy. PhD thesis, Universität Bremen

Schwarz R, Dörscher S, Al-Masoudi A, Benkler E, Legero T, Sterr U, Weyers S, Rahm J, Lipphardt B, Lisdat C (2020) Long term measurement of the ^{87}Sr clock frequency at the limit of primary cs clocks. Phys Rev Res 2(3). https://do.org/10.1103/PhysRevResearch.2.033242

Soffel M, Langhans R (2013) Space-time reference systems. Springer, Berlin, Heidelberg. https://doi.org/10.1007/978-3-642-30226-8

Sören D, Klose J, Palli SM, Lisdat C (2022) Experimental determination of the E2-M1 polarizability of the strontium clock transition. Phys Rev Res 5(1):Article id.L012013. https://doi.org/10.48550/ARXIV.2210.14727

Takamoto M, Ushijima I, Ohmae N, Yahagi T, Kokado K, Shinkai H, Katori H (2020) Test of general relativity by a pair of transportable optical lattice clocks. Nat Phot 14:411–415

Voigt C, Denker H, Timmen L (2016) Time-variable gravity potential components for optical clock comparisons and the definition of international time scales. Metrologia 53(6):1365–1383

Wu H, Müller J (2020) Towards an international height reference frame using clock networks. In: International Association of Geodesy Symposia, vol 2020(97), 2020. https://doi.org/10.1007/1345

Wu H, Müller J (2021) Clock networks and their sensibility to time-variable gravity signals, pp 21–1074. EGU

Xu D, Lee W-K, Stefani F, Lopez,O, Amy-Klein A, Pottie P-E (2018) Studying the fundamental limit of optical fiber links to the 10^{-19} level. Opt 26(8):9515–9527

Hybrid Geoid Modeling for the Kingdom of Saudi Arabia

Rossen S. Grebenitcharsky, Georgios S. Vergos, Sultan Al-Shahrani, Abdullah Al-Qahtani, Golubinka Iuri, Alrubayyi Othman, and Suliman Aljebreen

Abstract

A significant improvement in the accuracy and homogeneity has been achieved with the new gravimetric geoid model for the Kingdom of Saudi Arabia (KSA-Geoid21GRAV) w.r.t the previous Geoid models KSA2009, KSA2015 and KSA-Geoid17. The gravimetric geoid prediction was carried with the remove-compute-restore technique resulting in external absolute accuracies at the 10–11 cm level and relative accuracies at the 1–5 ppm. In this work, the estimation of the hybrid KSA-Geoid21 model is described. A hybrid deterministic and stochastic approach is used to model the residuals of the gravimetric model relative to available GNSS/Levelling geoid heights. Various parametric models ranging from simple north-south bias and tilt one to second and third degree polynomial models have been evaluated. After various tests a second order polynomial model was selected resulting in a 10.3 cm absolute difference of the adjusted residuals between the gravimetric KSA-Geoid21 geoid model and the GNSS/Levelling geoid heights. Following that, a stochastic modelling of the residuals after the fit has been carried out, resulting in errors relative to the GNSS/Levelling data at the 0.014 m level. Compared to the previous geoid model, KSA-Geoid2017, improved residuals to 75.2% of the benchmarks is found with a mean improvement at the 1.1 cm, while for the rest 24.8% a mean deterioration of 0.7 cm is found.

Keywords

GNSS/levelling validation · Hybrid geoid · Kingdom of Saudi Arabia · Parametric models · Stochastic modeling

1 Introduction

The new high-accuracy and resolution gravimetric geoid model for KSA, KSA-GEOID21GRAV (Vergos et al. 2023), is a gravimetric geoid model evaluated employing the Remove Compute Restore (RCR) approach (Barzaghi et al. 2018), XGM2019e (Zingerle et al. 2019) as a reference field and residual terrain model topographiccorrections from a global model (Rexer et al. 2018). It is based on more than 808,000 land and marine gravity data from two mainly sources (Arab American Petroleum Company – ARAMCO and General Authority for Surveying & Geospatial Information – GEOSA former GASGI), satellite altimetry data from the DTU18 model (Andersen and Knudsen 2019), and a new dataset of airborne gravity data covering almost the 68% of the KSA territory. All these data were pre-processed, evaluated and validated in terms of consistency with the IGSN71 as gravity reference system, KSA-GRF17 as geodetic reference frame, KSA-VRF14 as vertical reference frame and refer to the tide free system. A homogeneous database containing a total number of 2,010,766 land, airborne and shipborne

R. S. Grebenitcharsky · G. S. Vergos (✉) · S. Al-Shahrani · A. Al-Qahtani · G. Iuri · A. Othman · S. Aljebreen
General Authority for Surveying & Geospatial Information, Riyadh, Kingdom of Saudi Arabia
e-mail: vergos@topo.auth.gr

© The Author(s) 2023
J. T. Freymueller, L. Sánchez (eds.), *Gravity, Positioning and Reference Frames*,
International Association of Geodesy Symposia 156, https://doi.org/10.1007/1345_2023_215

gravity data has been used for the determination of the gravimetric geoid employing a classical FFT-based solution to evaluate Stokes' kernel function and a Wong-Gore modification (Wong and Gore 1969; Sideris 2013). Its overall agreement with GNSS/Levelling data from GEOSA reached a standard deviation of 13.6 cm level and relative accuracies at the 1–5 ppm over distances ranging from 10 to 2,000 km. In the frame of the determination of the final Hybrid geoid model for KSA, a hybrid deterministic and stochastic approach is followed employing 3,522 GAGSI GNSS/Leveling benchmarks (BMs), so as to provide a geoid model (KSA-GEOID21GASGI) appropriate for surveying and engineering applications as best fit to the BMs.

2   KSA-Geoid21 Hybrid Geoid Modeling

Following the initial validation of the final gravimetric geoid model, the next stage was to use a deterministic parametric model to reduce and remove biases and trends in the gravimetric model, relative to the GNSS/Levelling geoid heights at selected BMs. The determination of the hybrid geoid is in essence a geometric fit of the gravimetric geoid to the available GNSS/Levelling data, hence a geoid solution that best fits the latter and provides small residuals. It should be mentioned that KSA uses a new geopotential-based Vertical Reference Frame which is called Jeddah2014 and is tied to epoch 2014.75. In this work, the determination of the hybrid geoid model for the KSA is based on the high-quality GPS/Levelling data from GEOSA (former GASGI) (3,522 BMs). For the deterministic part of the fit, simple north-south and east-west bias and tilt models have been tested, as well as the classical 4- and 5-parameters transformation models (Tziavos et al. 2012; Vergos et al. 2014). Nevertheless, despite the fact that their estimated parameters practically have no physical meaning, the selected parametric models refer to second and third order polynomial ones, as the goal was to minimize the residuals to the GNSS/Leveling data as much as possible and let the stochastic part of the transformation model treat the remaining, unbiased, residuals. The observation equation of the differences between the gravimetric and GNSS/Levelling geoid height in this parametric LSC is given as (Moritz 1980):

$$\ell_i = \left(h_i^{GPS} - H_i\right) - N_i^{grav} = N_i^{GPS} - N_i^{grav}, \quad (1)$$

where ℓ_i denotes the observation, h_i^{GPS} the ellipsoidal height, H_i the orthometric height, N_i^{grav} the gravimetric geoid height and N_i^{GPS} the so-called GNSS geoid height. In matrix notation it becomes

$$b = Ax + s + v, \quad (2)$$

where, A is the design matric, x is the matrix of the unknowns, s denotes stochastic signal, v denotes the errors of the observations b. With Eq. (2) we can easily treat first the deterministic part to first absorb any systematic differences between the various types of heights and then estimate the stochastic residual signal with least-squares collocation. The unknown deterministic parameters of the transformation model are determined as:

$$\hat{x} = \left(A^T P A\right)^{-1} A^T P b, \quad (3)$$

where $\hat{x}$ denotes the adjusted unknowns and P is the weight matrix. In the next step, after the removal of the deterministic part, an appropriate covariance function is estimated and employing LSC the stochastic signal is estimated and the hybrid geoid heights are computed from the gravimetric geoid heights as a combination of stochastic and deterministic modeling:

$$N_{Hybrid} = N^{grav} + a_i^T \hat{x} + \hat{s}. \quad (4)$$

The deterministic part $a_i^T x$ depends on the chosen parametric model and in the case of the second order polynomial model becomes (Kotsakis and Katsambalos 2010)

$$\begin{aligned} a_i^T x = \ & x_0 + x_1(\varphi_i - \varphi_0)^0(\lambda_i - \lambda_0)^1 cos^1\varphi_i \\ &+ x_2(\varphi_i - \varphi_0)^1(\lambda_i - \lambda_0)^0 cos^0\varphi_i \\ &+ x_3(\varphi_i - \varphi_0)^1(\lambda_i - \lambda_0)^1 cos^1\varphi_i \\ &+ x_4(\varphi_i - \varphi_0)^2 \\ &+ x_5(\lambda_i - \lambda_0)^2 \end{aligned} \quad (5)$$

2.1   KSA-Geoid21GASGI Hybrid Geoid Determination

As already mentioned, the hybrid geoid is based on the KSA-Geoid21GRAV and a set of 3,522 GNSS/Levelling BMs over KSA. Before the practical determination of the transformation model a 3σ test has been performed to remove possible blunders in the BM database. During the 3σ test, 23 points have been removed so that after the 3σ test the std. of the differences between the gravimetric geoid and the GNSS/Levelling BMs reduced to 13.3 cm and the mean to −10.0 cm (see Table 1). In the practical evaluation of the various parametric models tested, their fit has been evaluated

Table 1 Statistics of geoid height differences between the gravimetric geoid model and GNSS/levelling BMs. Units [m]

Number of points	Max	Min	Mean	RMS	Std
Original 3,522 pts.	0.465	−0.563	−0.102	0.170	0.136
After 3σ 3,499 pts.	0.418	−0.421	−0.100	0.166	0.133
After 2nd order pol. ft	0.508	−0.447	0.000	0.097	0.097

Table 2 Relative differences with baseline length for the geoid model after the parametric fit. Units: [ppm]

	0–10	10–20	20–30	30–40	40–50	50–60	60–70	70–80	80–90	>100
Relative differences	7.542	1.973	1.480	1.237	1.118	1.043	0.967	0.923	0.849	0.182

in terms of the std. after the fit, the system condition number and coefficient of determination. For the simple NS-tilt and WE-tilt models the std. is at 10.8 cm and 13.0 cm, respectively, for the 4- and 5-parameter Helmert transformation models it reaches the 10.5 cm and 10.4 cm, and for the second and third order polynomial the 9.7 cm and 8.6 cm. As the goal was to model with the deterministic part the residuals and provide a smooth signal for the prediction with LSC, it was decided to use the second order polynomial model to treat the deterministic part. It provided both a reasonable reduction of the std. (from 13.3 to 9.7 cm), an adjusted coefficient of determination at the 0.467 level and a condition number of the system of normal equations at the 109.196. Note that the third order polynomial model may provide a smaller std. but the condition number was at the 1.4×10^6 which shows that the model results in over-parametrization, hence it was deemed as not appropriate.

To validate the adjusted residuals after the second order polynomial fit, the absolute and relative differences between the gravimetric geoid heights and GPS/Levelling geoid heights have been computed. 98.9% of the differences are lower than the $2cm\sqrt{dist(km)}$ error, 92.3% are lower

than the $1cm\sqrt{dist(km)}$ error and 71.4% are lower than the $0.5cm\sqrt{dist(km)}$ error. These statistics show the significant improvement in the GRAV-Geoid21 with most of the baseline differences (92%) being below the 1 cm error, showing that there are only a few exceptions with errors larger than 1 cm Kingdom-wide. Table 2 summarizes the relative differences as a function of baseline length for the adjusted gravimetric geoid, where relative accuracies smaller than 1.9 ppm are found for distances larger than 10–20 km and for shorter baselines the relative accuracy is at the 7.5 ppm level.

The next step for the determination of the hybrid KSA-Geoid21GASGI model was the stochastic treatment of the adjusted, with the deterministic second order polynomial model, residuals of the gravimetric geoid model. An empirical covariance function of the stochastic signal to be modelled (see Fig. 1) was estimated and to that a Gauss-Markov analytical covariance function models have been fitted, so that the auto- and cross-covariance matrices needed for the prediction of the stochastic signal of the hybrid geoid model using LSC will be carried out. Following Eq. (4) the hybrid geoid is determined combining the estimated determinis-

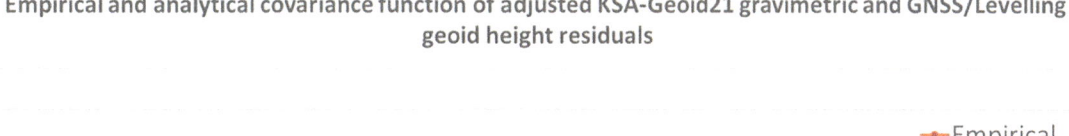

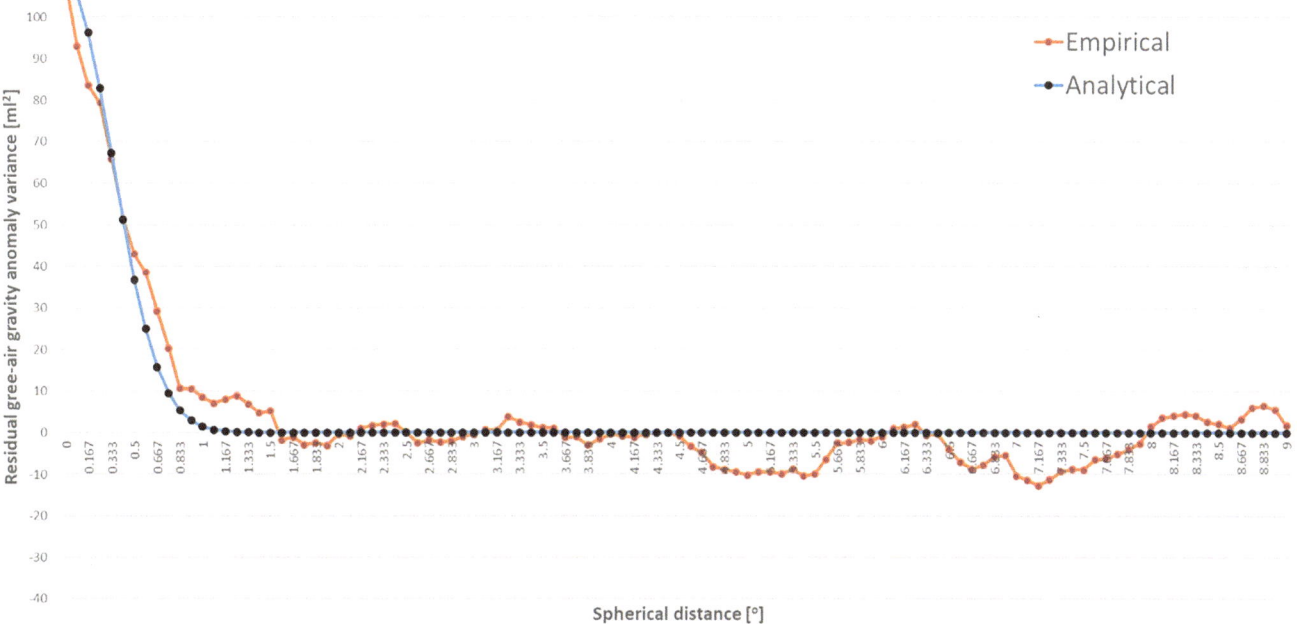

Fig. 1 Empirical and analytical covariance functions of the adjusted residuals between KSA-Geoid21 gravimetric geoid and GEOSA GNSS/levelling geoid heights

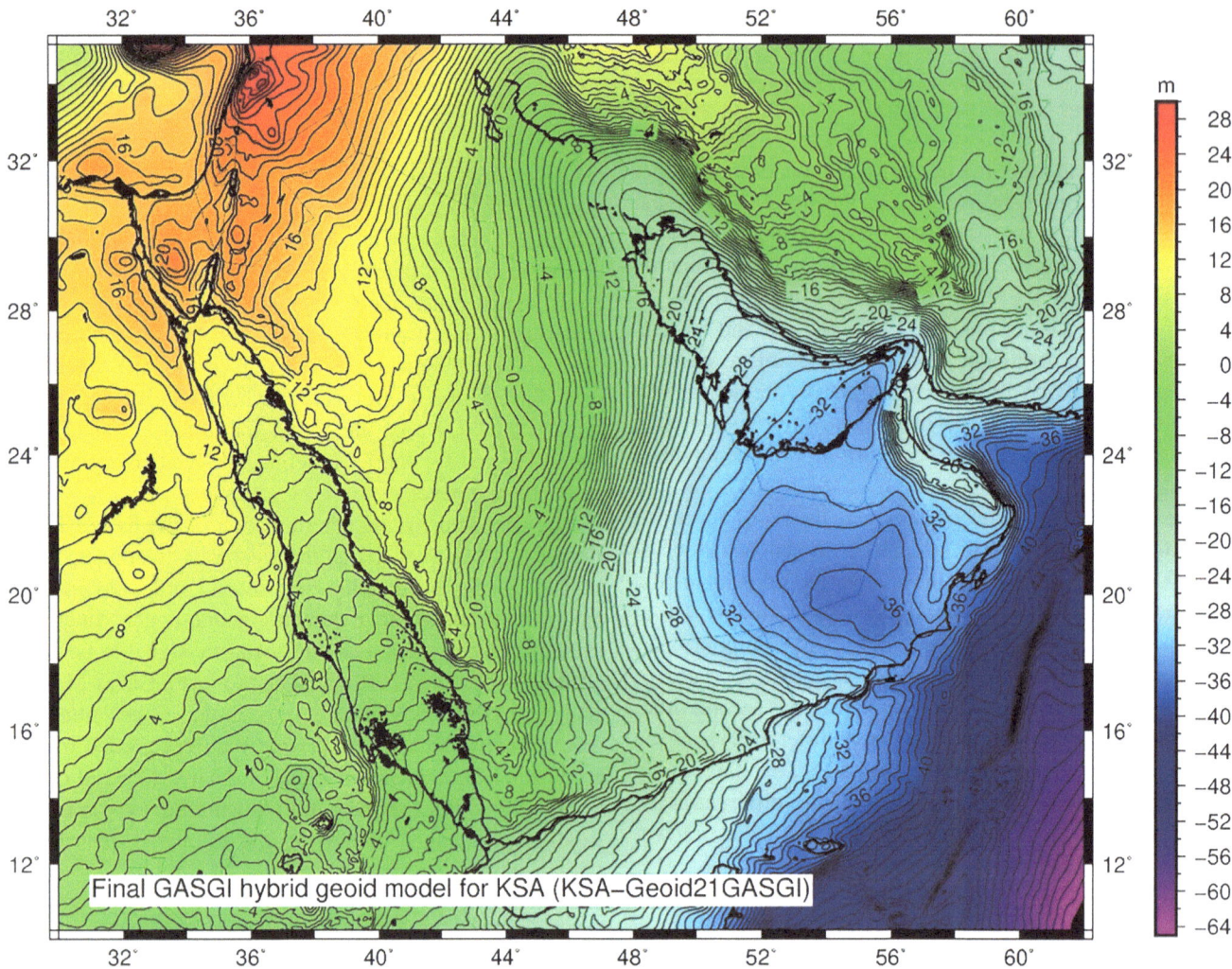

Fig. 2 The KSA-Geoid21GASGI hybrid geoid model

tic and stochastic modeled signals. Figure 2 presents the hybrid KSA-Geoid21GASGI model, which is to be used in accordance with the KSA VRF and GRF, while Fig. 3 depicts the hybrid geoid standard error. It provides a standard error of 0.199 cm (see Table 3) while its fit to the GASGI GPS/Levelling BMs has a zero mean a std. of 0.02 m.

To evaluate the possible improvement of the new KSA-Geoid21GASGI hybrid geoid model over the previous model KSA-GEOID17, an extended set of 17,528 GNSS levelling dataset comprising of observations over BMs from GEOSA (former GASGI), ARAMCO and MOMRA (Ministry of Municipalities and Rural Affairs) has been used. Over these BMs we have evaluated the level of Improvement/Deterioration of the new geoid model compared to KSA-GEOID17 based on the absolute values of the residuals to the GNSS/Levelling geoid heights. Figure 4 summarizes the results of this analysis where it can be seen that for 75.2% of the BMs there is an improvement, with a mean value at the

1.1 cm level, while for the rest 24.8% there is a deterioration of the difference, with a mean at the 0.7 cm level. The main improvement is found over the south-eastern part of the Kingdom where the ARAMCO BMs are situated, reaching 83.8% of the BMs. For the MOMRA BMs improvement is found for 74.8% of the BMs, while for the GEOSA BMs there is a mean improvement of 0.4 cm for 43% of the BMs and a mean deterioration of 0.6 mm for 57% of the BMs. The reason that for the GEOSA (former GASGI) BMs the improvement is not a significant as for the other two datasets is that these BMs have been used in the development of the KSA-GEOID17 model, which is a hybrid one as well, so it is expected to fit well.

A final evaluation test for the new hybrid geoid model was performed by acquiring new real time kinematic (RTK) data, both in network RTK and single-base modes depending on the network coverage, has been conducted. A total number of 149 BMs have been surveyed with the new hybrid geoid

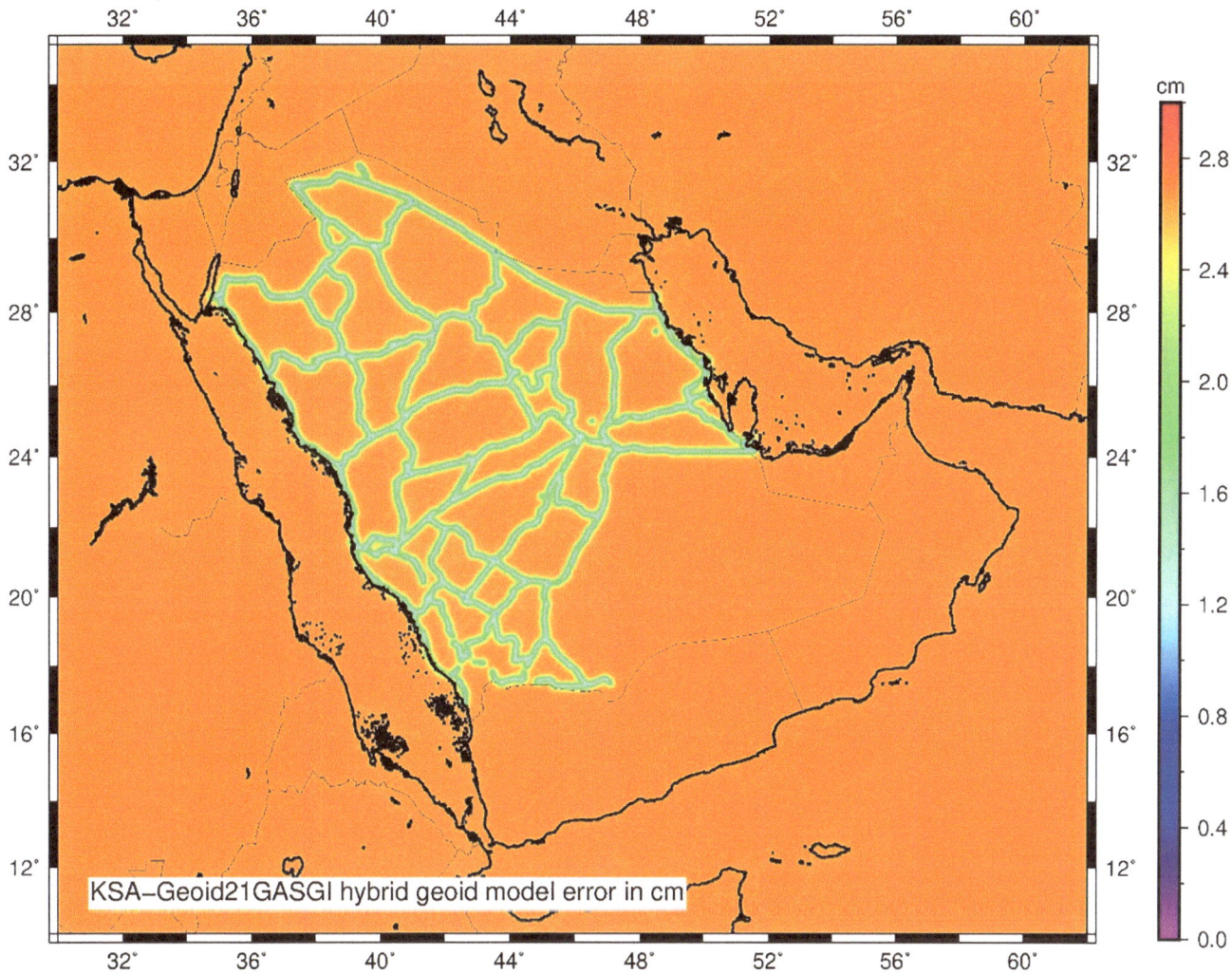

Fig. 3 The associated KSA-Geoid21GASGI hybrid geoid model standard error

Table 3 Statistics of the final KSA-Geoid21GASGI hybrid geoid models, its errors and differences to GASGI GPS/levelling and GRAV-Geoid21

	Max	Min	Mean	RMS	Std
$N^{\text{KSA-Geoid21GASGI}}$ [m]	29.253	−66.363	−10.415	22.258	19.670
$N^{\text{KSA-Geoid21GASGI}}$ error [cm]	2.660	0.670	2.615	2.623	0.199
$N^{\text{KSA-Geoid21GASGI}} - N^{\text{GRAV-Geoid21}}$ [m]	0.982	−0.987	0.077	0.296	0.286
$N^{\text{KSA-Geoid21GASGI}} - N^{\text{GNSS/Lev GEOSA}}$ [m]	0.194	−0.334	0.000	0.020	0.020

models providing residuals with a mean value of −2.3 cm and std. of 7.4 cm and the KSA-GEOID17 having a mean of −2.0 cm and a std. of 8.4 cm. The largest residuals are found, as expected, outside the coverage of the KSA positioning service, where network corrections in the form of a virtual reference station are not available and single-base RTK solutions are provided. Given that these results are achieved in RTK mode, hence the errors in ellipsoidal height determination are higher, the uniform quality of the hybrid KSA-Geoid21GASGI is confirmed.

3   Conclusions

In this work the estimation of the hybrid geoid model KSA-Geoid21GASGI is described. Based on the high-accuracy and resolution gravimetric geoid model for KSA, KSA-GEOID21, with external absolute accuracies at the 13.6 cm level, the hybrid KSA-Geoid21 model was estimated. This was based on a deterministic second-order polynomial parametric model to reduce and remove biases and trends in the gravimetric model relative to the GNSS/Levelling geoid heights followed by the estimation of the residual stochastic

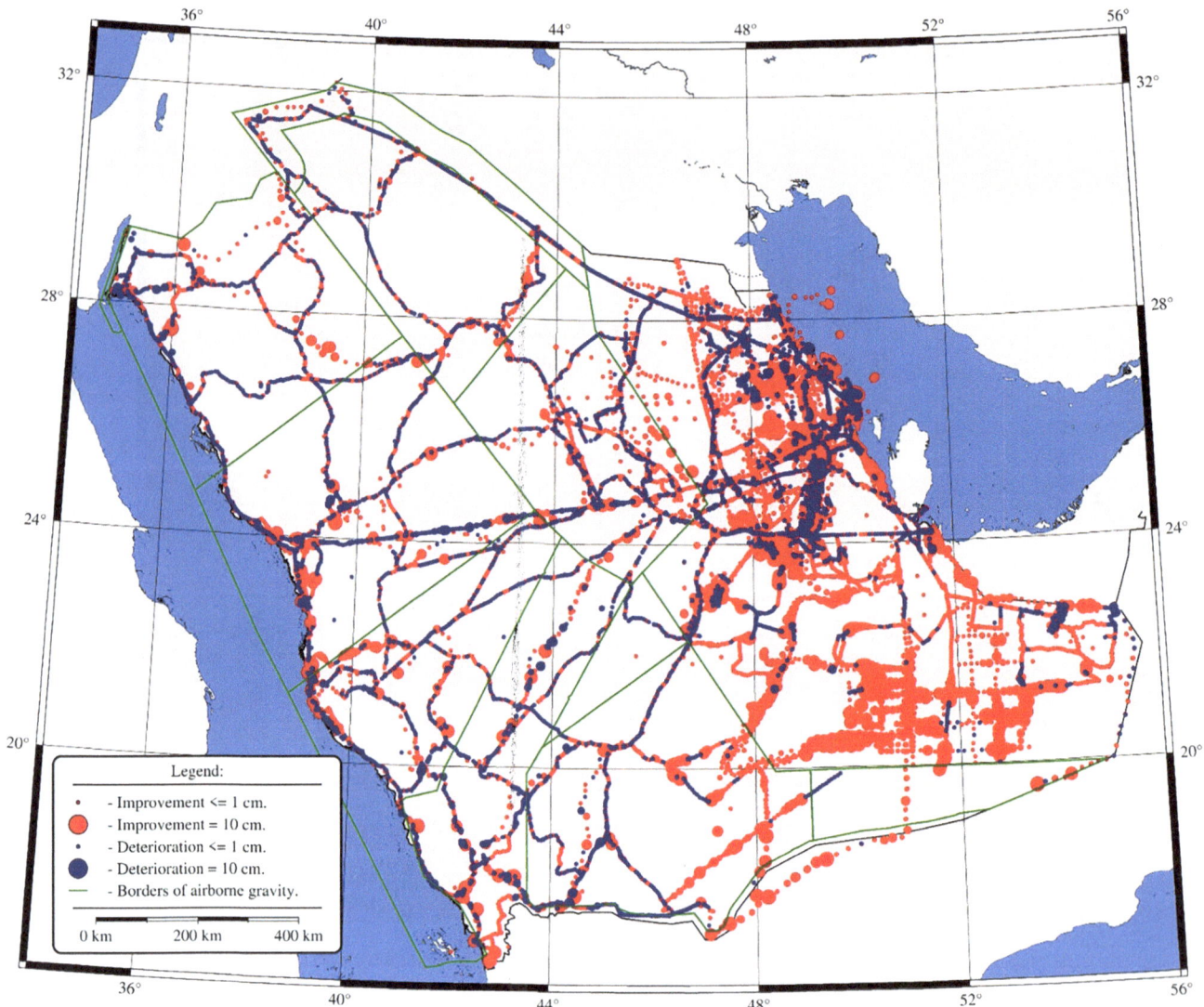

Fig. 4 Improvement/deterioration of the new KSA-Geoid21 GASGI hybrid geoid model compared to KSA-GEOID17 over the extended set of GNSS/levelling BMs

part with LSC. The hybrid geoid model reaches a standard error of 2.0 cm and relative accuracies of 1.9 ppm for distances larger than 10–20 km. Compared to the previous hybrid geoid model, KSA-GEOID17, it provides a mean improvement of 1.1 cm for 75.2% of the BMs and a mean deterioration of 0.7 cm for 24.8% of them. Finally, even in RTK mode, the hybrid geoid model gives a std. of 7.4 cm which is 1 cm better than that of the previous model.

References

Andersen OB, Knudsen P (2019) The DTU17 global marine gravity field: first validation results. In: Mertikas S, Pail R (eds) Fiducial reference measurements for altimetry, International Association of Geodesy Symposia, vol 150. Springer, Cham. https://doi.org/10.1007/1345_2019_65

Barzaghi R et al (2018) GEOMED2: high-resolution geoid of the Mediterranean. In: Freymueller J, Sánchez L (eds) International symposium on advancing geodesy in a changing world, International Association of Geodesy Symposia, vol 149. Springer, Cham. https://doi.org/10.1007/1345_2018_33

Kotsakis C, Katsambalos K (2010) Quality analysis of global geopotential models at 1542 GPS/levelling benchmarks over the Hellenic mainland. Surv Rev 42(318):327–344. https://doi.org/10.1179/003962610X12747001420500

Moritz H (1980) Advanced physical geodesy. Wichmann, Karlsruhe

Rexer M, Hirt C, Bucha B, Holmes S (2018) Solution to the spectral filter problem of residual terrain modelling (RTM). J Geod 92(6):675–690. https://doi.org/10.1007/S00190-017-1086-Y

Sideris MG (2013) Geoid determination by FFT techniques. In: Sansò F, Sideris M (eds) Geoid determination, Lecture notes in earth system sciences, vol 110. Springer, Heidelberg. https://doi.org/10.1007/978-3-540-74700-0_10

Tziavos IN, Vergos GS, Grigoriadis VN, Andritsanos VD (2012) Adjustment of collocated GPS, geoid and orthometric height observations in Greece. Geoid or orthometric height improvement? In:

Kenyon S, Pacino C, Marti U (eds) Geodesy for planet earth, International Association of Geodesy Symposia, vol 136. Springer, Heidelberg, pp 481–488

Vergos GS, Grigoriadis VN, Tziavos IN, Kotsakis C (2014) Evaluation of GOCE/GRACE global geopotential models over Greece with collocated GPS/levelling observations and local gravity data. International Association of Geodesy Symposia, vol 141, pp 85–92. https://doi.org/10.1007/978-3-319-10837-7_11

Vergos GS et al (2023) Development of the national gravimetric geoid model for the Kingdom of Saudi Arabia. In: Gravity geoid and height systems 2022, International Association of Geodesy Symposia (accepted for publication)

Wong L, Gore R (1969) Accuracy of geoid heights from modified Stokes kernels. Geophys J R Astron Soc 18:81–91. https://doi.org/10.1111/j.1365-246X.1969.tb00264.x

Zingerle P, Pail P, Gruber T, Oikonomidou X (2019) The experimental gravity field model XGM2019e. GFZ Data Serv. https://doi.org/10.5880/ICGEM.2019.007

Almost-Instantaneous PPP-RTK Without Atmospheric Corrections

Andreas Brack, Benjamin Männel, and Harald Schuh

Abstract

Ambiguity resolution enabled precise point positioning (PPP-RTK) can provide fast, potentially even instantaneous, centimeter-level positioning results, given that the phase ambiguities are correctly resolved. Without external ionospheric corrections, a time-to-first-fix the ambiguities of around 30 min is often reported for GPS-only solutions. In this contribution we investigate the capabilities of almost-instantaneous PPP-RTK without any a-priori ionospheric information. The key aspects are the mean square error-optimal best integer-equivariant estimator, a multi-GNSS solution using GPS, Galileo, BDS, and QZSS, and a proper weighting of the satellite clock and bias corrections with their inverse covariance matrix in order to obtain realistic observation models. Real data experiments with dual-frequency observations show that centimeter-level horizontal positioning errors are reached within one and two epochs in 87.6% and 99.7% of the cases, thereby demonstrating that almost-instantaneous PPP-RTK without atmospheric corrections is indeed possible with the current constellations.

Keywords

Best integer-equivariant estimation · Integer ambiguity resolution · Multi-GNSS · Precise point positioning (PPP) · Real-time kinematic (RTK)

1 Introduction

In this contribution we provide an analysis of the capabilities of *almost-instantaneous* ambiguity resolution enabled precise point positioning (PPP-RTK) using only *a few epochs* of GNSS observations. With the high precision of the carrier phase observations, centimeter-level positioning results are immediately obtained once the phase ambiguities are correctly resolved. A main obstacle for fast and reliable ambiguity resolution are the ionospheric delays in the user's GNSS observations. A time-to-first-fix the ambiguities of around 30 min is generally reported for GPS-only solutions (Geng et al. 2011; Zhang et al. 2019), which can to some extent be shortened when combining systems (Li and Zhang 2014; Geng and Shi 2017; Li et al. 2018). Faster solutions are also possible when external ionospheric corrections are provided (Teunissen et al. 2010; Banville et al. 2014), but these have to be at the level of at most a few centimeters for a clear gain in terms of the convergence time (Psychas et al. 2018). Such a precision is currently not possible with global ionospheric models but requires corrections from nearby reference stations, which limits the field of applications. We therefore focus on the case without a-priori ionospheric information.

A typical example for kinematic GPS L1/L2 PPP-RTK is shown in Fig. 1 using data recorded at the station PERT

A. Brack (✉) · B. Männel
GFZ German Research Centre for Geosciences, Potsdam, Germany
e-mail: brack@gfz-potsdam.de; benjamin.maennel@gfz-potsdam.de

H. Schuh
GFZ German Research Centre for Geosciences, Potsdam, Germany

Chair of Satellite Geodesy, Technische Universität Berlin, Berlin, Germany
e-mail: schuh@gfz-potsdam.de

J. T. Freymueller, L. Sánchez (eds.), *Gravity, Positioning and Reference Frames*,
International Association of Geodesy Symposia 156, https://doi.org/10.1007/1345_2023_196

73

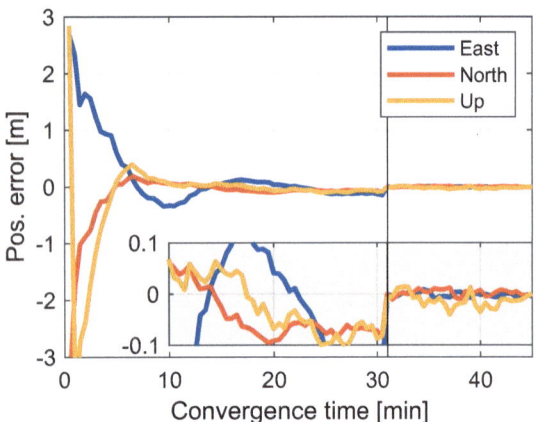

Fig. 1 GPS L1/L2 positioning errors of the station PERT for a kinematic PPP-RTK example during April 1, 2022. The time of ambiguity fixing is indicated by the black vertical line

in Perth, Australia, during April 1, 2022. While satellite clock and bias corrections are applied, no corrections for the atmospheric delays are used, so that tropospheric and ionospheric delays have to be estimated. The ambiguities are fixed once the failure rate drops to below 0.1%, indicated by the black vertical line after slightly more than 30 min. The ambiguity-float solution reaches the sub-meter level after several minutes, but the ambiguity-fixed solution is directly at the centimeter level.

The considered key aspects to obtain a similar performance within only a few epochs are (1) the mean square error (MSE)-optimal best integer-equivariant (BIE) estimator, introduced in Teunissen (2003), which does not 'fix' the ambiguities to integers but rather weights different candidates, (2) a multi-GNSS solution using GPS, Galileo, BDS, and QZSS, and (3) a proper weighting of the satellite clock and bias corrections in order to obtain realistic observation models. Simulations and real data analyses are used to demonstrate the impact of these three aspects. We show that centimeter-level horizontal positioning errors are reached within one and two epochs in 87.6% and 99.7% during an exemplary day.

2 Multi-GNSS PPP-RTK: Experimental Setup and Formal Analysis

The multi-GNSS PPP-RTK performance is analyzed using one day of simulated and real 30 s GPS (G) L1/L2, Galileo (E) E1/E5a, BDS (C) B1/B3, and QZSS (J) L1/L2 data in the area of Perth, Australia, during April 1, 2022.

The single-system undifferenced, uncombined GNSS code and carrier phase observations $p_{r,f}^s$ and $\varphi_{r,f}^s$ between

the user receiver r and satellite s on frequency f are modeled as

$$E[p_{r,f}^s] = \mathbf{g}_r^{s,\mathrm{T}} \Delta \mathbf{x}_r + dt_r - dt^s + m_r^s \tau_r + \mu_f i_r^s$$
$$+ d_{r,f} - d_{\cdot,f}^s$$
$$E[\varphi_{r,f}^s] = \mathbf{g}_r^{s,\mathrm{T}} \Delta \mathbf{x}_r + dt_r - dt^s + m_r^s \tau_r - \mu_f i_r^s$$
$$+ \lambda_f (\delta_{r,f} - \delta_{\cdot,f}^s + a_{r,f}^s), \quad (1)$$

with the expectation operator $E[\cdot]$, the satellite-to-receiver unit vector $\mathbf{g}_r^s$, the incremental user coordinates $\Delta \mathbf{x}_r$, the receiver and satellite clock offsets dt_r and dt^s, the residual zenith tropospheric delay τ_r with the mapping function m_r^s, the ionospheric slant delay i_r^s with the coefficients $\mu_f = \lambda_f^2/\lambda_1^2$ depending on the wavelengths λ_f, the frequency specific receiver and satellite code biases $d_{r,f}$ and $d_{\cdot,f}^s$, the respective phase biases $\delta_{r,f}$ and $\delta_{\cdot,f}^s$, and the carrier phase integer ambiguities $a_{r,f}^s$.

Most GNSS parameters as given in (1) cannot be determined in an absolute sense, but only as linear combinations with other parameters. The external satellite clock and phase bias corrections $d\tilde{t}^s$ and $\tilde{\delta}_{\cdot,f}^s$ are defined as

$$d\tilde{t}^s = dt^s + d_{\mathrm{IF}}^s - dt_1 - d_{1,\mathrm{IF}}$$
$$\tilde{\delta}_{\cdot,f}^s = \delta_{\cdot,f}^s - (d_{\mathrm{IF}}^s - \mu_f d_{\mathrm{GF}}^s - d_{1,\mathrm{IF}} + \mu_f d_{1,\mathrm{GF}})/\lambda_f$$
$$- \delta_{1,f} - a_{1,f}^s, \quad (2)$$

i.e., the satellite clock corrections also contain the clock offset of the reference receiver and ionosphere-free (IF) combinations of the code biases, and the satellite phase bias corrections contain IF and geometry-free (GF) combinations of the code biases as well as phase biases and ambiguities of the reference receiver. The corrections are either assumed deterministic or are computed by a single reference station on an epoch-by-epoch basis, for which the station NNOR (88.5 km distance to PERT) is used in the real data experiments.

The 'rover' station PERT is assumed kinematic with no constraints on the relative movement. After removing the PPP-RTK corrections (2) from (1), the estimable versions of its parameters are given by

$$d\tilde{t}_r = dt_{1r} + d_{1r,\mathrm{IF}}$$
$$\tilde{i}_r^s = i_r^s + d_{r,\mathrm{GF}} - d_{\mathrm{GF}}^s$$
$$\tilde{\delta}_{r,f} = \delta_{1r,f} - (d_{1r,\mathrm{IF}} - \mu_f d_{1r,\mathrm{GF}})/\lambda_f + a_{1r,f}^1$$
$$\tilde{a}_{r,f}^s = a_{1r,f}^s - a_{1r,f}^1, \qquad s \neq 1, \quad (3)$$

with $(\cdot)_{1r} = (\cdot)_r - (\cdot)_1$. The code biases $d_{r,f}$ and $d^s_{,f}$ are absorbed by the clock and ionosphere parameters via their IF and GF combinations, and $\Delta \mathbf{x}_r$ and τ_r are directly estimable. In a multi-GNSS solution, the receiver clock offset $d\tilde{t}_r$ and phase biases $\tilde{\delta}_{r,f}$ are estimated per constellation, and a separate pivot satellite is chosen for the ambiguity parameters $\tilde{a}^s_{r,f}$. We note that the residual tropospheric zenith delay τ_r, using the global mapping function (Boehm et al. 2006), as well as the ionospheric slant delays $\tilde{i}^s_r$ are estimated at the user receiver and are assumed unlinked in time, so that the results are valid for any ionospheric activity.

Figure 2 shows the average formal ambiguity-float positioning precision of the east component with the very weak single-epoch, single-station corrections (solid lines) and with the strongest possible, i.e., deterministic, corrections (dashed lines). Although the benefit of combining multiple systems is significant, we cannot expect centimeter-level results within a few epochs even in the four-system case with deterministic corrections. The ambiguity-fixed solutions, on the other hand, would already provide values of below 1 cm even in the GPS-only case after one epoch and with single-station corrections. The average times-to-first-fix presented in Table 1, however, show that even in the best case of a multi-GNSS solution with deterministic corrections, more than seven minutes are needed. The fixing criterion is an integer bootstrapping failure rate of 0.1% or lower (Teunissen 1998). Combining systems generally implies lower failure rates for ambiguity resolution and should lead to shorter fixing times. At the same time, rising satellites – which occur

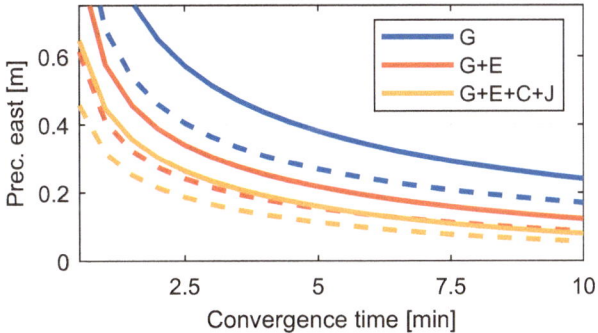

Fig. 2 Average formal ambiguity-float kinematic PPP-RTK positioning precision of the east component with single-epoch, single-station corrections (solid) and with deterministic corrections (dashed)

Table 1 Average time-to-first-fix in [min] for kinematic PPP-RTK with single-epoch, single-station corrections and with deterministic corrections. The fixing criterion is an integer bootstrapping failure rate of 0.1% or lower

	Single-stat.	Determ.
G	23.9	15.6
G+E	13.6	7.6
G+E+C+J	17.3	7.8

more often with more systems – cause additional parameters and extend the convergence time. In our analysis, the first aspect dominates when switching from GPS-only to the two-system case, whereas when switching from the two-system to the four-system case the second aspect has a larger impact, cf. Table 1.

3 PPP-RTK with Best Integer-Equivariant Estimation

An alternative for the ambiguity-float and ambiguity-fixed solutions is given by the BIE estimator (Teunissen 2003). Let $\hat{\mathbf{a}} \in \mathbb{R}^n$ and $\mathbf{Q}_{\hat{\mathbf{a}}} \in \mathbb{R}^{n \times n}$ be the float solution of the ambiguity vector $\mathbf{a} \in \mathbb{Z}^n$ and its covariance matrix. For normally distributed data, the BIE ambiguity estimates $\bar{\mathbf{a}}$ are the weighted sum of integers

$$\bar{\mathbf{a}} = \sum_{\mathbf{z} \in \mathbb{Z}^n} \mathbf{z} \frac{\exp\left(-\frac{1}{2} \|\hat{\mathbf{a}} - \mathbf{z}\|^2_{\mathbf{Q}_{\hat{\mathbf{a}}}}\right)}{\sum_{\mathbf{u} \in \mathbb{Z}^n} \exp\left(-\frac{1}{2} \|\hat{\mathbf{a}} - \mathbf{u}\|^2_{\mathbf{Q}_{\hat{\mathbf{a}}}}\right)}. \quad (4)$$

When implementing (4), the infinite sums are replaced by sums over the finite set of integers contained within an ellipsoidal region around $\hat{\mathbf{a}}$. The BIE positioning solution follows from the conditional least-squares estimator assuming the ambiguities given by $\bar{\mathbf{a}}$. The BIE results are *MSE-optimal*, meaning that they are always at least as good as the ambiguity-float or any ambiguity-fixed solution in that sense. The BIE estimator automatically adapts to the strength of the underlying model – without the need to define a fixing criterion. It is identical to the ambiguity-float solution for very poor precision of $\hat{\mathbf{a}}$ and converges to the ambiguity-fixed solution for very high precision of $\hat{\mathbf{a}}$ (Teunissen 2003). Further, as the BIE results are MSE optimal, they can serve as a benchmark for analyzing the theoretically best possible performance of any GNSS model.

An extension of the BIE principle for elliptically contoured distributions is provided in Teunissen (2020), and a sequential scalar approximation of the BIE estimator is proposed in Brack et al. (2014). A performance analysis of the BIE estimator for single-baseline RTK positioning is given in Odolinski and Teunissen (2020) for low-cost receivers and in Yong et al. (2022) for smartphone receivers.

In order to gain some insight into the basic properties of the BIE estimator, we consider a simulated kinematic GPS+Galileo PPP-RTK example with single-station corrections. The horizontal positioning errors after six epochs are shown in Fig. 3 for 10,000 samples together with their root mean square (RMS) errors. The ambiguity-float solution (gray) is normally distributed with an uncertainty at the few-decimeter level. The ambiguity-fixed solution using the integer least-squares estimator is at the sub-centimeter level

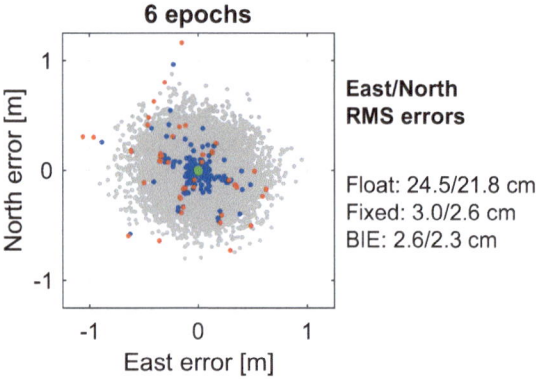

Fig. 3 Simulated horizontal positioning errors for kinematic GPS+Galileo PPP-RTK after six observation epochs with single-epoch, single-station corrections. The ambiguity-float solution is shown in gray, the ambiguity-fixed solution in green and red for correct and incorrect ambiguity estimates, and the BIE solution in blue

with correct ambiguity estimates (green) and can otherwise have large errors (red). The BIE solution (blue) is less likely to result in very large errors than the ambiguity-fixed solution, but also has a smaller probability of very small positioning errors. It is generally more concentrated around the true position than the ambiguity-fixed solution, which is also reflected by the smallest RMS errors of 2.6 cm and 2.3 cm for the east and north components.

Figure 4 shows the average simulated RMS east positioning error of the considered kinematic PPP-RTK positioning example for the first ten minutes after initialization using different systems with single-station and deterministic

corrections. As already observed in Fig. 2, the ambiguity-float solutions cannot provide centimeter-level results within such a short convergence time. For the GPS+Galileo case, centimeter-level positioning results are obtained with the ambiguity-fixed and BIE estimators after slightly more than five minutes with single-station corrections and after around three minutes with deterministic corrections. In the four-system case, sub-decimeter results are obtained within one minute (two epochs) and sub-centimeter results within one and a half minutes (three epochs) with single-station corrections, which can both be reduced by around half a minute with deterministic corrections. The BIE results are always RMS-optimal. It is noted that although the ambiguity-fixed and BIE RMS errors are often very close, the error characteristics of both estimators can still be quite different, cf. Fig. 3.

From the above simulation results we can expect centimeter-level horizontal PPP-RTK results with four systems within only a few observation epochs. Real-data PPP-RTK results of the rover station PERT with satellite clock and phase bias corrections from the station NNOR are shown in Fig. 5 for the 24 h of April 1, 2022, using one and two observation epochs. The horizontal RMS positioning errors of the BIE solution (shown in blue) are at the one-decimeter level after one epoch and at the centimeter level

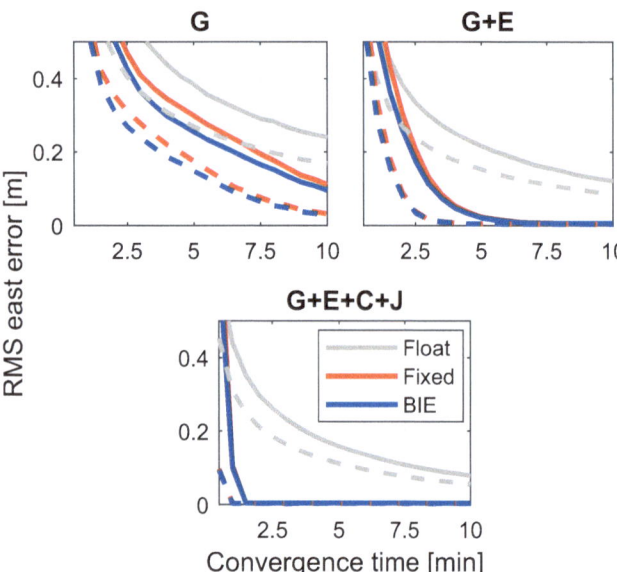

Fig. 4 Average simulated RMS east positioning error for kinematic PPP-RTK with single-epoch, single-station corrections (solid) and with deterministic corrections (dashed)

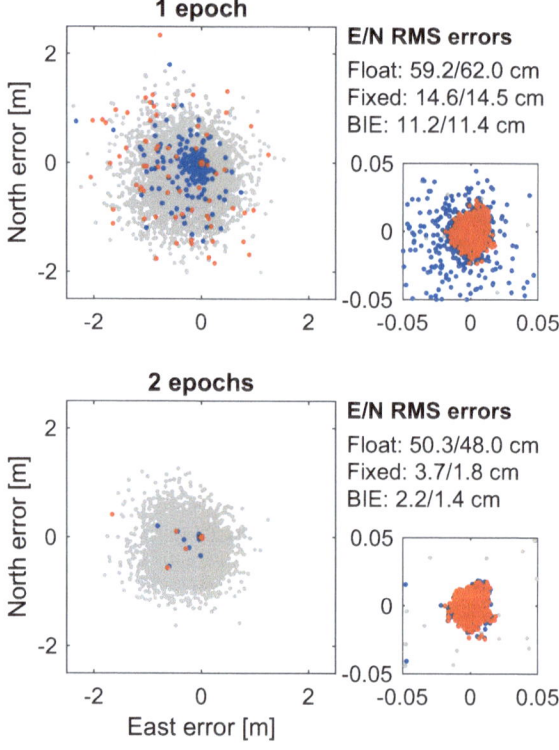

Fig. 5 Horizontal positioning errors of the station PERT for kinematic GPS+Galileo+BDS+QZSS PPP-RTK with corrections from the station NNOR. The ambiguity-float solution is shown in gray, the ambiguity-fixed solution in red, and the BIE solution in blue

already after only two epochs. A positioning error of less than 3 cm for the horizontal components is obtained in 87.6% and 99.7% of the cases, respectively. The corresponding ambiguity-fixed solutions show larger RMS errors caused by incorrect ambiguity estimates, but also have a higher probability of very small positioning errors, as can be seen in the zoom plot for one epoch.

4 Neglecting the Uncertainty of the PPP-RTK Corrections

So far, the PPP-RTK corrections have been applied to the user observations together with their full covariance information. In this way, the user obtains a realistic description of his stochastic observation model, and the corrected observations are weighted with their actual inverse covariance matrix in the least-squares adjustment, leading to minimum-variance parameter estimates with a realistic description of their precision. Neglecting the uncertainty of the corrections can, therefore, not only result in an increased failure rate when fixing the ambiguities, but also in unrealistic *formal* success rates as computed from the precision of the float ambiguity estimates (Psychas et al. 2022). The latter is particularly problematic, as a user might have too much confidence that the ambiguities can be resolved correctly, while in fact the success probability could be quite poor.

In the context of BIE ambiguity estimation, we face a similar problem: As neglecting the uncertainty of the corrections can have an impact on both $\hat{a}$ and its covariance matrix, suboptimal weights of the integer candidates might be obtained when computing the BIE ambiguity estimates $\bar{a}$ in (4) and the MSE-optimality of the positioning solution might be lost.

Figure 6 shows the magnitude of the three-dimensional PPP-RTK errors of the station PERT using two consecutive observation epochs with the BIE estimator, where the uncertainty of the corrections from the station NNOR is either included by means of their full covariance matrix as before (red), or completely neglected (blue). We can see that neglecting the uncertainty of the corrections generally leads to larger positioning errors, most notably around 1 h 40 min with an increase of more than 3 m. The corresponding empirical RMS positioning errors are given in Table 2 for the east, north, and up components, and show an increase of up to 67% when neglecting the uncertainty of the corrections.

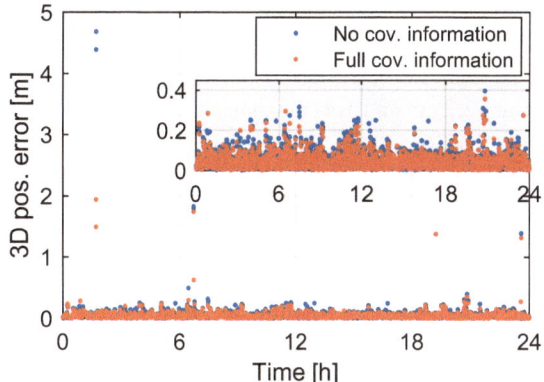

Fig. 6 Three-dimensional BIE positioning errors of the station PERT for kinematic GPS+Galileo+BDS+QZSS PPP-RTK using two consecutive epochs. The precision of the corrections from the station NNOR is fully considered (red) or completely neglected (blue)

Table 2 Empirical BIE east, north, and up RMS positioning errors of the station PERT for kinematic GPS+Galileo+BDS+QZSS PPP-RTK using two epochs in [cm]. The precision of the corrections from the station NNOR is fully considered or completely neglected

	Full cov.	No cov.	Increase
East	2.2	3.7	65%
North	1.4	1.8	27%
Up	8.5	14.1	67%

5 Conclusion

An analysis of the PPP-RTK performance with the current GNSS constellations in the absence of atmospheric corrections was provided. As ionospheric delay parameters have to be estimated in this case, fast and reliable ambiguity resolution is difficult, as was demonstrated in the beginning of this contribution. In order to achieve *almost-instantaneous* centimeter-level results, the use of the BIE estimator in a multi-GNSS solution was proposed. As the BIE positioning results are MSE-optimal, they can also be interpreted as the limits of the positioning performance of a given model.

PPP-RTK examples with a different selection of systems were analyzed through simulations, where rather weak single-epoch, single-station corrections and 'perfect' deterministic corrections were applied. The results showed that centimeter-level RMS positioning errors within a few (one to three) epochs can indeed be achieved when combining all four considered GNSS, even with single-station corrections.

An analysis of real GNSS data from the station PERT with corrections from NNOR confirmed these results. The empirical east and north RMS positioning errors after two epochs are 2.2 cm and 1.4 cm when combining GPS, Galileo, BDS, and QZSS data.

It was further demonstrated how the user positioning performance is degraded when neglecting the uncertainty of the PPP-RTK corrections, caused by unrealistic assumptions on the user's stochastic observation model. A significant increase of the RMS positioning errors was observed, reaching 67% for the up component.

Besides the BIE estimator, another alternative to conventional ambiguity fixing in weak models is *partial ambiguity resolution*. As demonstrated in Brack et al. (2021) for multi-GNSS single-baseline RTK positioning, it enables similar convergence times to reach centimeter-level results when ionospheric delays are estimated.

A more detailed version of this study is published in Brack et al. (2023).

Acknowledgements RINEX observation data were provided by the International GNSS Service (Johnston et al. 2017) and are available at https://cddis.nasa.gov/archive/gnss/data; GFZ multi-GNSS orbit products (Deng et al. 2017) are available at ftp://ftp.gfz-potsdam.de/GNSS/products/mgex. This support is gratefully acknowledged.

References

Banville S, Collins P, Zhang W, et al (2014) Global and regional ionospheric corrections for faster PPP convergence. J Inst Navig 61(2):115–124. https://doi.org/10.1002/navi.57

Boehm J, Niell A, Tregoning P, et al (2006) Global Mapping Function (GMF): a new empirical mapping function based on numerical weather model data. Geophys Res Lett 33(7). https://doi.org/10.1029/2005GL025546

Brack A, Henkel P, Günther C (2014) Sequential best integer-equivariant estimation for GNSS. Navigation 61(2):149–158. https://doi.org/10.1002/navi.58

Brack A, Männel B, Schuh H (2021) GLONASS FDMA data for RTK positioning: a five-system analysis. GPS Solut 25(1):9. https://doi.org/10.1007/s10291-020-01043-5

Brack A, Männel B, Schuh H (2023) Two-epoch centimeter-level PPP-RTK without external atmospheric corrections using best integer-equivariant estimation. GPS Solut 27(1):12. https://doi.org/10.1007/s10291-022-01341-0

Deng Z, Nischan T, Bradke M (2017) Multi-GNSS rapid orbit-, clock-& EOP-product series. GFZ data services. https://doi.org/10.5880/GFZ.1.1.2017.002

Geng J, Shi C (2017) Rapid initialization of real-time PPP by resolving undifferenced GPS and GLONASS ambiguities simultaneously. J Geod 91(4):361–374. https://doi.org/10.1007/s00190-016-0969-7

Geng J, Teferle FN, Meng X, et al (2011) Towards PPP-RTK: ambiguity resolution in real-time precise point positioning. Adv Space Res 47(10):1664–1673. https://doi.org/10.1016/j.asr.2010.03.030

Johnston G, Riddell A, Hausler G (2017) The international GNSS service. In: Springer handbook of global navigation satellite systems. Springer, pp 967–982. https://doi.org/10.1007/978-3-319-42928-133

Li P, Zhang X (2014) Integrating GPS and GLONASS to accelerate convergence and initialization times of precise point positioning. GPS Solut 18(3):461–471. https://doi.org/10.1007/s10291-013-0345-5

Li X, Li X, Yuan Y, et al (2018) Multi-GNSS phase delay estimation and PPP ambiguity resolution: GPS, BDS, GLONASS, Galileo. J Geod 92(6):579–608. https://doi.org/10.1007/s00190-017-1081-3

Odolinski R, Teunissen PJG (2020) Best integer equivariant estimation: Performance analysis using real data collected by low-cost, single-and dual-frequency, multi-GNSS receivers for short-to long-baseline RTK positioning. J Geod 94(9):91. https://doi.org/10.1007/s00190-020-01423-2

Psychas D, Verhagen S, Liu X, et al (2018) Assessment of ionospheric corrections for PPPRTK using regional ionosphere modelling. Meas Sci Technol 30(1):014,001. https://doi.org/10.1088/1361-6501/aaefe5

Psychas D, Khodabandeh A, Teunissen PJG (2022) Impact and mitigation of neglecting PPP-RTK correctional uncertainty. GPS Solut 26(1):33. https://doi.org/10.1007/s10291-021-01214-y

Teunissen PJG (1998) Success probability of integer GPS ambiguity rounding and bootstrapping. J Geod 72(10):606–612. https://doi.org/10.1007/s001900050199

Teunissen PJG (2003) Theory of integer equivariant estimation with application to GNSS. J Geod 77(7–8):402–410. https://doi.org/10.1007/s00190-003-0344-3

Teunissen PJG (2020) Best integer equivariant estimation for elliptically contoured distributions. J Geod 94(9):82. https://doi.org/10.1007/s00190-020-01407-2

Teunissen PJG, Odijk D, Zhang B (2010) PPP-RTK: Results of CORS network-based PPP with integer ambiguity resolution. J Aeronaut Astronaut Aviat A 42(4):223–230. https://doi.org/10.6125/JoAAA.20101242(4).02

Yong CZ, Harima K, Rubinov E, et al (2022) Instantaneous best integer equivariant position estimation using Google Pixel 4 smartphones for single- and dual-frequency, multi-GNSS short-baseline RTK. Sensors 22(10):3772. https://doi.org/10.3390/s22103772

Zhang B, Chen Y, Yuan Y (2019) PPPRTK based on undifferenced and uncombined observations: theoretical and practical aspects. J Geod 93(7):1011–1024. https://doi.org/10.1007/s00190-018-1220-5

Multi-GNSS Tomography: Case Study of the July 2021 Flood in Germany

Karina Wilgan, Hugues Brenot, Riccardo Biondi, Galina Dick, and Jens Wickert

Abstract

Due to climate change, intensive storms and severe precipitation will continue to happen, causing destructive flooding. In July 2021, a series of storms with prolonged rain episodes took place in Europe. Several countries were affected by severe floods following that rainfall, causing many deaths and material damage. Thus, a good understanding and forecasting of such events are of uttermost importance. This study highlights the interest of multi-GNSS tomography for the 3D modelling of the neutral atmosphere refractivity. The tropospheric parameters have been retrieved for the July 2021 flood in Germany from two tomographic solutions with different constraining options using either GPS-only or multi-GNSS estimates. Our investigations show that the stand-alone solution (especially the multi-GNSS) is producing more patterns of refractivity, and is temporally more stable. We compare the tomographic results with external observations such as radiosondes and GNSS radio-occultations from Metop-A & -B satellites. The results show that tomography is producing wetter conditions than the reference. However, we can see the precursor information of the initiation of deep convection in the ground-based GNSS technique.

Keywords

Deep convection · GNSS tomography · Multi-GNSS · Severe weather events

1 Introduction

GNSS tomography is a technique that unwraps a simple integrated signal into a 3D distribution of the atmosphere parameters, usually related to water vapor (Flores et al. 2000;

K. Wilgan (✉) · J. Wickert
Technische Universität Berlin (TUB), Berlin, Germany

German Research Centre for Geosciences (GFZ), Potsdam, Germany
e-mail: karina.wilgan@gfz-potsdam.de

H. Brenot
Royal Belgian Institute for Space Aeronomy (BIRA-IASB), Brussels, Belgium

R. Biondi
University of Padova, Padova, Italy

G. Dick
German Research Centre for Geosciences (GFZ), Potsdam, Germany

Seko et al. 2000; Gradinarsky and Jarlemark 2004; Champollion et al. 2005). The method is based on the inverse Radon transform (Fiddy 1985), which states that a continuous field can be successfully reconstructed from integrated observations providing an infinite number of observations penetrating the field from an infinite number of angles. Due to the geometrical constraints such as one-way communication between satellite and receiver, availability of visible satellites only above the receiver, and very limited number of side observations, the tomography system is ill-conditioned and ill-posed (Troller et al. 2006), which evokes many research questions.

The idea of GNSS tomography originated in the early 2000s (Flores et al. 2000). In the traditional voxel approach, the tropospheric parameters, i.e. the refractivity or water vapor density, are obtained from the GNSS Slant Tropospheric Delay (STD) products on a 3D grid (voxels). Many methodological enhancements have been introduced. Some included adding supplementary data from external sources into the functional model (e.g., Bender et al. 2011a; Rohm et al. 2014), some new parametrizations (e.g., Perler et al. 2011; Brenot et al. 2019). Improvements are expected by using multi-GNSS (Bender et al. 2011b). The recent studies focus on function-based tomography, instead of voxel-based (e.g., Haji-Aghajany et al. 2020; Forootan et al. 2021).

In this study, we focus on the voxel-based tomography using multi-GNSS STD retrievals for a part of Germany that was affected by severe rainfall and flooding in July 2021. We have retrieved the total refractivity using Singular Value Decomposition method, with a novel iterative approach. We show the comparisons of the tomography-based total refractivity from different strategies with the reference data.

2   Data and Meteorological Conditions

We retrieve the tomography solutions for the period of July 10–18, when the severe rainfall and devastating floods in Europe occurred. The rain episodes started between July 6 and 12. Additional heavy precipitation on July 13–15 along with the slow-moving pressure system led to destructive flooding (Puca et al. 2021). In Germany, the most affected regions were North Rhine-Westphalia and Rhineland-Palatinate, especially in the district of Ahrweiler. In Cologne, the rain gauges indicated 154 mm of rainfall for July 14, the day of the highest rainfall. More detail on the meteorological conditions can be found in Wilgan et al. (2023). Figure 1

shows the chosen tomography area, indicating the GNSS stations and their GPS (G), GLONASS (R) and Galileo (E) signals' capability. The GNSS data are calculated using the GFZ-developed software EPOS.P8 with 2.5 min temporal resolution for the 70 stations located between 6° and 10° longitude and 49° and 52° latitude. More details about the processing can be found in Wilgan et al. (2022).

Figure 1 also shows the location of the radiosonde (RS) station Essen, 10410 (near GNSS station EDZE), situated within the tomography region as well as the two radio-occultations (RO) from Metop-A&B satellites that occurred during our chosen period (July 15, 19:55 UTC and July 14, 17:07 UTC). Both RS and RO are used as reference data in this study. The GNSS RO can be used to retrieve the vertical properties of the atmosphere with high accuracy and high vertical resolution (Scherllin-Pirscher et al. 2011). Each GPS Receiver for Atmospheric Sounding (GRAS) on board of the Metop satellites (Luntama et al. 2008) provides more than 600 daily atmospheric profiles globally distributed and it is the only operational RO instrument at the moment. The ROs can be downloaded here: https://www.cosmic.ucar.edu/what-we-do/data-processing-center/data.

The a priori model for tomography and another reference is Numerical Weather Model (NWM) Icosahedral Nonhydrostatic (ICON) run by the German Weather Service (DWD). We have used the nested ICON-D2 version of the global model with the resolution of 0.02° × 0.02° with 65 vertical layers up to 20 km. The GNSS ZTDs and ROs are assimilated into the ICON global model, but not into the nested, regional model.

Fig. 1 The location of the tomography region with marked GNSS and radiosonde stations as well as the two radio-occultations

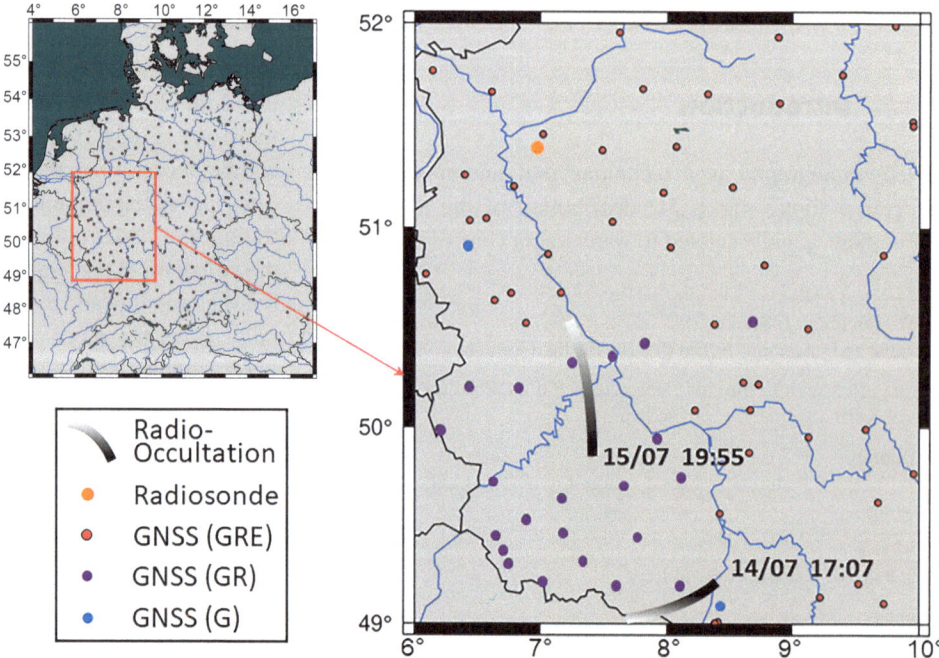

3 Strategy of GNSS Tomography

Located in western Germany (see Fig. 1), the tomography grid has a latitude × longitude horizontal resolution of $0.2° × 0.3°$ ($21 × 22$ km^2; $15 × 14$ elements). With 15 vertical levels, from 0 km above the sea level, every km until 15 km, the number of tomography voxels is 3,150. The temporal resolution of tomography matches the 2.5 min resolution of the GNSS data. We retrieve the total refractivity with the GNSS tomography principle, i.e., using the GNSS STDs. The STD can be related to the total refractivity N_{tot} using the equation:

$$STD = 10^{-6} \int N_{tot} ds \cong 10^{-6} \sum N_{tot} \Delta s. \quad (1)$$

The tomographic model m can be represented as:

$$m = m_0 + \left(G^t C_d^{-1} G + C_m^{-1}\right)^{-1} G^t C_d^{-1} (d - Gm_0), \quad (2)$$

where d is the data (GNSS STDs), G the geometrical matrix ($15 × 14 × 15$ voxels), m the model solution (calculated using Singular Value Decomposition), m_0 a priori model (forecasts from the ICON-D2), C_d the covariance operator of the data and C_m covariance operator of the a priori model.

The solutions are calculated using an iteration process, which stops when the absolute bias between previous and new retrievals is under 1% (convergence to the final solution). C_d characterizes the confidence in the data and C_m the confidence in the a priori model. In this study, we test estimates of $C_d = (STD * coeff\_C_d)^2$ with $coeff\_C_d = 10\%$, 15%, 20%, 25% or 30%, and $C_m = (N_{ap} * coeff\_Cm)^2$ with $coeff\_Cm = 90\%$, 85%, 80%, 75% or 70%. N_{ap} is the refractivity from the m_0 a priori model. The interest of using multi-GNSS in tomography is to improve the geometrical representation by increasing the number of forced voxels (the ones that tomography retrieves, i.e., with STDs crossing the voxels). In this study, for the G solution, the number of forced voxels is 70% (2,205 voxels) and it is improved to 74% (2,331 voxels) and 76% (2,394 voxels) by using GR and GRE, respectively.

We have used two types of tomographic solutions: constrained and stand-alone. In the constrained solution, we take the hourly a priori information from the ICON-D2, while in the stand-alone solution, ICON-D2 is used only to initiate the tomography, and then a priori values are taken from the previous tomography retrievals (TRs). On average, three iterations are needed for the constrained solution and only one iteration is required for the stand-alone solution.

4 Results

This section shows the results of the tomography retrievals. First, we compare different solutions with each other and then the TRs to the reference ICON-D2, RS and RO data.

4.1 Tomography Cross-Section

We present the total refractivity values obtained using the constrained and stand-alone solutions. Figure 2 shows the results using different GNSS signals: G, GR and GRE for a sample date and height of 1.5 km and Fig. 3 the time evolution of the two TRs for a fixed altitude and longitude.

We can see in Fig. 2 that the three constrained solutions are similar, while the three stand-alone solutions show stronger differences with more patterns. Especially the GRE solution shows more variability, compared to the G and GR solutions, which are closer to each other. However, if we have considered solutions for the consecutive times (Fig. 3), we can see that the constrained solutions show a lot more time variability as they try to move from the a priori ICON-D2 to the converge solution (closer to the stand-alone results), while the stand-alone solutions are smoother. In the above comparisons, a set-up of $coeff\_C_d = 10\%$ and $coeff\_C_m = 90\%$ is used. These parameters indicate how much confidence we have in the data and the a priori model, respectively, and can be modified. Figure 4 shows five different set-ups of the covariance parameters for the GRE stand-alone solutions. We can see that the set-ups differ from each other. The higher the $coeff\_C_d$ values, the lower the refractivity obtained with this solution. More detailed analyses are in the comparisons with RS and RO chapters.

4.2 Comparisons with ICON-D2

In the next step, we compare the TRs to the reference ICON-D2 data. Please note that these comparisons are not independent, because ICON data is used as a priori to calculate the tomography solutions. Figure 5 shows the total refractivity fields from ICON and TRs for the GRE solution on 13 July 2021, at 08:00 UTC and for the altitude of 1.5 and 2.5 km. For the constrained solution, the Root-Mean-Square Error (RMSE) is 11.4 ppm (12.5 ppm for the ICON datasets from July 10–18, 2021), and 15.7 ppm (17.9 ppm) for the stand-alone solution.

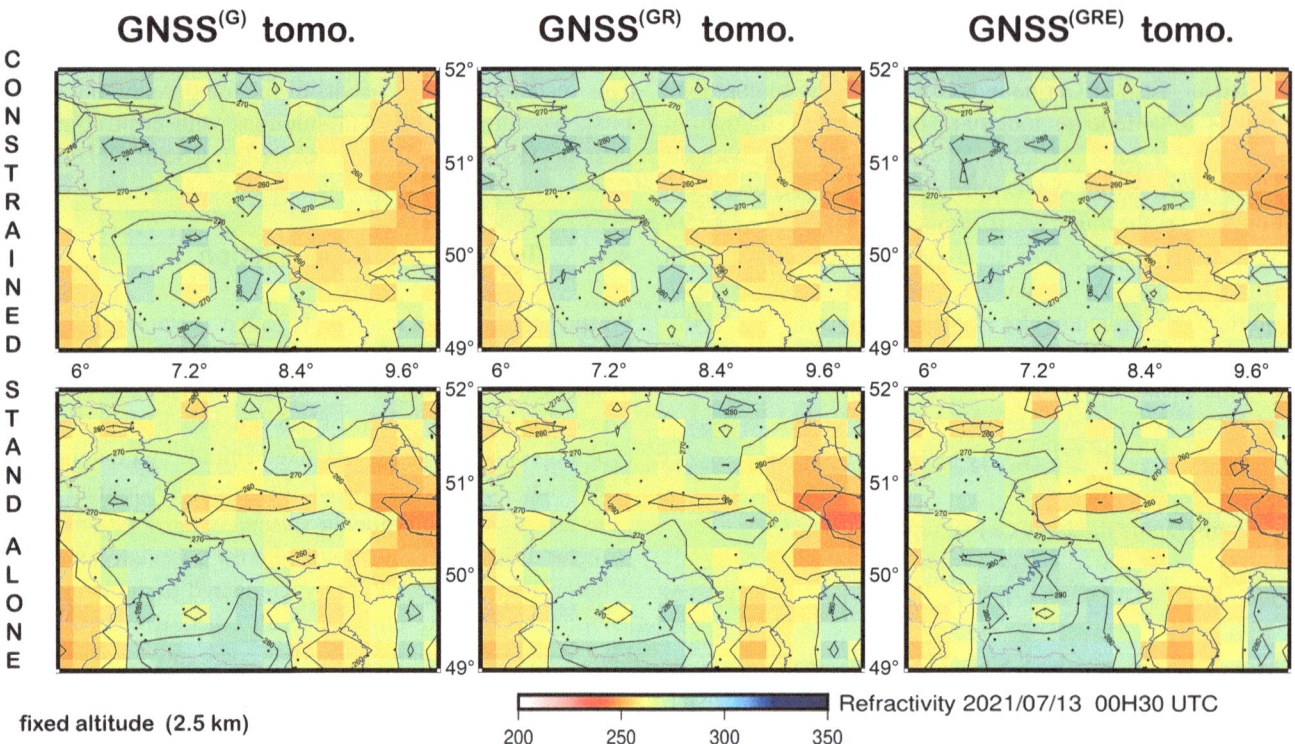

Fig. 2 Total refractivity from the constrained (top) and stand-alone (bottom) TR (G, GR and GRE solutions, from left to right) on 13 July, 00:30 UTC, for a fixed altitude of 2.5 km a.s.l.

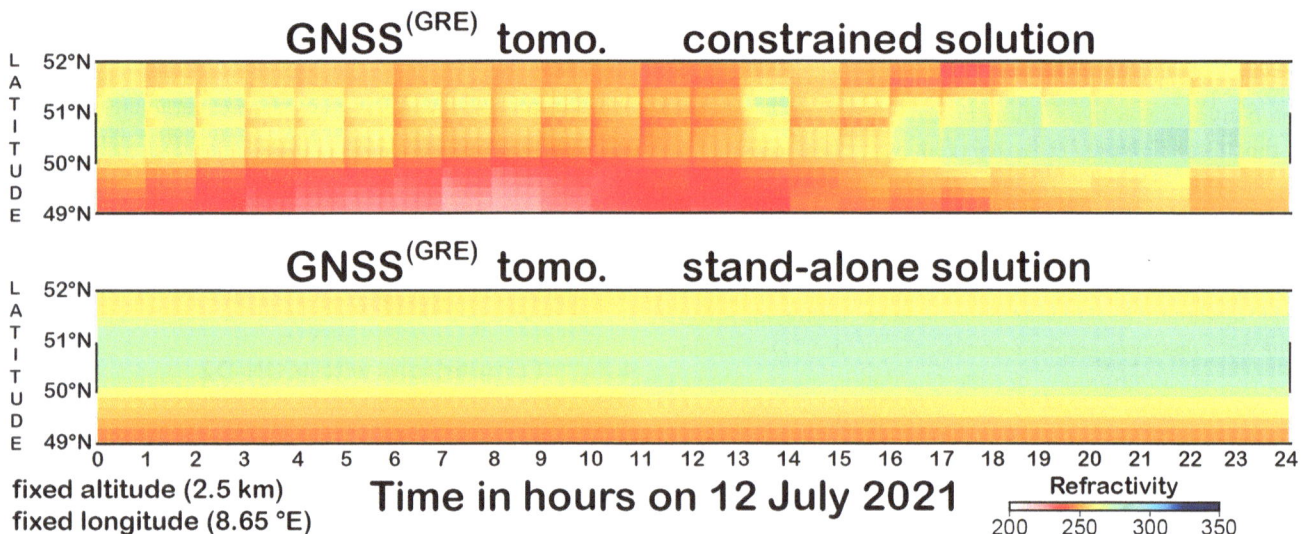

Fig. 3 Total refractivity from the constrained (top) and stand-alone (bottom) TR (GRE solution) on 12 July, from 00:00 to 24:00 UTC, for a fixed altitude (2.5 km) and a fixed longitude (8.65°E)

As shown in Fig. 5, the TRs are producing wetter conditions than ICON data. Moreover, the constrained solution is 40% closer to ICON than the stand-alone, which is not surprising, as we have used ICON as the a priori for the constrained solution. However, the two TRs are still closer to each other than to ICON, with a RMSE of 8.4 ppm (10.2 ppm) for the G solutions, and 6.3 ppm (8.0 ppm) for the GRE solutions. Moreover, closer to the ground (1.5 km vs 2.5 km) and when deep convection took place on July 13 northeastwards of the Grand Duchy of Luxembourg, Fig. 5 shows more structured refractivity fields, as there, the water vapor content and thus refractivity is higher and more variable. Such pattern is not seen by ICON, even though it offers more detailed fields, as the resolution of the model is 0.02°, which is 10 times larger than TRs.

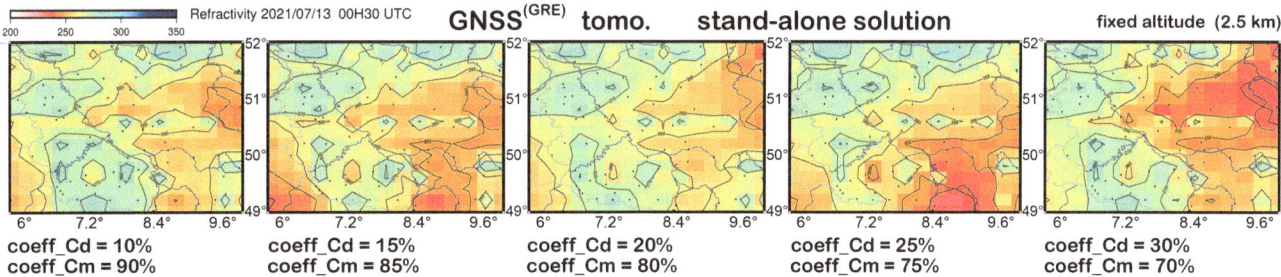

Fig. 4 The evaluation of using different covariance values coeff_C_d = 10%, 15%, 20%, 25%, 30% while coeff_C_m = 90%, 85%, 80%, 75%, 70%. Results are shown for the stand-alone GRE TR

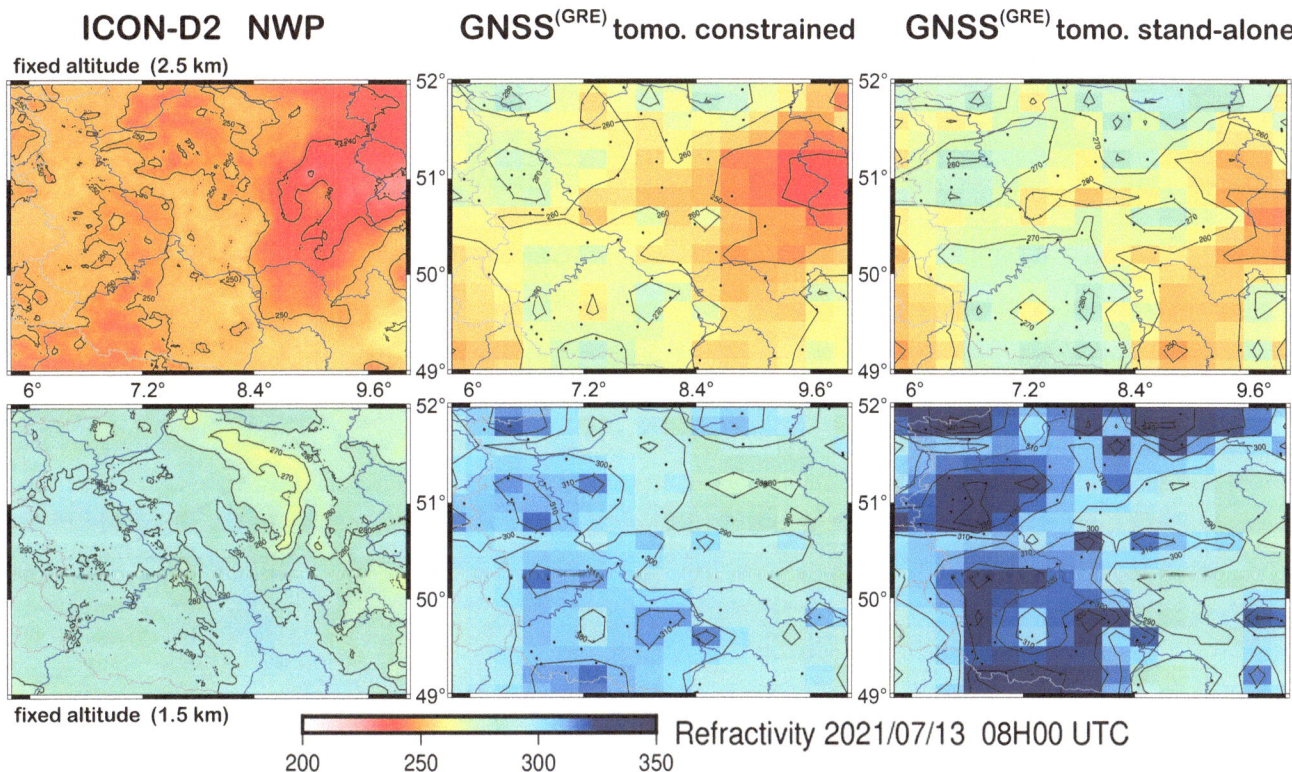

Fig. 5 Comparison of ICON (left), GRE tomography constrained (middle) and GRE tomography stand-alone (right) for July 13 (08:00 UTC), height 2.5 km (top) and 1.5 km (bottom)

4.3 Comparisons with Radiosonde Data

Another reference data in this study is radiosonde. There is one RS station 10410 located in the north-east of the chosen area, in Essen (see Fig. 1). Figure 6 shows the total refractivity values from the RS, ICON and TR from G and GRE solutions for a sample date of July 13, 0:00 UTC. The RMSE RS-ICON is of 5.2 ppm (4.3 ppm for the 18 radiosondes from July 10–18, 2021).

As seen in Fig. 6, RS and ICON data are closer to each other than to the TRs, meaning that tomography produces wetter conditions than the reference data. For the constrained retrievals, both G and GRE solutions are very similar (RMSE of 1.2 ppm for the 18 radiosondes), but, there are some differences for the stand-alone solution (RMSE of 4.7 ppm), where

GRE is closer to the reference data (RMSE of 15.7 ppm against 17.3 ppm for the G solution). In the bottom panel, we see the impact of using different covariance operators for the stand-alone GRE solution. The variant with *coeff_C_d* = 20% and *coeff_C_m* = 80% is closer to the RS on the ground level (15% decrease of the RMSE with respect to the solution with *coeff_C_d* = 10% and *coeff_C_m* = 90%), while with *coeff_C_d* = 30% and *coeff_C_m* = 70% is the closest for the middle layers (45% decrease of the RMSE).

4.4 Comparisons with Radio-Occultations

In the next step, the TRs are compared to the RO data (two profiles; see Fig. 1). Figure 7 shows the total refractivity

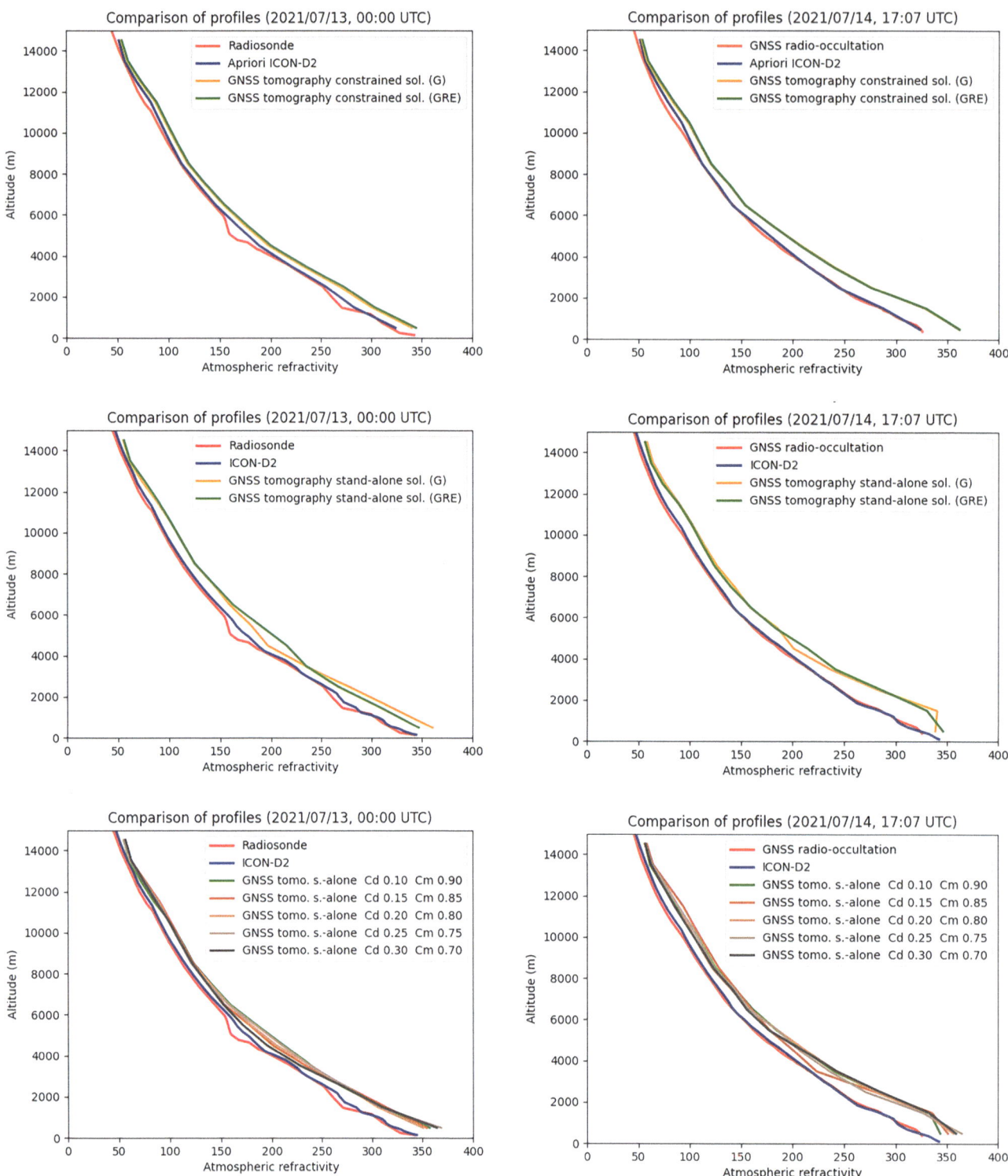

Fig. 6 Reference RS and ICON-D2 data vs. TRs for the constrained (top), stand-alone (middle) and stand-alone solution with different covariance operators (bottom) solutions

Fig. 7 Reference RO and ICON-D2 data vs. TRs for the constrained (top), stand-alone (middle) and stand-alone solution with different covariance operators (bottom) solutions

from RO, ICON and TRs for July 14, 17:07, one of the occultations occurrences.

As shown in Fig. 7, for the RO we have a similar situation to RS: the RO and ICON are close to each other (RMSE of 3.1 ppm for the two ROs), while the TRs are producing wetter conditions (RMSE of about 21 ppm for the G/GRE constrained/stand-alone solutions). Here, we see an improvement for the stand-alone GRE solution for the layer close to the ground, i.e., under 3 km with a 28% decrease of the RMSE, however, a 25% increase of the RMSE is observed for the middle layers, i.e., between 4 and 8 km. From the covariance parameters, the closest to RO is $coeff\_C_d = 10\%$ and $coeff\_C_m = 90\%$ for the lowest layers and $coeff\_C_d = 15\%$ and $coeff\_C_m = 85\%$ for the middle layers (26% decrease of the RMSE with respect to the solution with $coeff\_C_d = 10\%$ and $coeff\_C_m = 90\%$), so slightly different options than for RS.

5 Conclusions

We showed the first results of multi-GNSS tomography for a severe precipitation and flooding event in July 2021. We presented a new retrieval algorithm with an iteration process for stand-alone and constrained tomography solutions based on G, GR and GRE data. The two types of TRs differed between each other, especially in space, where the stand-alone solution was smoother, while the constrained solution tried to converge to the a priori data, here taken from ICON. The GRE solution was the best fit, as it showed more patterns in the obtained total refractivity. Using the multi-GNSS also retrieved more forced voxels. The TRs were compared with reference ICON, RS and RO data. In general, the TRs tended to produce wetter conditions compared to the reference data, which was, however, in line with the previous findings. During the phase of the initiation of deep convection on July 13, 2021, TRs show high values of total refractivity northeastwards of the Grand Duchy of Luxembourg (see Fig. 5), which is not seen by ICON-D2 NWM and could be substantial information to be considered in an assimilation system.

Moreover, we checked the impact of different covariance operators on the tomography retrievals. We reached a better agreement with the reference data for some of the variants. TRs show wetter estimates for the lower layers (between 0 and 3 km) than reference external solutions. As the impact of GNSS ground-based data is stronger for the lower layers than for the middle layers (between 3 and 5 km), we suggest using a low covariance coefficient for the data ($coeff\_C_d = 10\%$) and a high covariance coefficient for the a priori model ($coeff\_C_m = 90\%$). However, this requires having good a priori estimates. To improve the quality of TRs, we think a mixed strategy/solution can be implemented, which combines the use of conservative covariance for the lower layers and less conservative coefficients for the middle layers (e.g., with $coeff\_C_d = 20\%$ and $coeff\_C_m = 80\%$).

Acknowledgements This study was performed under the framework of the Deutsche ForschungsGemeinschaft (DFG) project AMUSE, grant no. 418870484 and ALARM H2020 SESAR, grant no. 891467. It is also a part of the IAG WG 4.3.6. We thank Michael Bender from the DWD for providing outputs from the ICON-D2 model and valuable inputs. The GNSS data are provided by the GFZ, IGS and SAPOS networks.

References

Bender M, Dick G, Ge M, Deng Z, Wickert J, Kahle HG, Raabe A, Tetzla G (2011a) Development of a GNSS water vapour tomography system using algebraic reconstruction techniques. Adv Space Res 47:1704–1720

Bender M, Stosius R, Zus F, Dick G, Wickert J, Raabe A (2011b) GNSS water vapour tomography—expected improvements by combining GPS, GLONASS and Galileo observations. Adv Space Res 47(5):886–897

Brenot H, Rohm W, Kačmařík M, Möller G, Sá A, Tondaś D et al (2019) Cross-comparison and methodological improvement in GPS tomography. Remote Sens 12(1):30

Champollion C, Masson F, Bouin MN, Walpersdorf A, Doerflinger E, Bock O, Van Baelen J (2005) GPS water vapour tomography: preliminary results from the ESCOMPTE field experiment. Atmos Res 74(1-4):253–274

Fiddy MA (1985) The radon transform and some of its applications. Opt Acta 32(1):3–4

Flores A, Rufifni G, Rius A (2000) 4D tropospheric tomography using GPS slant wet delays. Ann Geophys 18:223–234

Forootan E, Masood D, Saeed F, Ali SK (2021) A functional modelling approach for reconstructing 3 and 4 dimensional wet refractivity fields in the lower atmosphere using GNSS measurements. Adv Space Res 68(10):4024–4038

Gradinarsky L, Jarlemark P (2004) Ground-based GPS tomography of water vapor: analysis of simulated and real data. J Meteorol Soc Jpn 82(1B):551–560

Haji-Aghajany S, Amerian Y, Verhagen S (2020) B-spline function-based approach for GPS tropospheric tomography. GPS Solut 24(3):88

Luntama JP, Kirchengast G, Borsche M, Foelsche U, Steiner A, Healy S, von Engeln A, O'Clerigh E, Marquardt C (2008) Prospects of the EPS GRAS mission for operational atmospheric applications. Bull Am Meteorol Soc 89(12):1863–1875

Perler D, Geiger A, Hurter F (2011) 4D GPS water vapor tomography: new parameterized approaches. J Geod 85:539–550

Puca S, Brocca L, Panegrossi G et al (2021) A slow moving upper-level low brought devastating floods to parts of north west Germany and other parts of western Europe in July 2021. https://www.eumetsat.int/devastating-floods-western-europe. Published 26.07.2021

Rohm W, Zhang K, Bosy J (2014) Limited constraint, robust Kalman filtering for GNSS troposphere tomography. Atmos Meas Tech 7(5):1475–1486

Scherllin-Pirscher B, Steiner AK, Kirchengast G, Kuo YH, Foelsche U (2011) Empirical analysis and modeling of errors of atmospheric profiles from GPS radio occultation. Atmos Meas Tech 4(9):1875–1890

Seko H, Shimada S, Nakamura H, Kato T (2000) Three-dimensional distribution of water vapor estimated from tropospheric delay of GPS data in a mesoscale precipitation system of the Baiu front. Earth Planets Space 52:927–933

Troller M, Geiger A, Brockmann E, Bettems JM, Bürki B, Kahle HG (2006) Tomographic determination of the spatial distribution of water vapor using GPS observations. Adv Space Res 37(12):2211–2217

Wilgan K, Dick G, Zus F, Wickert J (2022) Towards operational multi-GNSS tropospheric products at GFZ Potsdam. Atmos Meas Tech 15(1):21–39

Wilgan K, Dick G, Zus F, Wickert J (2023) Tropospheric parameters from multi-GNSS and numerical weather models: case study of severe precipitation and flooding in Germany in July 2021. GPS Solut 27(1):1–17

Quantum Diamond Magnetometry for Navigation in GNSS Denied Environments

X. Wang, W. Li, B. Moran, B. C. Gibson, L. T. Hall, D. A. Simpson, A. N. Kealy, and A. D. Greentree

Abstract

Satellite-based navigation is a transformational technology that underpins almost all aspects of modern life. However, there are environments where global navigation satellite systems (GNSS) are not available, for example undersea or underground, and navigation that is robust to GNSS outages is also required for resilient systems. Here we explore the potential for quantum diamond magnetometers as aids to obtain external position fix for navigation in global navigation satellite systems (GNSS)-denied environments. Diamond magnetometers offer high sensitivity and low measurement noise. We demonstrate this by simulating external position fix from the magnetic field measurements with a geographical data map using the probabilistic multiple hypotheses map matching filter with probabilistic data association for data mapping.

Keywords

Diamond magnetometry · Expectation maximisation · Map matching · Multiple hypotheses tracker · Probabilistic data association · Total magnetic intensity data map

X. Wang (✉) · W. Li
School of Science, RMIT University, Melbourne, VIC, Australia
e-mail: xuezhi.wang@rmit.edu.au

B. Moran · A. N. Kealy
School of Science, RMIT University, Melbourne, VIC, Australia

FEIT, University of Melbourne, Melbourne, VIC, Australia

B. C. Gibson
ARC CNBP, RMIT University, Melbourne, VIC, Australia

L. T. Hall
School of Chemistry, University of Melbourne, Melbourne, VIC, Australia

D. A. Simpson
School of Physics, University of Melbourne, Melbourne, VIC, Australia

A. D. Greentree
School of Science, RMIT University, Melbourne, VIC, Australia

ARC CNBP, RMIT University, Melbourne, VIC, Australia

1 Introduction

In GNSS-denied environments, platform navigation performance is dominated by the accuracy of onboard inertial sensors. Even with high end inertial sensors, which exhibit extremely low bias and drift, it is not possible to avoid the build-up of navigation errors over long time frames (Titterton and Weston 2004). Removing these accumulated navigation errors is therefore crucial for navigation accuracy (Groves 2013). This removal, or correction, is achieved using one or more aiding sources that provide positional information, i.e. a position fix.

Geophysical map matching is an effective method for localisation and navigation where GNSS is not available; such as underwater, urban, or hostile environments (Tyren 1982; Tuohy et al. 1996; Kamgar-Parsi and Kamgar-Parsi 1999; Goldenberg 2006; Wang et al. 2022a,b; Li et al. 2022). Although conceptually simple, map matching with geophysical maps suffers from map measurement ambiguity issues. First, the geophysical measurements themselves are

J. T. Freymueller, L. Sánchez (eds.), *Gravity, Positioning and Reference Frames*,
International Association of Geodesy Symposia 156, https://doi.org/10.1007/1345_2023_218

degraded by sensor noise so the measurements will not match the map exactly. Second, the measurements may match multiple points within the map as the map-lookup process is a scalar to vector mapping. Third, the location where the measurement was acquired is of course uncertain. Finally, the map itself suffers from finite spatial and signal resolution.

Here we consider magnetometry aided inertial navigation with total magnetic intensity (TMI) maps. The method include a probabilistic data association (PDA) approach to address the measurement ambiguity problem, and a probabilistic multiple hypothesis tracker (PMHT) for the map matching localisation using geophysical data maps. We show that the PDA method provides an effective way to map a field measurement into geolocation, but also enables a quantitative analysis of localisation error with respect to the magnetometer noise levels for a given TMI reference map. Furthermore, we implement a magnetometry aided INS using this method to determine the relationship between magnetometer noise levels and navigation performance.

Diamond magnetometry is a rapidly developing field, with potential applications for navigation Frontera et al. (2018). Sensitivity of diamond sensors is rapidly increasing, with additional techniques poised to transition from research to practical systems, meaning that it is timely to explore the potential of existing and future diamond sensors. Techniques designed to improve sensitivity include isotopic enrichment (Balasubramanian et al. 2009), portability through embedding in optical fibers (Ruan et al. 2018; Bai et al. 2020; Filipkowski et al. 2022), and laser threshold magnetometry (Jeske et al. 2016; Dumeige et al. 2019; Hahl et al. 2022).

2   INS Aiding via Map Matching

Aided INS can be described by a recursive Bayesian filtering system, where the system prediction is given by the onboard INS, and system update from measurements from external aiding sources. Figure 1 illustrates a generic aided INS with aiding from a map matching system. The INS is initialised from known parameters and at time k propagates the navigation state $\mathbf{X}_{\mathrm{INS},k|k-1}$ based on the earth surface motion model and inertial measurements ($\mathbf{f}_b\ \boldsymbol{\omega}_b$). The global position measurements, estimated from map matching, are assumed to be Gaussian distributed with mean $\hat{\boldsymbol{x}}^s$ – the estimated sensor location where s is taken and covariance Σ^s, and are incorporated into the system via an integration filter to update the navigation state $\mathbf{X}_{\mathrm{INS},k|k}$. For simplicity, we denote the navigation state at time k as $\mathbf{X}_k \in \mathbb{R}^n$: this comprises the components of vehicle kinematic state (position and velocity) expressed in the geographical coordinates, vehicle attitude (roll, pitch and yaw), and inertial sensor bias

terms. Based on inertial sensor measurements, navigation state is

$$\mathbf{X}_k = \mathbf{F}_{\mathrm{INS}}(\mathbf{X}_{k-1}, \boldsymbol{\omega}_b, \mathbf{f}_b) + \mathbf{w}_k, \qquad (1)$$

where the function $\mathbf{F}_{\mathrm{INS}}(\cdot)$ signifies the mechanization of INS which involves the prior navigation state $\mathbf{X}_{k-1}$, the measurements of accelerometer $\mathbf{f}_b$ and gyroscope $\boldsymbol{\omega}_b$ at k, respectively (Titterton and Weston 2004), where $\mathbf{w} \sim \mathcal{N}(0, \boldsymbol{Q})$ accounts for system process noise including the errors from accelerometer and gyroscope $\mathcal{N}(\mathbf{a}, \boldsymbol{B})$ signifies a Gaussian distribution with mean vector $\mathbf{a}$ and covariance matrix $\boldsymbol{B}$.

At each aiding update time k, the aiding position measurement is coupled into the navigation state via

$$\boldsymbol{y}_k = \mathbf{H}\mathbf{X}_k + \mathbf{v}_k. \qquad (2)$$

where $\mathbf{H}$ is a constant matrix and $\mathbf{v} \sim \mathcal{N}(0, \boldsymbol{R})$ is a Gaussian zero-mean noise term modeling the measurement errors.

The INS aiding problem is to find the posterior density $p(\mathbf{X}_k \mid \boldsymbol{y}_{1:k})$ based on the sequence of measurements $\boldsymbol{y}_{1:k}$ from aiding sources.

3   Probabilistic Multiple Hypothesis Map Matching

The probabilistic multiple hypothesis map matching involves probabilistic data association (PDA) for data mapping from TMI signal domain to vehicle position domain, and a batch based multiple hypothesis tracking algorithm to iteratively optimise the estimated vehicle trajectory.

For magnetometer measurement s_k at time k, the measurement model is

$$s_k = s_k^o + \nu_k, \qquad (3)$$

where s_k^o is the ground truth value and ν_k a noise term covering imperfect sensor measurements, assumed to be Gaussian distributed i.e., $\nu \sim \mathcal{N}(0, \sigma^2)$.

Following (3), we consider a set of candidate measurements from a single measurement s_k, one of which is the true sensor measurement. Let $Z_m = \{\boldsymbol{z}_i,\ i = 1, \cdots, n\}$ denote the set of possible map locations corresponding to s_k. We assume that at time k, the location of magnetometer, which takes the magnetic intensity measurement s_k, is a Gaussian random variable with mean $\boldsymbol{x}_k^s$ and covariance matrix Σ_k^s. Then, the location of true magnetic intensity measurement $\boldsymbol{z}_i,\ i = 1, \cdots, n$ should satisfy (Chi-Square Test)

$$(\boldsymbol{z}_i - \boldsymbol{x}^s)(\boldsymbol{\Sigma}^s)^{-1}(\boldsymbol{z}_i - \boldsymbol{x}^s)' \leq \gamma, \qquad (4)$$

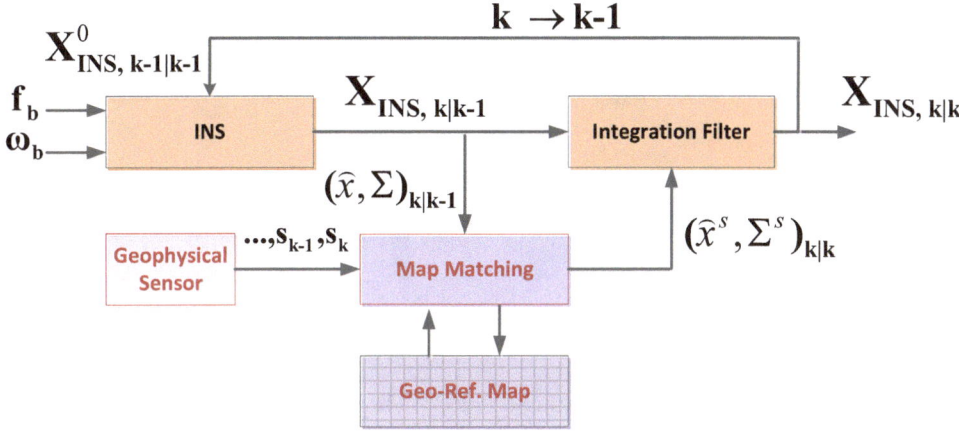

Fig. 1 Generic single recursion map matching aided inertial navigation system

Fig. 2 Collection of candidate signal locations $\{z_i^s, i = 1, 2, \cdots, n\}$ obtained via (4) based on knowledge of predicted vehicle position x_{INS} from INS, and sensor noise level

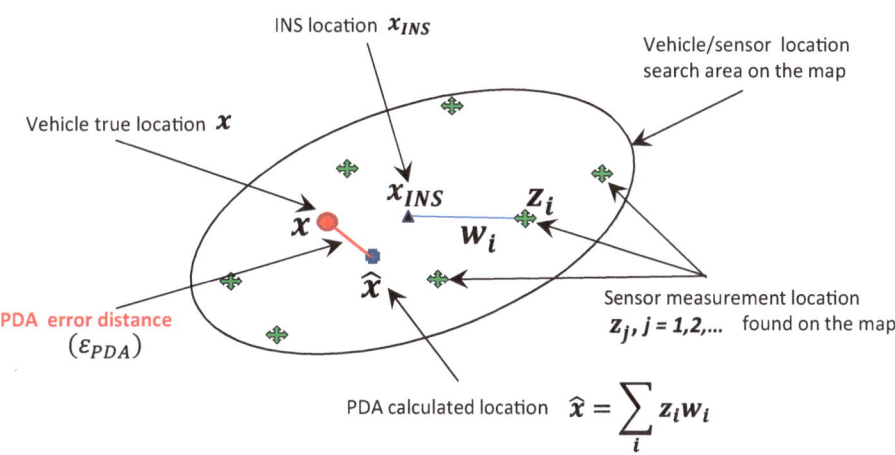

where γ is a probability threshold. This determines an ellipsoid on the data map containing the magnetometer location with a certain level of confidence. We refer to this area as a search window. Figure 2 illustrates the data PDA mapping process. A finite set of potential locations for signal s on the map can be obtained via (4).

The probability weight of each candidate location z_i is proportional to the geometric distance between z_i and the window centre x^s (i.e., x_{INS}). The probability weight can be found as

$$w_i = p(z_i \mid x^s) \left[\sum_{j=1}^{n} p(z_j \mid x^s) \right]^{-1}, \qquad (5)$$

where $p(z_i \mid x^s) \sim \mathcal{N}(z_i - x^s, R_i(\sigma))$, and $R_i(\sigma)$ is the associated variance which is a function of the signal noise variance, or in other words, signal-to-noise ratio (SNR). Thus, the mean $\bar{z}$ and variance $\bar{R}$ of PDA solution for the map location on magnetic intensity measurement s_k are given by

$$\bar{z} = \sum_{i=1}^{n} w_i z_i. \qquad \bar{R} = \sum_{i=1}^{n} w_i \left[R_i(\sigma) + (z_i - \bar{z})(z_i - \bar{z})' \right]. \qquad (6)$$

Using PDA, the map matching quality can be characterised the PDA error distance ε_{PDA}, defined as the Euclidian distance between the true magnetometer location and the location estimated via PDA, i.e., $\varepsilon_{PDA} = \|\hat{x}_{PDA} - x^s\|$.

Our simulations demonstrate that for a fixed resolution map, the measurement taken from a high sensitivity/low noise magnetometer will result in small PDA error distance. The TMI map used in the simulation is downloaded from Australia (2023). As shown in Fig. 3a, the actual data grid size is 85×85 metres. The simulation is carried out in the area surrounded by the green solid line rectangle. For every sensor noise level, 1000 samples are drawn randomly in the area, which are treated as the mean of sensor locations. The values of sensor location covariance Σ^s and probability threshold γ are chosen such that a search window approximately 6.8 km² is formed for collecting candidate measurement locations. PDA error distances are then calculated as a function of sensor noise level σ and map grid size.

Figure 3b shows the plot of PDA error distances versus sensor noise levels with original TMI map and the 5 and 10 times downsampled TMI maps. The plots show that localisation error decreases with improving magnetometer sensitivity until a localisation floor is reached. This floor is a function of the map resolution, with higher resolution

Fig. 3 (**a**) The total magnetic intensity map used in the simulation from Australia (2023) with superimposed platform travel trajectory (yellow line). (**b**) Comparison of PDA error distances vs. sensor noise levels in the TMI map at original, 5 and 10 times downsampled data grids

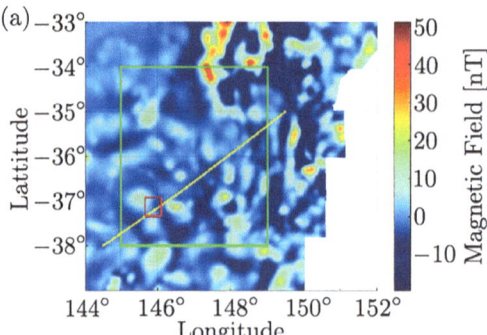
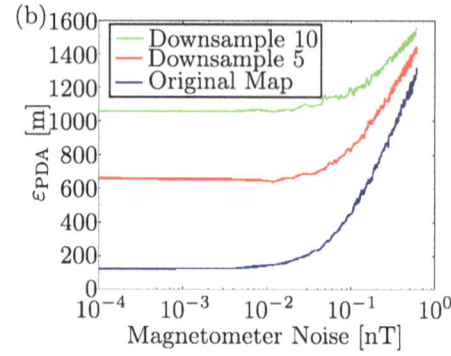

providing a lower floor. The implication is that for a finite resolution map there is a magnetometer sensitivity below which no improvement is expected.

We use Map Feature Variability (MFV) as a measure of data variation sparsity of the geophysical data map. The MFV at a data point i on a data map is defined as $\mathcal{C}_i = \frac{1}{n}\sum_j^n (s_{x_i} - s_{x_j})^2$, $\forall\, x_j \in$ search window, $x_j \neq x_i$. In a map matching based INS aiding, the value of $\mathcal{C}_i^{-1}$ may be used to weight the estimated sensor location covariance to provide additional parameter that locally describes quality of the data map used.

Figure 4c shows an example of the normalised map feature variability over the map area (Fig. 4a). For reference, we also show the original TMI map in Fig. 3a, and the PDA error distance maps for sensor noise levels $\sigma = 0.015$ nT and 0.15 nT in Fig. 3b and d, respectively.

The Map matching localisation problem, shown in Fig. 1, is solved using the probabilistic multiple hypothesis tracker based map matching (PMHT-MM) proposed in (Wang et al.

2023). It iteratively estimates the current vehicle location from a batch of measurements processed by the probabilistic multiple hypothesis map matching method introduced in Sect. 3 under the vehicle dynamic constraints.

4   Navigation Experiment

The simulation scenario is a constant velocity vehicle travelling along the surface of the earth at a fixed height of 100 m from $[-38°, 144.5°]$ to $[-35°, 150°]$ (i.e., from Melbourne to Sydney) and at a ground speed of 22 m/s. The entire journey takes more than 3.6 hours and navigation is conducted by an onboard INS in GNSS denied environment. The inertial sensors (both accelerometer and gyroscope) used in the INS are precision grade with errors specified according to Jekeli (2005), with measurement frequency of 1 Hz and are assumed to be well calibrated before the journey starts. We assume that a low noise magnetometer is onboard to take

Fig. 4 TMI map quality analysis in the red rectangle area shown in Fig. 3. (**a**) TMI map shown inside the red rectangle in Fig. 3a. (**c**) Map feature variability. (**b**) PDA error distance for $\sigma = 0.015$ nT. (**d**) PDA error distance for $\sigma = 0.15$ nT

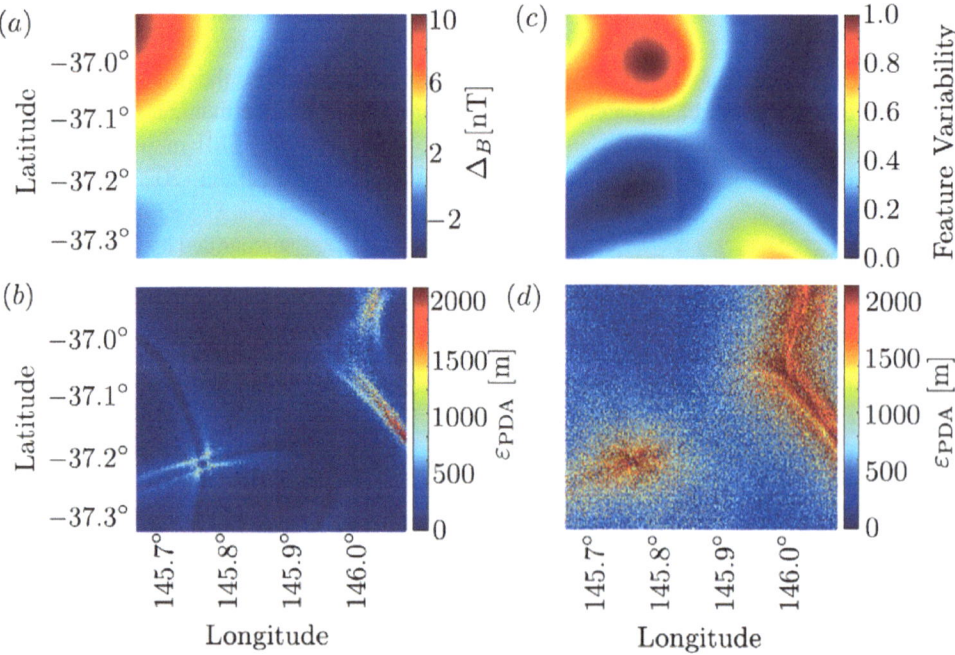

Fig. 5 Comparison of RMS position errors of the aided INS for magnetometer measurement noise levels 0.015 nT (red) and 0.15 nT (blue), along with INS-only case (black)

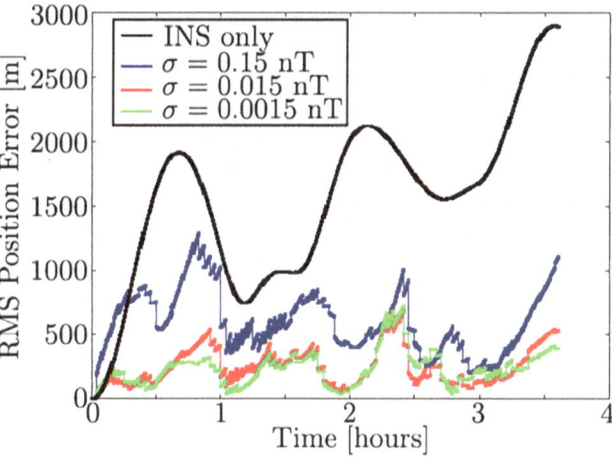

magnetic intensity measurement at an interval of every 10 seconds. The PMHT-MM algorithm works with a batch of 30 magnetic intensity measurements at a time in an aiding interval 300 seconds.

In this experiment, two noise levels for the magnetometer are considered: (1) $\sigma = 0.0015$ nT, which is to model a magnetometer of very high precision; (2) $\sigma = 0.15$ nT, which represents the level of sensitivity of magnetometers that are commercially available. We plot the vehicle root-mean-squared (RMS) position errors in Fig. 5 along with case of INS-only without aiding. The results were averaged from 100 Monte Carlo runs for each of cases.

The simulation results in Fig. 5 show that:

- when the noise level of magnetometer is 0.0015 nT, the INS with magnetometry aiding can achieve an average of RMS position error of 250 m; the RMS position error doubles if the noise level is 10 times larger at 0.15 nT.
- magnetometry INS aiding is robust with 100% success rate with a batch length (i.e., the number of magnetometer measurements to be processed in a batch) 30 at each time in this simulation. If a lower batch length or a high noise level magnetometer is used, the RMS position error increases and the magnetometry INS aiding will not be completely reliable.
- magnetometry INS aiding is able to remove position drift, as indicated by the INS-only case, which is accumulated over time due to the imperfection of inertial sensors.

5   Conclusions

In this paper, we describe a probabilistic method for map matching localisation based on magnetometry and total magnetic intensity maps. We demonstrated the effectiveness of the magnetometry map matching localisation via simulation. The magnetometry map matching removes accumulated position drift in the INS, that arises in the absence GNSS

positioning. Simulation results verified the robustness and effectiveness of the proposed algorithm, particularly, the aiding precision improves with increasing magnetometer sensitivity, until the quality of the magnetic map limits precision.

Funding This work is partially supported by the Australian Army Quantum Exploit funding.

References

Australia Government (2023) Geophysical Archive Data Delivery System. https://portal.ga.gov.au/persona/gadds

Bai D, Huynh MH, Simpson DA, Reineck P, Vahid SA, Greentree AD, et al. (2020) Fluorescent diamond microparticle doped glass fiber for magnetic field sensing. APL Mater 8(8):081102

Balasubramanian G, Neumann P, Twitchen D, Markham M, Kolesov R, Mizuochi N, et al. (2009) Ultralong spin coherence time in isotopically engineered diamond. Nat Mater 8(5):383–387

Dumeige Y, Roch JF, Bretenaker F, Debuisschert T, Acosta V, Becher C, et al. (2019) Infrared laser threshold magnetometry with a NV doped diamond intracavity etalon. Opt Express 27(2):1706–1717

Filipkowski A, Mrózek M, Stępniewski G, Kierdaszuk J, Drabińska A, Karpate T, et al. (2022) Volumetric incorporation of NV diamond emitters in nanostructured F2 glass magneto-optical fiber probes. Carbon 196:10–19

Frontera P, Alessandrini S, Stetson J (2018) Shipboard calibration of a diamond nitrogen vacancy magnetic field sensor. In: 2018 IEEE/ION Position, Location and Navigation Symposium (PLANS), Monterey, CA, USA, 2018, pp. 497–504. https://doi.org/10.1109/PLANS.2018.8373418

Goldenberg F (2006, April) Geomagnetic navigation beyond the magnetic compass. In: Proceedings of IEEE/ION PLANS, pp 684-694

Groves P (2013) Principles of GNSS, inertial, and multisensor integrated navigation systems. Artech House, Boston

Hahl FA, Lindner L, Vidal X, Luo T, Ohshima T, Onoda S, et al. (2022) Magnetic-field-dependent stimulated emission from nitrogen-vacancy centers in diamond. Sci Adv 8(22):eabn7192

Jekeli C (2005) Navigation error analysis of atom interferometer inertial sensor. Navigation 52(1):1–14

Jeske J, Cole JH, Greentree AD (2016) Laser threshold magnetometry. J Phys 18(1):013015

Kamgar-Parsi B, Kamgar-Parsi B (1999) Vehicle localization on gravity maps. In: Gerhart GR, Gunderson RW, Shoemaker CM (eds) Unmanned ground vehicle technology. SPIE. https://doi.org/10.1117/12.354447

Li W, Gilliam C, Wang X, et al (2022) Gravity aided navigation using viterbi map matching algorithm. https://arxiv.org/abs/2204.10492v1

Ruan Y, Simpson DA, Jeske J, Ebendorff-Heidepriem H, Lau DWM, Ji H, et al. (2018) Magnetically sensitive nanodiamond-doped tellurite glass fibers. Sci Rep 8(1):1268

Titterton DH, Weston JL (2004) Strapdown inertial navigation technology, 2nd edn. The Institution of Electrical Engineers, London

Tuohy S, Patrikalakis N, Leonard J, et al (1996) Map based navigation for autonomous underwater vehicles. Int J Offshore Polar Eng 6(01):278–289

Tyren C (1982) Magnetic anomalies as a reference for ground-speed and map-matching navigation. J Navigat 35(2):242–254

Wang B, Ma Z, Huang L, et al (2022a) A filtered-marine map-based matching method for gravity-aided navigation of underwater vehicles. IEEE/ASME Trans Mechatron 27(6):1–11.

Wang X, Gilliam C, Kealy A, et al (2022b) Probabilistic map matching for robust inertial navigation aiding. ArXiv abs/2203.16932

Wang X, Gilliam C, Kealy A, Close J, Moran B (2023) Probabilistic map matching for robust inertial navigation aiding. NAVIGATION J Instit Navigat 70(2):navi.583

Feasibility of CSAC-Assisted GNSS Receiver Fingerprinting

Qianwen Lin and Steffen Schön

Abstract

Interference and jamming of Global Navigation Satellite System (GNSS) signals can induce inaccurate Position, Velocity and Time (PVT) information, resulting in crucial integrity and even security issues. The poor stability and accuracy of the GNSS receivers' internal clocks, i.e. quartz oscillators, additionally impact the situation by hindering the detection of spoofing signals. High-precision atomic clocks are used to enhance PVT results, however, their bulk, weight and energy consumption constrain their deployment scenarios. Miniature atomic clocks (MAC) present a promising alternative that trades off between frequency stability and size/weight limitations of an atomic clock.

This paper investigates the potential of chip-scale atomic clocks (CSAC) as external clocks of GNSS receivers for fingerprinting the receivers in both static and dynamic environments. Fingerprinting is characterized by the clock's physical behavior expressed by Allan Deviation (ADEV) or Time Interval Error (TIE), both of which relate to the clocks' frequency stability. Thus, unique receiver clock features serve as clock fingerprints. The optimal combinations of features are explored by three feature extraction methods. We gathered GNSS data in diverse scenarios, consisting of a four-day static experiment, a car and a flight experiment as well as the corresponding static experiment for comparison. Results indicate that CSAC-aided fingerprinting is feasible in static conditions, achieving an overall accuracy (OA) of 90% across the three methods. One of the three methods is proven effective to handle clock fingerprinting in dynamic conditions, but yielded a comparably lower OA than in static conditions.

Keywords

Allan deviation (ADEV) · Chip-scale atomic clock (CSAC) · GNSS interference · Machine learning · Receiver clock fingerprinting

1 Introduction

Ensuring signal authenticity is critical for Global Navigation Satellite System (GNSS) users, particularly intentional attacks to GNSS receivers are feasible during the signal transmission or when transmitting GNSS data to location-based applications (Borio et al. 2017). In the field of Wireless Local Area Networks (WLAN), wireless device fingerprinting is a considerably viable strategy to address the issue of signal authenticity. The fingerprints/signatures derived from the device-specific metrics are generated to identify individual devices or separate different devices (Xu et al. 2015; Polak and Goeckel 2015).

GNSS receiver fingerprinting has accordingly been investigated to preliminarily discriminate receivers in static scenarios (Borio et al. 2016). The motivation is that the receiver

Q. Lin (✉) · S. Schön
Institut für Erdmessung, Leibniz Universität Hannover, Hannover, Germany
e-mail: lin@ife.uni-hannover.de; schoen@ife.uni-hannover.de

© The Author(s) 2023
J. T. Freymueller, L. Sánchez (eds.), *Gravity, Positioning and Reference Frames*,
International Association of Geodesy Symposia 156, https://doi.org/10.1007/1345_2023_221

clock errors show certain controlled behaviour, i.e. frequency stability, thanks to the clocks' physical properties. Borio et al. (2017, 2016) indicated that a combination of three clock-specific features enables the separation of a few geodetic receivers from a few mass-market receivers. Note that, the Temperature-Compensated Crystal Oscillator (TCXO) embedded in GNSS receivers has limited long-term frequency stability and is highly sensitive to environmental factors like accelerations and vibrations (Jain and Schön 2020). Thus, the reliability of deriving clock-related features of an internal clock to fingerprint receivers is not guaranteed. In turn, high-precision atomic clocks show stable performance. A comprehensive overview of the foundational principles of quartz oscillators and atomic clocks is available in Teunissen and Montenbruck (2017). By equipping GNSS receivers with precise clocks like miniature atomic clocks (MAC) or chip-scale atomic clocks (CSAC), the vertical accuracy of the positioning results can be significantly improved, and even navigation using three satellites can be realized (Clock Coasting (Sturza 1983; Weinbach and Schön 2011; Krawinkel and Schön 2014)). It has been proven that the holdover performance of a CSAC-aided GNSS receiver, i.e. the recovery time from signal outages, is always stable and better than a normal receiver when time lapse exceeds 1 min (Fernández et al. 2017).

This paper investigates the potential of GNSS receiver fingerprinting in static and dynamic conditions by utilizing CSACs as receivers' external clocks. The clock-specific features for characterizing fingerprinting are presented in Sect. 2. Section 3 proposes the approaches adopted to feature extraction. Several GNSS measurements in various scenarios are then collected in Sect. 4. Finally, the feasibility of fingerprinting using clock-specific features is analyzed in Sect. 5, followed by the conclusion summarized in Sect. 6.

2   Clock-Derived Features

Clocks' fingerprints derive from unique clock-related features tied to frequency stability, demonstrating how an instant frequency adheres to its nominal frequency over time, thus reflecting clocks' unique physical behaviour. Consequently, 13 such features, consistent with those in Borio et al. (2017, 2016), are extracted from the characteristics of the metrics summarized in the following.

2.1   Allan Deviation (ADEV)

ADEV, the standard deviation of the first differences of fractional frequency values, is the most common way to measure frequency stability in time domain. The overlapping ADEV further improves original ADEV by utilizing all possible sample combinations overlapped to each other, leading to better estimate confidence (Riley 2008). Equation 1 is the way to calculate overlapping ADEV $\sigma_y(\tau)$. y represents the fractional frequency samples determined from the receiver clock drift, and N is the samples' total number. The sampling interval τ, also known as averaging time, is the multiplication of averaging factor n and sampling rate T_s ($\tau = n \cdot T_s$).

$$\sigma_y^2(\tau) = \frac{1}{2n^2(N-2n+1)} \sum_{j=1}^{N-2n+1} \left(\sum_{i=j}^{j+n-1} (y_{i+n} - y_i) \right)^2 \quad (1)$$

Figure 1a gives an example of overlapping ADEV for various oscillators. Apparently the Rubidium frequency standard (*SRS PRS10*) has the best frequency stability for both short and long term. The advantage of CSACs is the stable performances at long-term averaging time, while the sta-

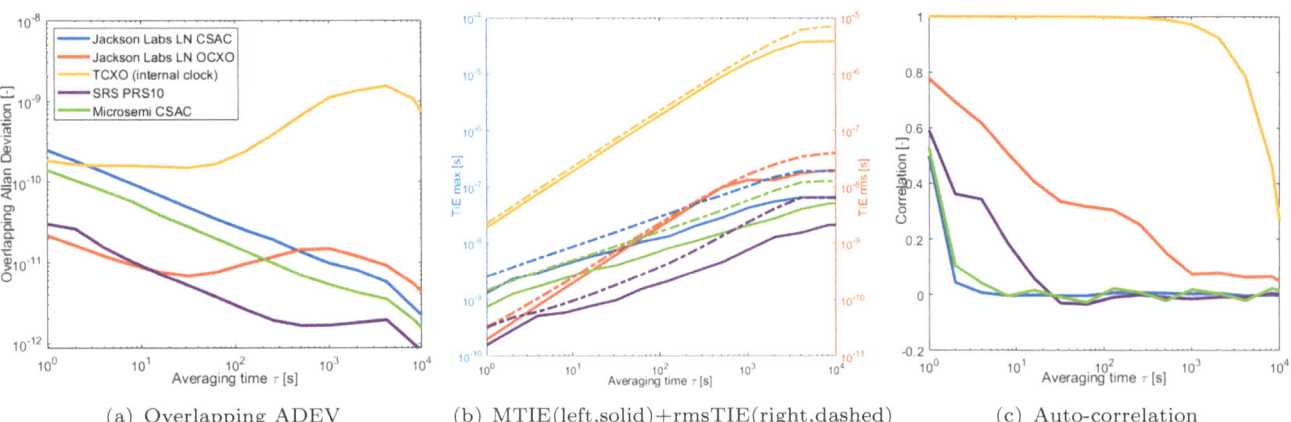

(a) Overlapping ADEV   (b) MTIE(left,solid)+rmsTIE(right,dashed)   (c) Auto-correlation

Fig. 1 Metrics of frequency stability for oscillators including CSACs (*Jackson Labs CSAC* (blue), *Microsemi CSAC* (green), *Stanford Research Systems PRS10* (high-precision) (purple)), quartz oscillators (*Jackson Labs OCXO* (orange), *TCXO* (yellow)). The metrics are derived from static GNSS data collected for Project *VENADU-A2* (Krawinkel and Schön 2014). (**a**) Overlapping ADEV. (**b**) MTIE(left,solid)+rmsTIE(right,dashed). (**c**) Auto-correlation

bility of the quartz oscillators becomes comparably unstable. However, the OCXO post-filter shows good short-term stability. Based on the oscillators' different performances, the candidate features, noted as OA_x, are decided to be the values at short-term $1s$ OA_1, $30s$ OA_{30} and their slope OA_{slope}. The minimum value OA_{min} and the averaging time τ_{min} of OA_{min} are also supposed to be useful for differentiating the clocks.

2.2 Time Interval Error (TIE)

Time Interval Error (TIE) is another measure of a clock's time errors. It describes time error variations through a time interval τ starting from the time point t_0, defined as Eq. 2. TE means time errors, referring to the differences between instantaneous times and its ideal times (Bregni 2002). It can be calculated by the integral of frequency errors ($\sum_{i=0}^{n} y_i T_s$, n: sample lag with similar meaning to m mentioned in Eq. 1) (Borio et al. 2016). In this way the maximum TIE ($MTIE$, Eq. 3) and the root mean square of TIE (TIE_{rms}, Eq. 4) are meaningful for characterizing a clock's stability behaviours. Different from the measures determined by averaging data samples, $MTIE$ refers to variations of the peak values of TIE within a time period T, as described in Eq. 3 (Bregni 2002).

$$TIE(\tau) = TE(t_0 + \tau) - TE(t_0) \tag{2}$$

$$MTIE(\tau) = \max_{t_0=0}^{T-\tau}\left(\max_{t=t_0}^{t_0+\tau} TE(t) \quad \min_{t=t_0}^{t_0+\tau} TE(t) \right) \tag{3}$$

$$TIE_{rms}(\tau) = \sqrt{\frac{1}{N-n} \sum_{t_0=1}^{N-n} TIE(\tau)^2} \tag{4}$$

From Fig. 1b, we can see the $MTIE$ and TIE_{rms} curves of each oscillator are less distinctive than $\sigma_y(\tau)$. The curves rise along averaging time with very similar slopes, especially for CSACs. Nevertheless, the values at $1s$, $30s$ and their slope describe generally the clocks' behaviours. Thus the features $MTIE_1$, $MTIE_{30}$, $MTIE_{slope}$, $rmsTIE_1$, $rmsTIE_{30}$ and $rmsTIE_{slope}$ are extracted as the potential features of fingerprinting.

2.3 Correlation Between Time Series

In Polak and Goeckel (2015), Borio et al. (2016), the autocorrelation of normalized frequency errors is utilized to produce features as fingerprints of oscillators. The autocorrelation curves of several oscillators from short to long time intervals are shown in Fig. 1c. It is noticeable that the internal clock generates high-correlated time series ($\sim$1) until the interval increases to $\sim$ $10^3 s$. Another quartz oscillator (orange) shows also correlations, whereas the time series of CSACs decorrelate themselves quickly within $\sim$30s. To choose specific characters for separating different clocks, the candidate features are decided as correlation at time intervals of $20s$ R_{20} and of $60s$ R_{60}.

3 Feature Extraction

Feature extraction is a way to exploit practical features for fingerprinting. Attention should be paid on the reduction of feature dimension because features can be redundant (Xu et al. 2015). The idea is to form a feature set by either creating important new features from, or directly selecting several essential features among, the candidate features. Three related machine learning approaches are proposed in this section, essentially recasting clock fingerprinting as a classification problem.

3.1 Pre-Processing Procedures

First of all, a series of pre-processes are implemented to acquire more precise frequency data, enabling candidate features to better reflect the clocks' real stability behaviours. Figure 2 outlines the pre-processing steps for GNSS raw data. The receiver clock parameter *Clock Drifts* is initially resolved by passing the Doppler observations through a Single Point Positioning (SPP) estimation. This raw frequency data then undergoes further processing: small gaps are filled, deterministic effects like frequency offsets or frequency drifts are subtracted, and outliers are removed.

To determine the minimum observation duration for extracting reliable features, the processed static data sampling in 1 Hz is divided into non-overlapping segments. The segment length increases from 20 min to 120 min, with 10 min increment. For kinematic data sampling in 10 Hz, the length of data segments starts from $30s$ with 1 min increment. Note that, a longer segment duration allows the extracted features more broadly representing the clocks' frequency stability but in our case induces a reduction of sample size due to the fixed total data duration. Conversely, a shorter segment duration presents the opposite situation. For instance, given a five-day measurement sampling in 1 Hz, the quantity of segments ranges between 360 and 60. Suppose

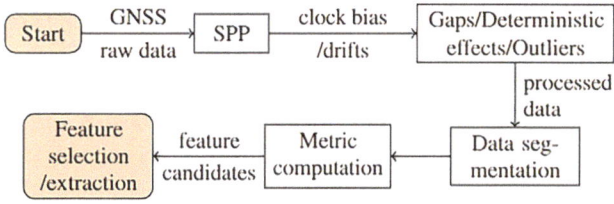

Fig. 2 Pre-processing flowchart of fingerprinting

five receivers are measuring simultaneously, the accumulated sample size n increases correspondingly, ranging from 1800 to 300. For each segment/sample, a feature vector of 13 features is subsequently computed. Hence, for each dataset of a specific observation duration, a feature matrix A is compiled for classification, accomplished by stacking feature vectors with the dimension of $n \times 13$.

3.2 Singular Value Decomposition and Support Vector Machine

A widely-used method for reducing data dimension is Principle Component Analysis (PCA) (Bishop and Nasrabadi 2006). It projects the features to a lower-dimensional space where the most important information is determined from dimensions with the greatest variances and meanwhile decorrelated because the dimensions are orthogonal to each other. This can be realized by Singular Value Decomposition (SVD) of feature matrix, where the right singular vector sorts the importance of information in columns of original data.

Equation 5 specifies this process. The columns C_i of the feature matrix A are firstly normalized by $\frac{C_i - \min(C_i)}{\max(C_i) - \min(C_i)}$ because of features' different units and magnitudes. Decomposing the normalized A, we obtain a matrix Σ containing the singular values arranged on the diagonal from large to small, or from important to unimportant. The right singular vector V consists of the columns $V_{i,13 \times 1}$ referring to the singular values in sequence. If the cumulative proportion of the first m singular values exceeds the empirical 95% threshold, we consider the first m columns of V contains sufficient information to describe A. Hence, the new feature matrix A' is derived by a multiplication in Eq. 5. Specifically, each new feature element is calculated by $\sum_{i=1}^{13} v_i a_i$, in which a_i is an element of the original feature vector $A_{i,1 \times 13}$ and v_i is an element of $V'_{i,13 \times 1}$, equivalent to a weight. Thus, each feature vector is characterized by m new generated features instead of 13 original features. However, the new features cannot be physically interpreted like the original features.

$$A_{n \times 13} = U_{n \times n} \cdot \Sigma_{n \times 13} \cdot V_{13 \times 13}^T$$
$$A'_{n \times m} = A_{n \times 13} \times \begin{bmatrix} V_1 & V_2 & \cdots & V_{\mathbf{m}} \end{bmatrix} \qquad (5)$$

Support Vector Machine (SVM) is a widely-used classifier, fitting for scenarios with sparse data samples, akin to our experimental situation. The essential of SVM is to discover an optimal hyperplane that simultaneously maximizes the vertical distances from the hyperplane to the planes formed by each class's support vectors. Support vectors signify each class's data samples nearest to the hyperplane. In our case, multi-class classification problems are considered due to the

experiments conducted usually with multiple clocks. This can be solved through *one-versus-the-rest* approach which trains multiple hyperplanes (Bishop and Nasrabadi 2006). Each trained hyperplane separates one and the rest classes.

Moreover, cross validation is performed ten times to train desired classifiers with robust generalization. Each iteration utilizes 90% randomly-selected samples to train a classifier which is subsequently tested by the remaining 10%. The testing samples are assigned to the classes yielding the highest scores or probabilities. Subsequently the feasibility of fingerprinting of this approach is assessed by calculating the overall accuracy (OA, different from OA_x above), precision and recall. Additionally, the optimal, i.e. minimum, observation duration is marked by superior OA.

3.3 Decision Tree

The key idea is to construct a binary tree by issuing a decision for each tree node, i.e. choosing the optimal attribute for partitioning the whole data samples. For multi-class classification problems, the optimal attribute for the root node should separate the dataset into two portions which have as few samples of the same category as possible. The same arrangement is in turn adopted to the two portions until all classes have been identified. This idea can be realized by measuring the information gain of each attribute for every node decision. The information gain is computed by comparing the change in entropy before and after partitioning the dataset. The entropy describes the degree of impurity or disorder in a dataset.

$$Gain(D, a) = Ent(D) - \sum_{v=1}^{V} \frac{|D_v|}{|D|} Ent(D_v)$$
$$Ent(D) = -\sum_{k=1}^{K} p_k log_2 p_k \qquad (6)$$

Equation 6 explains information gain mathematically, in which D, a and v represent a dataset, a feature/an attribute, and possible values of a feature, respectively. $\frac{|D_v|}{|D|}$ equals to a weight for v by means of the proportion of the positive samples. k and p_k denotes the classes and the proportion of samples of a class k (Wang et al. 2017). The greater the information gain, the greater the purity gain obtained by using feature a to segment the dataset D.

Similar to the approach SVD+SVM, ten-times cross validation is implemented for decision tree. During the training phase, 90% samples are used to train a decision tree. Several efficient features are selected during node decisions. Few features will be decided for the dataset with high purity and vice versa. The remaining samples test the decision tree, resulting in scores or probabilities of all classes, and quality measures.

3.4 Filtering Method

We adopt the filtering method successfully developed in Borio et al. (2017) to fingerprint various oscillators as a comparison with the approaches above. The candidate features are randomly assigned to groups with a capacity of three, i.e. $C_{13}^3 = 286$ combinations. Essentially, it defines a score function (Eq. 7) to rank these combinations, selecting the one with the highest score for fingerprinting. The score function is a ratio between the minimum inter-class (different classes i and j) distances and the maximum intra-class distances (class i). F denotes a feature subset, cf. Borio et al. (2017).

$$Score\ G(F) = \frac{\min_{i \neq j} d_{i,j}(F)}{\max_i d_i(F)} \rightarrow \max \qquad (7)$$

In our case, we utilize the filtering method during the training procedures of 10 times cross validation. Hence, the feature subset of the largest Score G is used in the testing procedure, in which the testing samples are characterized by the three features. The classification is done by comparing the Mahalanobis distances from the samples to the class centers, followed by an evaluation process.

4 Overview on Experiments

GNSS data documented in various scenarios are gathered to demonstrate the effectiveness of the approaches mentioned above (Table 1). Firstly a static experiment was executed on the institute's roof top in $1Hz$ sampling rate for four days (Krawinkel and Schön 2014). A fast-driving experiment consists of tracks along the route comprising a highway, an urban area in city Siegen, three tunnels and a small road with plaster, producing a ~1.5 hours dataset sampled in $10Hz$. A flight experiment was realized in Dortmund with the same equipment setup, yielding ~2.5 hours data in $10Hz$ sampling rate (Jain and Schön 2020).

Each GNSS receiver of the same type functions by either utilizing its built-in clock or connecting externally to a

miniature clock, cf. Table 1. For each kinematic experiment, a reference trajectory is created using a relative positioning approach, based on high-quality observations from Inertial Measurement Unit (IMU) of *IGI AEROcontrol* and GPS phase measurements. The operation setup of the same clocks is tested in static scenario.

5 Fingerprinting Results

5.1 Static Scenarios

Figure 3 gives an overview of the three approaches' performances with five clocks in static condition. Only OA is shown here because precision and recall have similar behaviours. First, we can notice the accuracy of the three approaches for datasets in all durations is larger than 99%, denoting the capability of the three approaches for fingerprinting the clocks. The missing part of less than 1% is resulted from one or two wrongly-classified samples visualized in Fig. 4. Additionally, for the necessary amount of features for fingerprinting, the feature dimension of SVD+SVM reduces from 13 to 7, while the other two methods require only three features. This implies that SVD+SVM is not as efficient as the other two approaches in this experiment context. It also accounts for its minimum observation duration of only $30min$ to achieve its best accuracy, as supposed to $50min$ required by the other two approaches.

The classification results of filtering method are shown in Fig. 4a. 814 samples are denoted in five colors in three

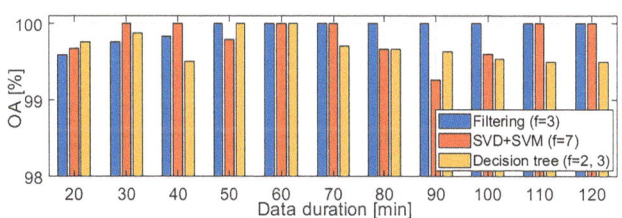

Fig. 3 OA of classification results for datasets with various observation periods. The datasets are obtained from static Exp1 and processed using the three proposed approached. f is the number of selected features

Table 1 Experiment data summary and description

	Sta. Exp1	Kin. Exp2	Kin. Exp3
Measured scenario	Fixed	Various conditions on road	Flight
Duration	~ 4 days	~ 1.5 hours	~ 2.5 hours
Sample rate	1 Hz	10 Hz	10 Hz
Reference trajectory		GPS+IGI Aerocontrol	GPS+IGI Aerocontrol
Receiver type	Javad Delta TRE_G3T	Javad Delta TRE_G3T	Javad Delta TRE_G3T
Clocks[a,b]	**1-3**, 6-7	**3-5**, 8	**4-5**, 7-8

[a]**CSAC**: 1. Microsemi CSAC SA.45s, 2. Jackson Labs LN (JLN) CSAC, 3. Standard Research Systems (SRS) PRS 10, 4. Microsemi MAC SA.35m, 5. Spectratime LCR 900
[b]Quartz: 6. JLN OCXO, 7. internal clock, 8. SRS SC 10

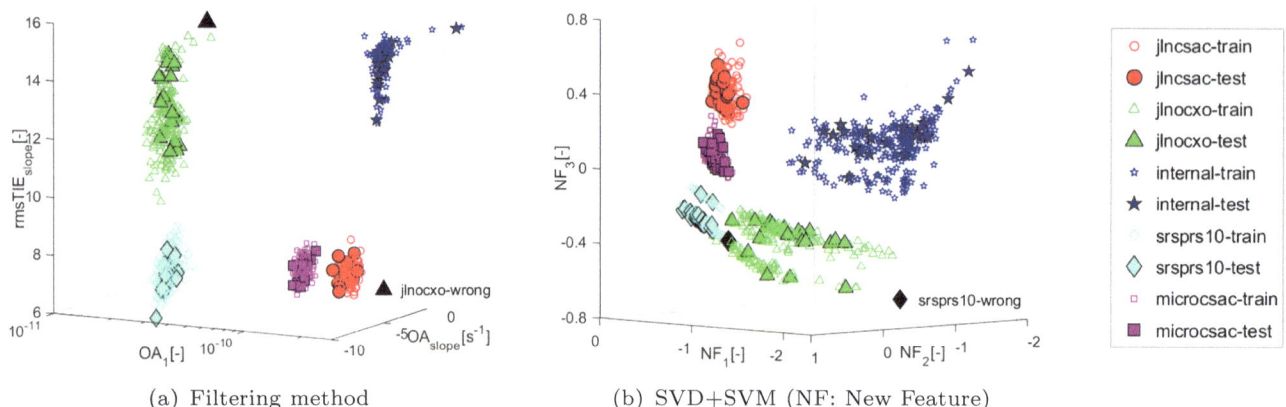

(a) Filtering method   (b) SVD+SVM (NF: New Feature)

Fig. 4 3D visualization of fingerprinting results for data of static Exp1. (**a**) Filtering method: obtained feature combination of OA_1, OA_{slope} and $rmsTIE_{slope}$, selected for frequency data of 30 min time length.

(**b**) SVD+SVM: first three of seven generated features, selected for frequency data of 20 min time length

dimensions of the selected features. The features of each sample are derived from a $30min$ observation dataset. Intuitively the clusters of five clocks are distinctive, demonstrating the feasibility of fingerprinting of this method. The CSAC clusters (red & magenta) are relatively concentrated and their 3D locations are close. This can be interpreted by the two CSACs' similar physical properties. In reverse, the distribution of quartz oscillators' samples (green & blue) are scattered, especially in the dimension of $rmsTIE_{slope}$, implying their instability. The class of high-precision clock (cyan) is additionally easy to be identified due to its distinct locations to others. Specifically, the sample marked by a black triangle is wrongly assigned, resulting from the shorter Mahalanobis distance of the sample to the blue cluster center instead of the green one.

The performance of SVD+SVM is displayed in Fig. 4b. 1222 samples are expressed by the first three of seven new features, originated from 20-min observation datasets. Similarly, the clusters are well separated to each other and CSACs in red and magenta symbols are especially straightforward to be distinguished. Besides that, the five clusters get closer comparing to those in Fig. 4a because seven features are in fact used in SVM to divide the classes. Admittedly, the new features derived via SVD lack a clear physical explanation, which makes it difficult to discover the efficiency of each feature candidate.

A different feature set, derived from the same 1222 samples, is decided by decision tree. Figure 5 presents how the trained tree distinguishes classes. Noticing its left column, OA_{30} distinctly isolates the internal clock (blue) which has the features markedly differing from others. In the middle column, $rmsTIE_{slope}$ further separates the green from the rest. Though both OA_{30} and OA_1 can separate the cyan from the magenta, the tree opts for OA_1 because it provides a greater separation ($\sim 6 \times 10^{-11}$ in 1-D distance)

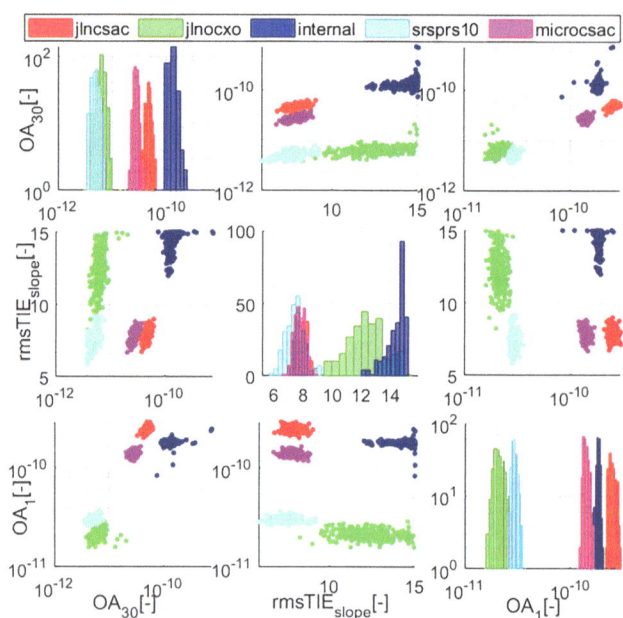

Fig. 5 3×3 scatter plots of fingerprinting results using method decision tree for data of static Exp1. Three features OA_{30}, $rmsTIE_{slope}$ and OA_1 are selected for frequency data of 20 min time length

between the two. The two CSACs (red and magenta) are again divided by OA_1. Finally, the effectiveness of a feature to distinguish classes can be concluded from the diagonal plots. Overlapping bars suggest the feature is not powerful for distinguishment, and vice versa.

In summary, CSACs are more identifiable by forming focused sample sets. Note that, although the selected features, minimum observation duration, and fingerprinting results may vary with each execution, fingerprinting in static scenarios via the three approaches are proven feasible, given appropriate features are chosen. Moreover, the filtering method and decision tree outperform SVD+SVM in efficiency due to fewer required features.

5.2 Dynamic Scenarios

Clocks employed in dynamic scenarios suffer from environmental factors like temperature variations, accelerations and vibrations etc. Especially the data is recorded in hard GNSS conditions like urban areas, and through flight maneuvers. The derived clock-related features are likely to mix with massive noise, undoubtedly complicating the clock fingerprinting process. Table 2 summarizes the classification results of three approaches in such scenarios. Principally, the OA of dynamic scenarios ($<$ 75%) is lower than static scenarios ($\sim$ 90%). The filtering method's comparably low OA indicates three features cannot adequately handle such complicated situation. SVD+SVM performs slightly better with more features required. Furthermore, decision tree has the most accurate classification results and generally needs few features. The large number required for selected features partly results from the high volume of samples, which increases the class confusion.

Figure 6 gives the classification results of decision tree using kinematic data segmented in 30 s. Note that, the

results are visualized in only three dimensions, whereas more than three are needed to get the OA shown in Table 2. Apparently the four clusters in the first two plots are not successfully distinguished, accompanying with many wrong-classified samples (black). This is also reflected in the confusion matrices of Fig. 6c. Nevertheless, the centroid of the magenta (internal clock) in Fig. 6b is clearly isolated from others, leading to a $\sim$ 100% classification precision of this class. In return, the sample sets of the rest classes are densely mixed because of the similar properties of high precision and stability. The four clusters in Fig. 6a are scattered except the green (LCR900) is slightly apart from others. This indicates again the comparable characteristics of the four clocks and the necessity of exploring more efficient features to distinguish them.

6 Conclusion

In this paper, we investigate the feasibility of receiver fingerprinting aided by CSACs in static and kinematic conditions. 13 features related to clocks' frequency stability are treated as fingerprints, derived from overlapping ADEV, TIE and autocorrelation. The approaches SVD+SVM and decision tree are adopted to determine practical features for fingerprinting. An existing filtering method is implemented for comparison.

The three approaches are proven effective to classify clocks in static scenarios with OA exceeding 99%. Besides, CSACs are advantageous for clock identification due to the extremely stable clock behaviour. For dynamic GNSS data, decision tree outperforms with over 70% OA, followed by SVD+SVM, while filtering method is not useful due to

Table 2 The number of selected features (#f) and OA of three approaches in two kinematic experiments and the corresponding static experiment

	Car Exp2		Flight Exp3		C. Static	
	#f	OA[%]	#f	OA[%]	#f	OA[%]
Filtering	3	37	3	63	3	87
SVD+SVM	8	48	7	68	9	96
D. Tree	1,4,5	71	2:9	74	2:9	87

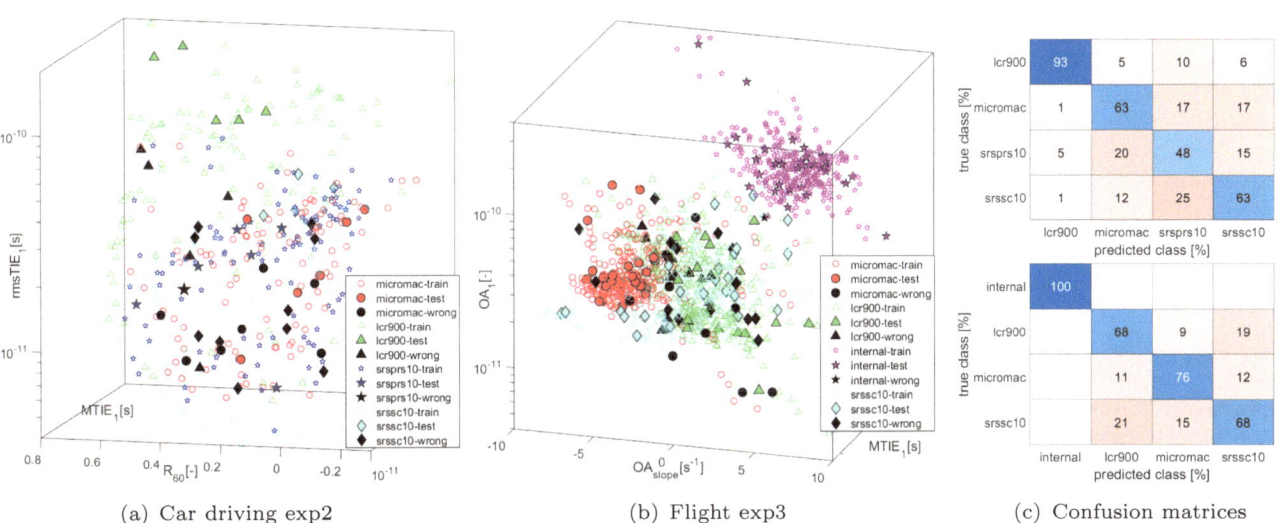

(a) Car driving exp2   (b) Flight exp3   (c) Confusion matrices

Fig. 6 Fingerprinting results of kinematic data segmented in 30s, via decision tree. (**a**) Car exp2: $\sim$440 samples visualized by first 3 of 5 selected features with OA $\sim$ 65%; (**b**) Flight exp3: $\sim$1400 samples visualized by first 3 of 9 selected features with OA $\sim$ 79%; (**c**) Confusion matrices of car exp2 (top) and flight exp3 (bottom), the numbers are rounded in percentage

insufficient selected features. Lastly, the amount of necessary features and min. observation duration for fingerprinting highly rely on the complexity of experiment data. Static data with less noise and deterministic effects requires normally few features and short observation periods.

Acknowledgements The work is part of the project **FIRST** (50NA2101), which is funded by the German Federal Ministry for Economics Affairs and Climate Action.

References

Bishop CM, Nasrabadi NM (2006) Pattern recognition and machine learning, vol. 4. Springer, New York

Borio D, Gioia C, Baldini G, Fortuny J (2016) GNSS receiver fingerprinting for security-enhanced applications. In: 2016 ION GNSS+, pp 2960–2970

Borio D, Gioia C, Cano Pons E, Baldini G (2017) GNSS receiver identification using clockderived metrics. Sensors 17(9):2120

Bregni S (2002) Synchronization of digital telecommunications networks. Wiley, UK

Fernández E, Calero D, Parés ME (2017) CSAC characterization and its impact on GNSS clock augmentation performance. Sensors 17(2):370

Jain A, Schön S (2020) Influence of receiver clock modeling in GNSS-based flight navigation: Concepts and experimental results. In: 2020 IEEE/ION PLANS, pp 208–218

Krawinkel T, Schön S (2014) Applying miniaturized atomic clocks for improved kinematic GNSS single point positioning. In: 2014 ION GNSS+, pp 2431–2439

Polak AC, Goeckel DL (2015) Wireless device identification based on RF oscillator imperfections. IEEE Trans on Inf Foren Secur 10(12):2492–2501

Riley WJ (2008) Handbook of frequency stability analysis. NIST special publication 1065. Boulder (2008)

Sturza MA (1983) GPS navigation using three satellites and a precise clock. Navigation 30(2):146–156

Teunissen PJ, Montenbruck O (2017) Springer handbook of global navigation satellite systems, vol 10. Springer, New York (2017)

Wang Y, Li Y, Song Y, Rong X, Zhang S (2017) Improvement of ID3 algorithm based on simplified information entropy and coordination degree. Algorithms 10(4):124

Weinbach U, Schön S (2011) GNSS receiver clock modeling when using high-precision oscillators and its impact on PPP. Adv Space Res 47(2):229–238

Xu Q, Zheng R, Saad W, Han Z (2015) Device fingerprinting in wireless networks: Challenges and opportunities. IEEE Commun Surv Tutor 18(1):94–104

On the Impact of GNSS Receiver Settings on the Estimation of Codephase Center Corrections

Yannick Breva ⓘ, Johannes Kröger ⓘ, Tobias Kersten ⓘ, and Steffen Schön ⓘ

Abstract

The role of codephase center corrections (CPC), also known as group delay variations (GDV), becomes more important nowadays, e.g. in navigation applications or ambiguity resolution. CPC are antenna dependent delays of the received codephase. They are varying with the angle of arrival of the signal at the GNSS antenna, i.e. with azimuth and elevation. CPC can be determined with a robot in the field with a similar approach as used for phase center corrections (PCC) for carrierphase measurements. The big challenge in the estimation of reliable CPC pattern is to deal with relatively noisy codephase observations compared to the correction magnitude. A better repeatability can be reached by reducing the overall codephase noise. One possibility to do this is to understand and improve the tracking loops of the receiver, especially the loop filters, within the calibration process. Due to highly dynamic stress caused by the fast robot motion, a perfect tracking of the GNSS signals is challenging. In this paper, a detailed look on the impact of different loop filter settings, like the noise bandwidth, the filter order or the use of an aided or unaided delay lock loop, on the time differenced single differences is done. To this end, an antenna calibration experiment was carried out, where, in addition to the hardware receivers, the IFEN Sx3 software receiver was used. The software receiver allows to change the settings in post-processing. The experiment shows, that the noise of the observations can be reduced by decreasing the noise bandwidth, but pattern information can be lost by using a bandwidth, which is too small. The trade-off between a small bandwidth and consequently less overall noise and the signal dynamics, caused by the fast robot motion, must be chosen carefully. At the end, an improvement in the pattern repeatability from 99.2 mm, using a hardware receiver, to 65.6 mm, using a software receiver with carefully chosen parameters, can be achieved.

Keywords

Absolute antenna calibration · Codephase center correction · Group delay variations · Loop filter

1   Introduction

For navigation applications or ambiguity resolution, codephase center corrections (CPC) – also known as group delay variations (GDV) – are becoming more important, like e.g. in ambiguity resolution with Precise Point Positioning. They are delays in the received codephase of the GNSS

Y. Breva (✉) · J. Kröger · T. Kersten · S. Schön
Leibniz University Hannover, Institut für Erdmessung, Hannover, Germany
e-mail: breva@ife.uni-hannover.de; kroeger@ife.uni-hannover.de; kersten@ife.uni-hannover.de; schoen@ife.uni-hannover.de

© The Author(s) 2023
J. T. Freymueller, L. Sánchez (eds.), *Gravity, Positioning and Reference Frames*,
International Association of Geodesy Symposia 156, https://doi.org/10.1007/1345_2023_206

signal and can have an impact on the ambiguity resolution in code-carrier linear combination (LC), e.g. Melbourne-Wübbena, when the magnitude of the respective antenna CPC is in the range of the LC wavelength (Kersten and Schön 2017). A concept to estimate CPC and also absolute phase center corrections (PCC) has been developed at the *Institut für Erdmessung* (IfE) – an antenna calibration facility accepted by the International GNSS Service (IGS) – in close cooperation with Geo++ (Menge et al. 1998; Wübbena et al. 2000; Böder et al. 2001) and is constantly improved and optimized (Kröger et al. 2021; Kersten et al. 2022). PCC are required to ensure a highly accurate position in GNSS applications, like precise point positioning. The estimation of PCC of an antenna under test (AUT) is done with a robot in the field. The estimation process is independent of the reference antenna's PCC; thus referred to as an *absolute calibration*. This approach is adapted for the estimation of CPC. IfE estimated CPC of the GPS L1 signal for low cost and geodetic antennas were shown in Kersten and Schön (2017), whereas Breva et al. (2019) presented first multi-GNSS CPC.

Related work on CPC/GDV comprises first CPC estimation with a robot done in 2008 (Wübbena et al. 2008), where a Kalman filter based on undifferentiated observations was used in a real-time process. At *Deutsches Zentrum für Luft- und Raumfahrt* (DLR) the electromagnetic behaviour of aeronautic antennas is estimated in an anechoic chamber. The group uses this information to calculate antenna dependent pseudorange errors as well as their GDV (Caizzone et al. 2019). Work at TU Dresden is based on code-minus-carrier linear combinations (CMC) to estimate satellite and receiver GDV together in a network approach (Wanninger et al. 2017; Beer et al. 2019). Wübbena et al. (2019) published absolute GDV for 36 antennas, Beer et al. (2021) were able to estimate absolute GDV for GNSS-satellite antennas with their CMC approach.

In the goal of estimating absolute CMC accurately and repeatedly, the noise of the codephase observations within the calibration process needs to be reduced significantly without deforming the CPC pattern. Therefore, Breva et al. (2022) presented an alternative data preprocessing strategy by using the empirical mode decomposition. Another opportunity for the noise reduction is to modify different receiver settings to ensure a stable tracking of the GNSS signals within the fast robot motion.

The present paper is structured as follows: after a brief overview about the antenna calibration and tracking loops of the GNSS receiver in Sect. 2, the impact of different settings of the GNSS tracking loops on codephase observations is studied in Sect. 3, which are used for the CPC estimation. In Sect. 4 the resulting CPC pattern with different receiver settings are shown as well as their repeatability. Section 5 concludes this paper.

2 Theoretical Background

2.1 Antenna Calibration at IfE

The *Institut für Erdmessung* uses an antenna calibration robot for estimating absolute phase and codephase center corrections. CPC are antenna dependent delays of the received codephase. They are varying with azimuth (α) and elevation (el) of the incoming satellite signal and are divided into a codephase center offset (CCO) and codephase center variation (CPV). The CPC can be calculated by

$$\begin{aligned} CPC(\alpha^k, el^k) = &- CCO \cdot \mathbf{e}(\alpha^k, el^k) \\ &+ CPV(\alpha^k, el^k) + r. \end{aligned} \tag{1}$$

Here, the CCO is projected onto the line-of-sight unit vector $\mathbf{e}$ towards the satellite k. The constant parameter r cannot be estimated without additional information. The rank deficit is removed by defining a certain datum ($CPC(z = 0) = 0$).

To estimate absolute CPC, an antenna under test is mounted on top of the robot nearby a reference antenna (see, Fig. 1). Each antenna is connected to a GNSS receiver (e.g. Septentrio PolaRx5TR), which are synchronized by an external frequency standard (Stanford Rubidium FS725). This setup forming a baseline of around 8 m, which allows

Fig. 1 Antenna calibration setup at IfE with the robot (foreground) and the nearby reference station (background)

calculating receiver-to-receiver single differences (SD). By time differencing the SD (ΔSD), almost all error sources are either cancelled out or reduced to a negligible magnitude. The antenna pattern information of the AUT are obtained by tilting and rotating the antenna around a fixed, certain point in space with the robot:

$$\Delta SD^k(t_i) = CPC^k_{AUT}(t_{i+1}) - CPC^k_{AUT}(t_i) + \epsilon \qquad (2)$$

It is noted that in addition to noise, also multipath (MP) are gathered in the parameter ϵ. Its amount highly depends on the antenna gain: MP at antennas with symmetric gain behaviour is cancelled out by rotating the antenna horizontally. In tilting robot sequences, the differences in MP between two epochs are still present and consequently contained in the ΔSD. Detailed investigations on MP within robot-based antenna calibration are currently done in the DFG MAE-STRO project. The time differenced single differences are used as the input for the estimation process based spherical harmonics. A detailed description of this approach can be found Kröger et al. (2021) and Kersten and Schön (2017).

2.2 Tracking Loops of GNSS Receiver

The task of GNSS receiver's tracking loops is to continuously track the GNSS signals received from the antenna and determine the aggregated observations like code, Doppler and carrier-phase for the navigation processing. In Fig. 2 (left) the main components of GNSS receivers are depicted. After amplifying the satellite signal, a down conversion to an intermediate frequency (IF) is achieved. Afterwards, the analog IF is converted via the analog-to-digital converter (ADC) to a digital IF signal. In the signal processing, the inphase and quadrature parts of the digital IF are correlated with the replica signals, analysed by the loop discriminators and filtered to steer the numerically controlled oscillator (NCO). The focus in this contribution is on the tracking loops, a detailed description of all main components can be

found e.g. in Häberling (2016) or Kaplan and Hegarty (2017) and will not be discussed here.

The basic principle of tracking loops are presented in the right of Fig. 2. They are located in the *signal processing* part of the GNSS receiver and starts right after the signal acquisition algorithm, where the GNSS signal is roughly located in the $\{\tau, f_D\}$ search space, where τ is the propagation time and f_D indicates the Doppler frequency of the satellite signal. The loops improve these rough estimates and continuously track changes in these parameters from that point forward (Misra and Enge 2006). Three kinds of tracking loops exist: The delay lock loop (DLL), the phase lock loop (PLL) and the frequency lock loop (FLL). Each loop consist of a *discriminator*, a *loop filter* and a *numerical or voltage controlled oscillator* (NCO, VCO). The main goal of the loops is to align the replica signals, generated by the local oscillator, to the incoming signals. Their outcomes are the code delay, frequency and phase of the satellite signal, which correspond to the codephase, carrier phase and Doppler observables in GNSS processing.

The replica signal $u_2(t)$ could be shifted in its code time, its phase or in its frequency to the incoming signal $u_1(t)$. The *discriminator* compares $u_2(t)$ and $u_1(t)$ to estimate possible delays or shifts, at which the code time delay $\Delta\tau$ is estimated by the DLL, the phase shift $\Delta\theta$ by the PLL and the frequency shift Δf by the FLL. The *discriminator* output signal $u_d(t)$, which depends on the used discriminator function (e.g. early-minus-late for DLL), passes the *loop filter* next. The filter reduce the noise in order to produce an accurate estimate of the original signal as output $u_f(t)$. A detailed description and analysis of loop filters and their parameters are depicted in Sect. 3. The last component of the loop is the *VCO/NCO*. This oscillator of a DLL creates a new replica signal $u_2(t)$ by slowing down or speeding up the clock that controls the speed of the replica code generator by the amount of $\Delta\tau$. In case of FLL or PLL, it is synchronizing the frequency and the phase of $u_f(t)$ with the frequency and phase of $u_1(t)$ by Δf and $\Delta\theta$. Afterwards, the delays and shifts are estimate again by the *discriminator*. When these parameters

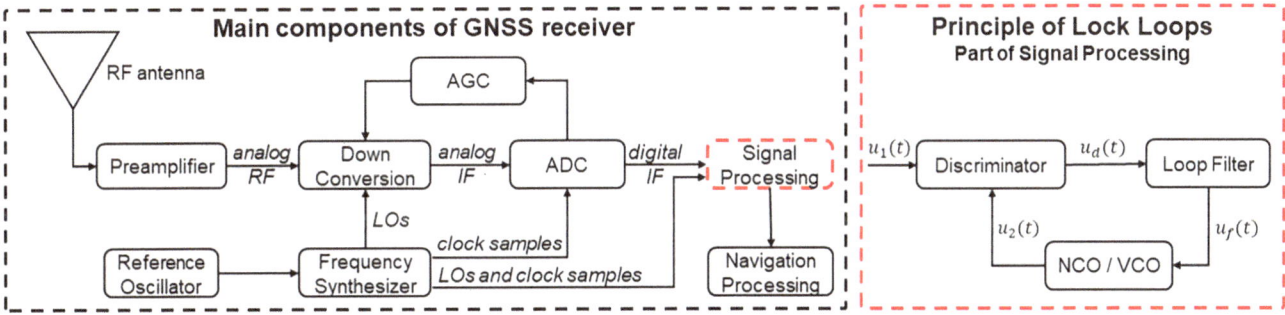

Fig. 2 (left) Overview of the GNSS receiver's main components and (right) a detailed look into the tracking loops, which are parts of the signal processing based on Häberling (2016)

are equal to zero or to a constant value, the loop is in a locked state.

The first operating tracking loop is the DLL. This loop provides the prompt correlation measurement required by the PLL and FLL. It must accurately estimate $\Delta\tau$ before the PLL begins to track. The FLL typically starts to operate, when the C/N_0 of the GNSS signals is too weak for a PLL operation.

3   Analysis of Different Receiver Settings

3.1   Practical Experiment

In order to analyse the impact of different receiver settings on the observations and the resulting CPC pattern, an antenna calibration experiment was set up. The goal of this experiment is to find the optimal receiver settings to optimize the tracking of codephase signals within the fast, challenging robot motion in a way that all CPC information are maintained.

Therefore, a Novatel antenna NOV703GGG.R2 NONE (S/N: 12420040) was mounted on the robot on 7th and 8th June 2022. Moreover, the Leica antenna LEIAR25.R3 LEIT (S/N: 9330001) was used as a reference. Each antenna was connected to one of two identical Septentrio PolaRx5TR GNSS receivers that are linked to the same external frequency standard FS725. In addition, the Novatel AUT was also connected to the IFEN software receiver. With this, a standard antenna calibration procedure was running with four individual sets (hereinafter called P_1 to P_4) with a duration of about 4 to 6 h (Table 1).

In general, hardware receivers have a limited amount of changeable settings for common users, like the bandwidth or order of the tracking loops. The directly changeable settings for the here used Septentrio receivers are the bandwidth of the DLL (default: 0.25 Hz) and the PLL (default: 15 Hz), as well as their coherent integration time (default: DLL 100 ms; PLL 10 ms). For further analysis, the default settings are used. The software receiver allows changing over 170 receiver settings in post-processing by using the same digitalized data stream. A list of all settings can be found in IFEN GmbH (2019). To enable a suitable solution, several receiver

settings are defined beforehand, like the correlator mode or correlator type. Here, the focus is on the impact of the DLL bandwidth and the DLL order on the time differenced single differences. Furthermore, the impact of an aided or unaided DLL, by using the default bandwidths for FLL (narrow: 1 Hz; wide: 10 Hz) and PLL (narrow: 9 Hz; wide: 50 Hz), is investigated.

3.2   Impact on Time Differenced Single Differences

To improve the estimation of CPC, less noisy codephase observations (ΔSD) are required. As mentioned in Sect. 2.2, the goal of loop filters is the noise reduction of the observations. Therefore, the noise bandwidth (B_N) and the filter order (FO) of the filter can be modified.

First, the impact of different DLL B_N on the ΔSD of the GPS C1C signal is presented in Fig. 3. In grey, the ΔSD observed with the Sx3 software receiver are presented for the first calibration set P_1. For a better comparison, the ΔSD observed with the classical Septentrio receivers are depicted in red. The parameters for both plots are the same, except the noise bandwidth. A B_N of 1 Hz is used in the left figure and a B_N of 0.25 Hz in the right figure. Obviously, decreasing the B_N leads to less noisy observations.

The second analysis belongs to the impact of different DLL filter orders. After Kaplan and Hegarty (2017) first order loop filter are sensitive to velocity stress, second order to acceleration stress and third order to jerk stress. In Fig. 4 the ΔSD with FO = 1 (left) and FO = 2 (right) are shown. By comparing both figures, the receiver sensitivity to robot motion (velocity) is clearly visible, because of the very noisy ΔSD from first order DLL. A DLL filter order of 2 and a B_N of 0.1 Hz leads to comparable ΔSD, observed with hardware receivers.

It should be noted, that previous studies consider an unaided DLL, so that the codephase tracking is done only by the DLL. An aided DLL uses information from the carrier tracking loops (FLL/PLL) to effectively remove the dynamic stress from the code loop. In this case, the aided DLL B_N can be as small as 0.005 Hz (Misra and Enge 2006). Figure 4 shows the differences between a second order unaided DLL (left) and a first order aided DLL (right) with a B_N of 0.05 Hz. The ΔSD of an unaided second order DLL with a small B_N are less noisy than the ΔSD acquired with the Septentrio receiver. The aided DLL leads to similar behaviour of software and hardware receiver, however with higher overall noise (Fig. 5).

Table 1 Duration of individual calibration sets P_1 to P_4

Set	P_1	P_2	P_3	P_4
GPS time	12:00–17:59	18:00–23:59	0:00–4:59	5:00–9:10
Duration [h]	6:00	6:00	5:00	4:10

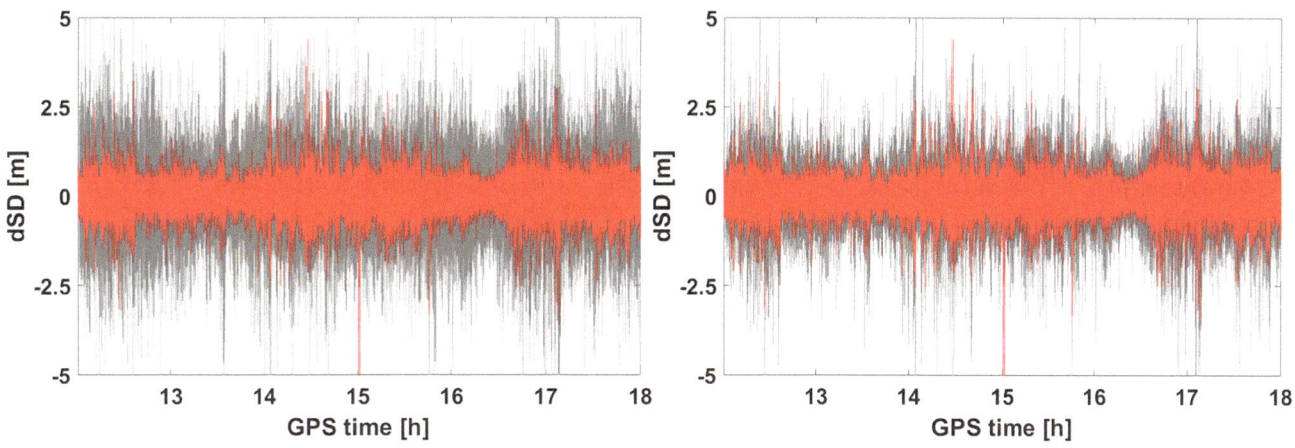

Fig. 3 Time differenced single differences of the GPS C1C signal of the first calibration set P_1. The ΔSD from the Septentrio receivers are presented in red and the ΔSD of the software receiver with different settings* are depicted in grey. *(left) $B_N = 1$ Hz, FO = 2, unaided DLL. (right) $B_N = 0.25$ Hz, FO = 2, unaided DLL

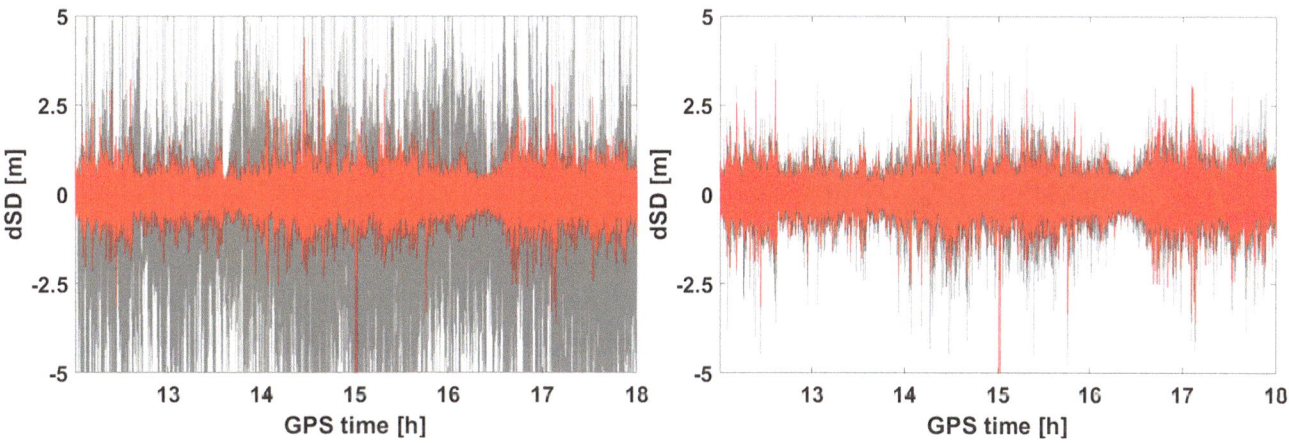

Fig. 4 Time differenced single differences of the GPS C1C signal of the first calibration set P_1. The ΔSD from the Septentrio receivers are presented in red and the ΔSD of the software receiver with different settings* are depicted in grey. *(left) $B_N = 0.1$ Hz, FO = 1, unaided DLL. (right) $B_N = 0.1$ Hz, FO = 2, unaided DLL

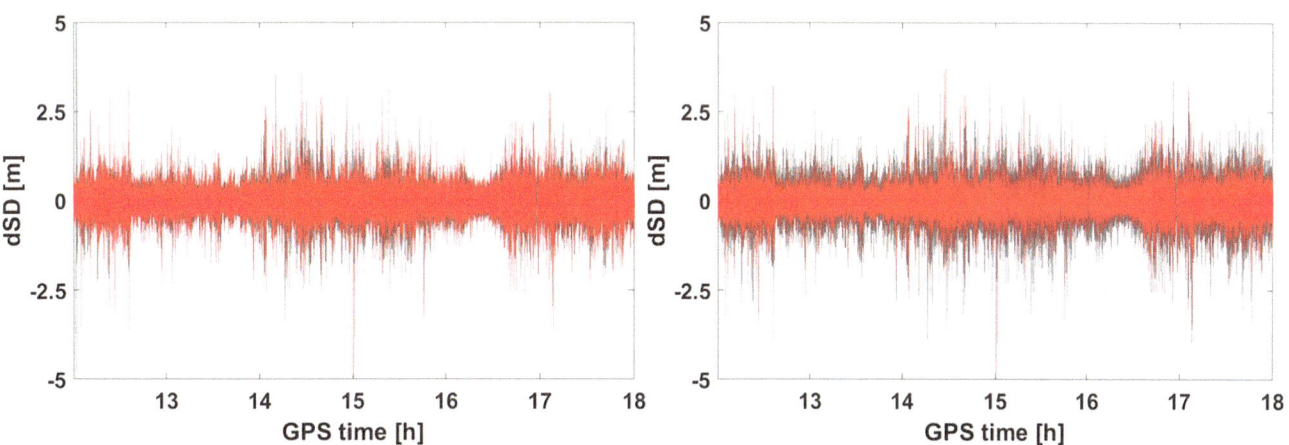

Fig. 5 Time differenced single differences of the GPS C1C signal of the first calibration set P_1. The ΔSD from the Septentrio receivers are presented in red and the ΔSD of the software receiver with different settings* are depicted in grey. *(left) $B_N = 0.05$ Hz, FO = 2, unaided DLL. (right) $B_N = 0.05$ Hz, FO = 1, aided DLL

4   Repeatability of Estimated CPC Pattern

Figures 6, 7, 8 show the CPC computation results for the hardware and the software receiver with different settings. The left side of Figs. 6, 7, 8 shows the estimated mean CPC pattern from the Novatel antenna for the GPS C1C signal, and the right side shows the absolute differences between two individual calibration sets P_1 to P_4 as a cumulative histogram. Here, only CPC values above 5° elevation in the antenna frame are considered. Each figure shows the results when using a different set of receiver settings:

- (1) Using Septentrio hardware receiver and default settings (Fig. 6).

- (2) Using Sx3 software receiver with a B_N of 0.05 Hz, FO of 1 and aided DLL (Fig. 7).
- (3) Using Sx3 software receiver with a B_N of 0.05 Hz, FO of 2 and unaided (Fig. 8).

The mean CPC pattern estimated with (1) and (2) have a very similar behaviour, however, the repeatability with (2) is worse with 95.4% of the differences below 111.72 mm, whereas 95.4% are below 99.2 mm for (1). This can be explained by the higher overall noise in the ΔSD calculated with (2) (Fig. 5, right). It can be assumed, that Septentrio receivers using an aided DLL. By using an unaided second order DLL with a small bandwidth (3), the repeatability of the antenna calibration sets are significantly improved (95.4% of the observations below 65.6 mm). However, the

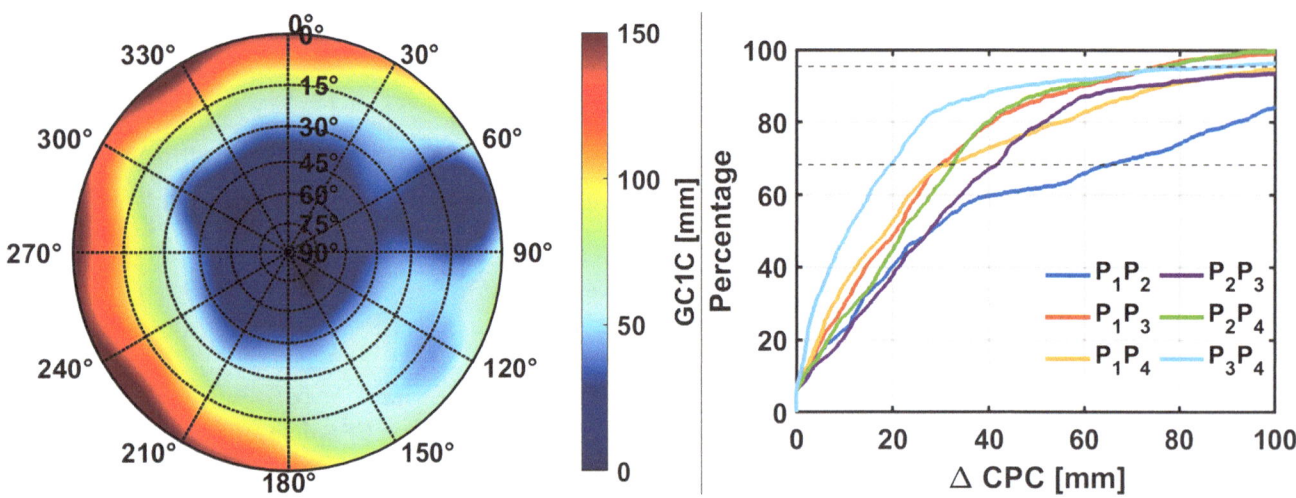

Fig. 6 (left) Estimated mean pattern of the Novatel antenna observed with (1)* hardware receiver and (right) absolute differences between estimated pattern (CPC ≤ 5°) from two different calibration sets (P_1–P_4) as cumulative histogram. *default settings

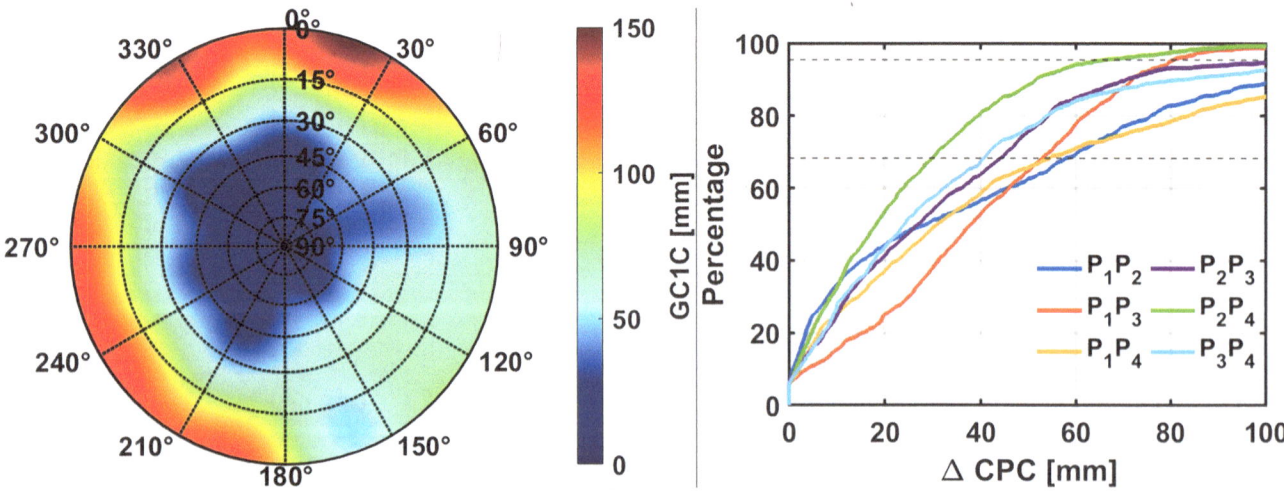

Fig. 7 (left) Estimated mean pattern of the Novatel antenna observed with (2)* software receiver and (right) absolute differences between estimated pattern (CPC ≤ 5°) from two different calibration sets (P_1–P_4) as cumulative histogram. *B_N of 0.05 Hz, FO of 1 and aided DLL

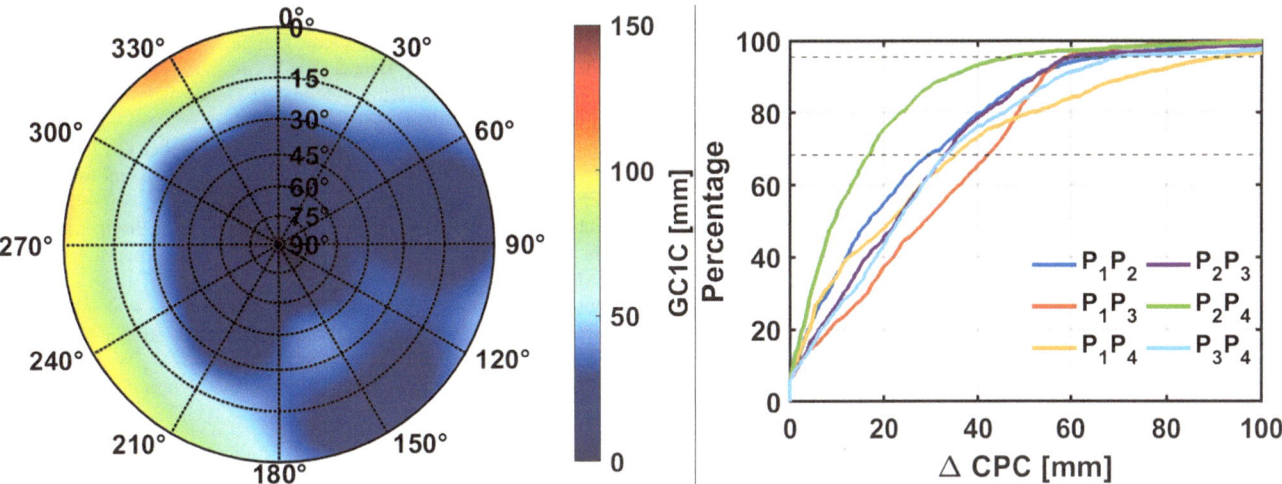

Fig. 8 (left) Estimated mean pattern of the Novatel antenna observed with (3)* software receiver and (right) absolute differences between estimated pattern (CPC $\leq 5°$) from two different calibration sets

$(P_1 - P_4)$ as cumulative histogram. *B_N of 0.05 Hz, FO of 2 and unaided DLL

estimated pattern differs from the pattern estimated with (1). This result shows very well the trade-off between noise performance and signal dynamics. Smaller bandwidth results in a smaller noise and accordingly better repeatability of the antenna calibration, when using an unaided DLL. But, a too small bandwidth leads to a different pattern, because the signal dynamics can no longer be tracked so well. By using an aided DLL, the signal dynamics are captured by the carrier tracking loops, which leads to stable tracking performance, but with higher noise and consequently a decrease in the repeatability. It should be noted, that the set duration of 4 to 6 h are very short for CPC estimation. Previous experiments with longer calibration duration, e.g. 12–14 h, show better repeatability.

5 Conclusion

In this contribution, the impact of different receiver settings on the antenna calibration was presented. In order to improve the CPC estimation, the overall noise of the observations needs to be reduced. One opportunity is to understand and modify the receiver tracking loop settings, especially the loop filters. Therefore, an experiment was carried out during a standard calibration of a Novatel antenna. Beside of the Septentrio hardware receiver, also the Sx3 software receiver from the IFEN company was used in a zero-baseline configuration. This experiment shows, that the interaction between noise bandwidth, filter order and the choice of an aided or unaided DLL plays an important role in the data acquisition of the antenna calibration. The trade-off between a small bandwidth and consequently less overall noise and the signal dynamics, caused by the fast robot motion, must be chosen carefully. For example, a small bandwidth in an

unaided DLL can increase the repeatability significantly, but the correctness of the CPC pattern can be lost. The default settings of the Septentrio hardware receivers have a good performance for the CPC estimation, however, a smaller bandwidth than the default 0.25 Hz can be chosen if an aided DLL is used.

Acknowledgements The work is funded by the Deutsche Forschungsgemeinschaft (DFG, German Research Foundation) - Project-ID 470510446 - MAESTRO.

References

Beer S, Wanninger L, Heßelbarth A (2019) Galileo and GLONASS group delay variations. GPS Solut **24**(1). https://doi.org/10.1007/s10291-019-0939-7

Beer S, Wanninger L, Heßelbarth A (2021) Estimation of absolute GNSS satellite antenna group delay variations based on those of absolute receiver antenna group delays. GPS Solut **25**(110), 10. https://doi.org/10.1007/s10291-021-01137-8

Böder V, Menge F, Seeber G, Wübbena G, Schmitz M (2001) How to deal with station dependent errors, new developments of the absolute field calibration of PCV and phase-multipath with a precise robot, 2166–2176. Institute of Navigation (ION)

Breva Y, Kröger J, Kersten T, Schön S (2019) Estimation and validation of receiver antenna codephase variations for multi GNSS signals. In: 7th International Colloquium on Scientific and Fundamental Aspects of GNSS

Breva Y, Kröger J, Kersten T, Schön S (2022) Estimation and validation of codephase center correction using the empirical mode decomposition. International Association of Geodesy Symposia

Caizzone S, Circiu M-S, Elmarissi W, Enneking C, Felux M, Yinusa K (2019) Antenna influence on global navigation satellite system pseudorange performance for future aeronautics multifrequency standardization. Navigation **66**(1), 99–116. https://doi.org/10.1002/navi.281

Häberling S (2016) Theoretical and practical aspects of high-rate gnss geodetic observations. PhD thesis, ETH Zürich

IFEN GmbH (2019) SX3 navigation software receiver - user manual

Kaplan ED, Hegarty CJ (2017) Understanding GPS/GNSS principles and applications, 3rd edn. GNSS technology and application series. Artech house, Norwood

Kersten T, Kröger J, Schön S (2022) Comparison concept and quality metrics for GNSS antenna calibrations: Cause and effect on regional GNSS networks. J Geodesy **96**(7), 48. https://doi.org/10.1007/s00190-022-01635-8

Kersten T, Schön S (2017) GPS code phase variations (CPV) for GNSS receiver antennas and their effect on geodetic parameters and ambiguity resolution. J Geodesy **91**(6), 579–596. https://doi.org/10.1007/s00190-016-0984-8

Kröger J, Kersten T, Breva Y, Schön S (2021) Multi-frequency multi-GNSS receiver antenna calibration at IfE: concept - calibration results - validation. Advances in Space Research (ASR). https://doi.org/10.1016/j.asr.2021.01.029

Menge F, Seeber G, Völksen C, Wübbena G, Schmitz M (1998) Results of the absolute field calibration of GPS antenna PCV, 31–38. Institute of Navigation (ION)

Misra P, Enge P (2006) Global positioning system: signals, measurements, and performance, 2nd edn. Ganga-Jamuna Press, Lincoln

Wanninger L, Sumaya H, Beer S (2017) Group delay variations of GPS transmitting and receiving antennas. J Geodesy **91**(9), 1099–1116. https://doi.org/10.1007/s00190-017-1012-3

Wübbena G, Schmitz M, Menge F, Böder V, Seeber G (2000) Automated absolute field calibration of GPS antennas in real-time, 2512–2522. Institute of Navigation (ION)

Wübbena G, Schmitz M, Propp M (2008) Antenna group delay calibration with the geo++ robot-extensions to code observable. In: IGS Analysis Workshop, Poster, June, pp 2–6

Wübbena G, Schmitz M, Warneke A (2019) Geo++ absolute multi frequency GNSS antenna calibration. In: Presentation at the EUREF Analysis Center (AC) Workshop, October 16–17, Warsaw, Poland (2019). http://www.geopp.com/pdf/gppcal125euref19p.pdf

Quality Control Methods for Climate Applications of Geodetic Tropospheric Parameters

Marcelo Santos, Jordan Rees, Kyriakos Balidakis, Anna Klos, and Rosa Pacione

Abstract

We have been analyzing the zenith total delay (ZTD) time series provided by six REPRO3 International GNSS Service (IGS) Analysis Centers (ACs), namely, COD, ESA, GFZ, GRG, JPL, and TUG, to compare their long-term trends. Long-term here means 20 years or longer. About thirty stations have been selected globally for this purpose. The estimated ZTD time series have gone through a process of homogenization using ERA-5 derived ZTDs as reference. The homogenized data is then averaged to daily values to minimize potential influences coming from different estimation strategies adopted by individual Analysis Centers as well as to mitigate the inherent autocorrelation. Similar averaging is applied to the ERA-5 ZTDs. Two combinations, using weighted mean and (a robust) least median of squares, are being generated from the six homogenized ACs. The combinations serve as quality control to each ACs. Analysis of the trends generated from each one of the seven ZTD time series is performed looking at their similarities in both time and frequency domains. This paper showcases the methodology and early results as presented during the second International Symposium of Commission 4: Positioning and Applications. Early results are based on station ALBH in Canada, showing an inter-AC scatter is 0.47 mm/decade for the trends, 0.11 mm for the annual amplitudes, and 0.29° for the annual phase.

Keywords

Climate · Combination · GNSS · Water vapor · ZTD

1   Introduction

Starting as a revolutionary theoretical possibility (Bevis et al. 1992), ground based Global Navigation Satellite System (GNSS) has turned into a contributor to weather forecast through assimilation of zenith total delays (ZTD) into numerical weather prediction (NWP) models of meteorological services [e.g., The UK Met Office (Bennitt and Jupp 2012), and others (Mascitelli et al. 2021)]. The collection of GNSS observations dates to the mid-90s with a growing number of stations distributed in permanent global and local networks being established since then. As a reference, we can take the year 1994, the start of IGS as an operational entity (Johnston et al. 2017), as the initial epoch of the continuous data

M. Santos (✉) · J. Rees
Department of Geodesy and Geomatics Engineering, University of New Brunswick, Fredericton, NB, Canada
e-mail: msantos@unb.ca

K. Balidakis
GFZ German Research Centre for Geosciences, Earth System Modelling, Potsdam, Germany

A. Klos
Military University of Technology, Warsaw, Poland

R. Pacione
e-Geos/ASI/Centro di Geodesia Spaziale (CGS), Matera, Italy

© The Author(s) 2023
J. T. Freymueller, L. Sánchez (eds.), *Gravity, Positioning and Reference Frames*,
International Association of Geodesy Symposia 156, https://doi.org/10.1007/1345_2023_233

collection. As time series grow longer so does the potential contribution of ground GNSS to climate. Such potential was one of the topics of an important COST Action (European Cooperation in Science and Technology) project (Bock et al. 2018). Essential questions follow: Are we, as a geodetic community, ready to contribute to climate? Are time series of GNSS-generated tropospheric parameters, ZTD, zenith wet delay (ZWD) and gradients good representation of long-terms for climate studies? Are there defined models and procedures for a dedicated estimation of such parameters? Are there established mechanisms for quality control? A reminder that VLBI (Very Long Baseline Interferometry) technique, even though not continuous, started earlier than GNSS.

This paper discusses the methodology and data sets used in the study currently going on within IAG JWG C.2: Quality control methods for climate applications of geodetic tropospheric parameters, as well as presents an early result based on station ALBH. In it, we define the concept of climate normals and the importance of long-term series, we revisit the growing importance of GNSS for meteorology and climate and present the overall strategy used in the research. Results are then presented and discussed, and lessons learned conclude the paper.

2   Climate Normals and the Importance of Long Terms

Climate scientists consider an average in weather taken over a 30 year-period, known as climate normals, as enough to evaluate climatological variables including temperature and precipitation, for a particular site. Under a stationarity assumption, the climate normal is long enough to smooth out year-to-year interannual fluctuations and short enough to represent climatic trends (WMO 2007).

The integrated water vapour (IWV), defined as the total mass of water vapor along a cross-section of 1 m^2 from the station till the topmost layer of the atmosphere, is an important meteorological quantity because it is related to changes in the temperature and the formation of clouds. This and the integrated precipitable water vapor (IPWV), or just precipitable water (PW) can be derived from GNSS estimates of ZWD.

Figure 1 illustrates the importance of long terms. Trends in IWV were computed based on ERA5 (ECMWF 2019) data (1979.0–2019.0). Given the time series, we vary the first and last epoch when estimating the trend together with some seasonal harmonics (the minimum duration set to 10 years) with a step of 1 month. The colour bar indicates the variation in trend values. Two red empty circles connected by a red line indicate where the climate normal falls within the 30-year period, varying from the '1979–2009 climate normal' to the

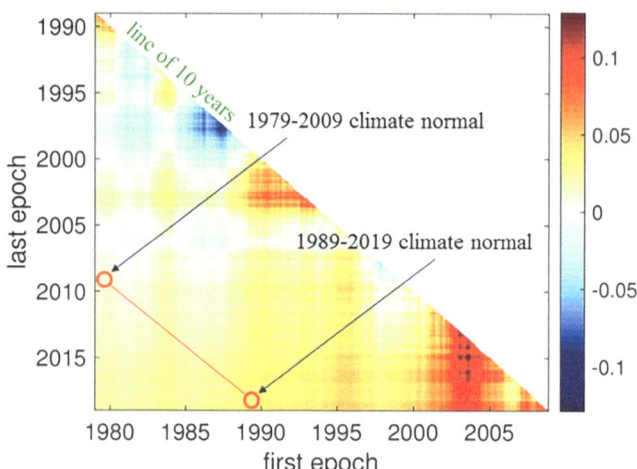

Fig. 1 IWV trend variations (kg/m^2/a) at Wettzell, Germany from ERA5 data

'1989–2019 climate normal' trends. The figure indicates that if we compute the normals using a shorter time span, such as 10 years, the trends values would be different. It is worthy of note that IWV trend variations as a function of the data span are more prominent should the seasonal component be ignored in the estimation process, especially if the data span deviates significantly from an integer factor if the dominant period inherent in the signal (typically the annual cycle).

The World Meteorological Organization (WMO) recommends 30-year normals and decadal updates, making it usual to find intermediary values (WMO 2007). For example, NOAA (National Oceanic and Atmospheric Administration) provides annual/seasonal normals, monthly normals, daily normals and even hourly normals. Climate normals can be computed in different ways (WMO 2011) but the data needs to go through stringent quality control and be made "self-homogeneous." To be sure (from a statistical viewpoint) that a small trend is valid, one might need even more than 30 years (Alshawaf et al. 2018).

Figure 2 shows a climograph. A climograph is a graphical representation of climatic parameters, such as monthly average temperature and precipitation, at a certain location. It is used for a quick view of the climate of a location. This climograph for station Addison, in Alabama, US, shows monthly normal values of precipitation, and temperature (minimum, medium and maximum).

Considering that the IGS started in 1994, in 2024 it will be 30 years. Are we ready?

3   Ground-Based GNSS for Meteorology: From Zero to Hero

Ground-based GNSS for meteorology has come a long way. In the beginning of the GNSS era, the meteorological com-

ADDISON, AL US

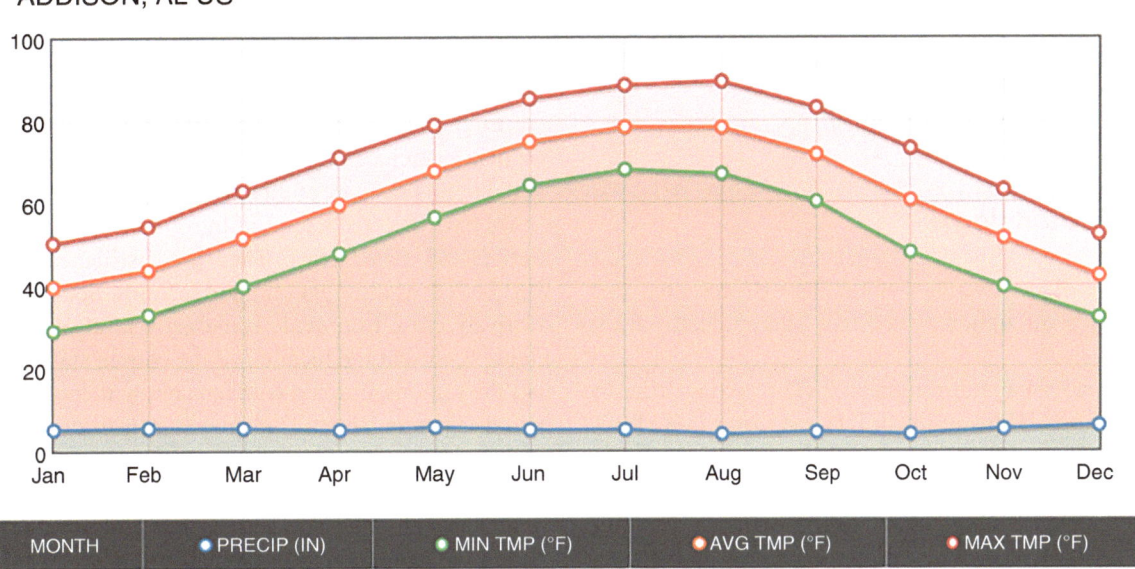

Fig. 2 Climograph from Station Addison, AL, USA, 1991–2020 (Courtesy, NOAA)

munity saw it with caution, sometimes with disbelief. In the early 1990s, an unmentioned meteorologist even stated, "There is no information of any quality that GPS can provide to weather analysis or forecast."

Nearly 30 years later, the consideration given to GNSS is very different. For example, during the December 2017 EGVAP Expert Meeting, Owen Lewis, from the UK Met Office, referred to ground based GNSS as the one providing the "second best observation impact among the various types of observations."

The GNSS-derived information that has found major use by the meteorological services for weather forecast is ZTD. The reason is simple to understand. As we are aware, from ZWD we can derive IWV and PW. However, the computation from ZWD to IWV requires water vapour-weighted mean temperature of the air column above the GNSS station, which depends on the vertical profile of temperature and humidity, at times not easily available.

In terms of climate, atmospheric water vapour is of great significance, as it is the major greenhouse gas. Therefore, the importance of its accurate, long-term monitoring and evaluation of trends and variability, potentially serving as independent benchmarks to climatological models. Climate scientists would love to have longer trends derived from GNSS, but also shorter trends, which could be used for assimilation and validation of climate models.

In the study that follows, we deal only with ZTD.

4 Research Questions

There are four underlying research questions we would like to tackle.

"Anyone can generate ZTD trends. How reliable are they?" One may think that it is, or that it tends to be as computational advances, easy to compute a long time series and derive trends from that. Let us work under this hypothesis. If so, anyone could feed GNSS trends to the climate community. But it is not just a question of estimating trends, but estimating trends that contain the proper information. This fosters another question.

"Can we define metrics to ascertain the quality of long-term trends provided to the climate community?" The word metrics here is used meaning standards of evaluation. That could involve from the treatment of input data, processing models and proper ways to determine trends, in such a way that they are meaningful to the climate community.

As far as processing modes are concerned, precise point positioning (PPP) seems to be a very attractive way to deal with the problem at hand. For example, the IGS tropospheric products are generated via a dedicated PPP processing. At the same time, the IGS Analysis Centers are in a continuous processing effort, and each one of them provide solutions that can be either used independently or cross-evaluated as a tool of quality control. We could, or probably, should, use the ZTD resulting from the IGS Analysis Centers, even considering that not all provide that, but a good number of them do that anyway, or per request as in the case of the REPRO3. This fact brings the fourth questions. "Are there advantages of combining ZTD estimates over not combining them? Is there any 'loss of information' if they are combined?"

Therefore, just looking from the ZTD time series derived by the IGS, either its tropospheric product or solutions generated by individual Analysis Centers, it is possible to estimate trends from each one of them. "Would there be difference in trends derived from them?" If so, that may have implication for feeding information to the climate community, either for validation or assimilation of models.

This WG wants to investigate that.

5   Methodology

A summary of the overall methodology is presented first, followed by a discussion on the only station dealt with so far, ALBH.

We take advantage of estimated ZTD resulting from the REPRO3 effort. The inclusion of ZTD estimates in REPRO3 followed from a request from the IGS to the Analysis Centers and represents a great opportunity for this study. A few problems arise though, namely, not all Analysis Centers provide ZTD, they do not use the same processing strategies and their output follow a different rate. At the end, six ZTD solutions were made available. In the sequel, each Analysis Center is represented by its three-letter code, followed by their own output rate: COD, 1 h; ESA, 1 h; GFZ, 1 h; GRG, 2 h; JPL, 100 s; TUG 5 m.

Dependency on the availability of REPRO3 was a delaying factor for the practical start of the work of this WG, but the wait was worthwhile.

The choice of using ZTD solutions resulting from REPRO3 brings a few challenges. The first one is that not all Analysis Centers provided ZTD, which narrowed down the number of ZTD time series to six. Another issue is that the Analysis Centers apply different strategies to their data processing. And, finally, their ZTD output rate is also different.

Besides REPRO3 ZTD estimates, we also used ERA-5 extracted ZTD, serving as a trustful independent reference. We plan to include the IGS ZTD product into the analysis as well, but it is not part of this paper.

We have selected a total of thirty-nine stations with long-term GNSS time series. They are distributed around the world to cover different climatic regimes. Some stations are relatively close to provide some extra level of comparison. Figure 3 present the location of the chosen stations. Table 1 lists the stations, ordered with the ones with longer operation period appearing first. The time shown discounts eventual interruptions or gaps during the period.

Each ZTD time series is then subjected to a process of homogenization (Klos et al. 2022; Van Malderen et al. 2020), using ERA-5 as reference. Homogenization is an important step to derive trends.

Following the homogenization, daily mean values are produced. The reduction to daily mean values is an attempt to accommodate the differences in strategies that each Analysis Center use. The daily mean values are computed using a simple weighted average using the ZTD standard deviation in the process. When available, the IGS final ZTD product in REPRO3 will go through the same process.

The next step in the methodology is to perform the combination among the six daily-averaged homogenized ZTD time series. We envision this process being done in two different ways, as a weighted mean of by computing the

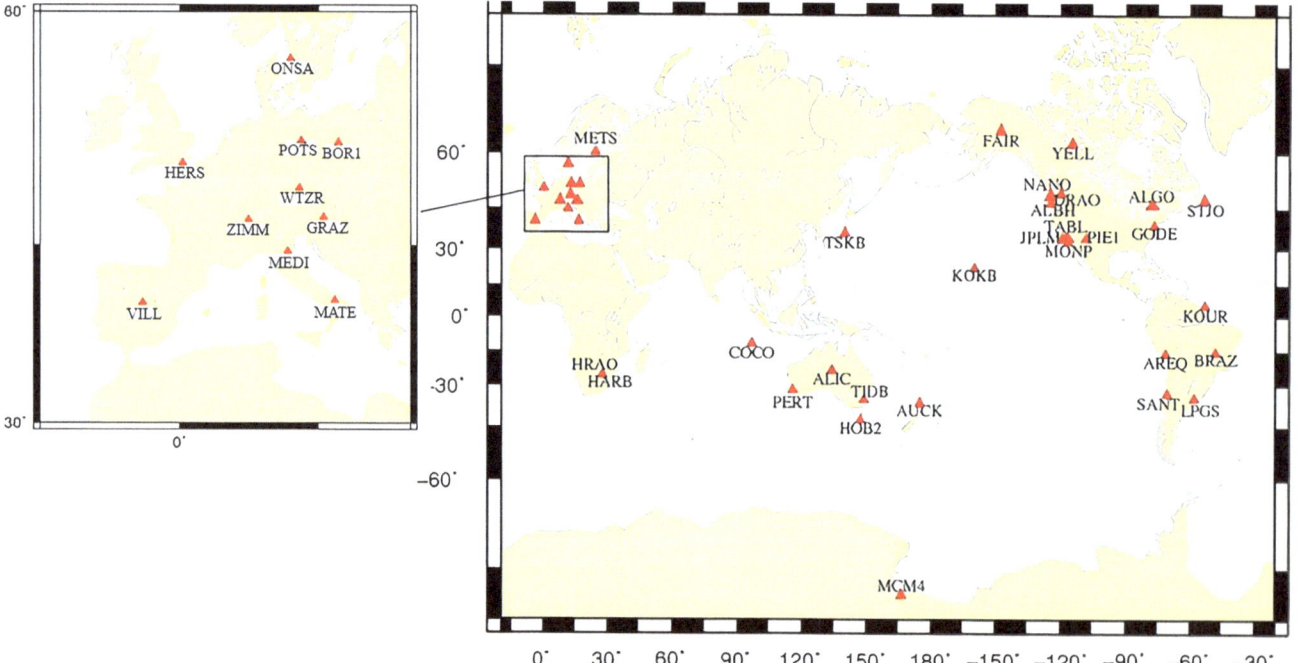

Fig. 3 Selected stations

Table 1 List of selected stations ordered by duration of operation (discounted gaps)

Station	Latitude (°)	Longitude (°)	Years
DRAO	49.3	−119.6	26.973
YELL	62.5	−114.5	26.971
ALBH	48.4	−123.5	26.938
ZIMM	46.9	7.5	26.867
MATE	40.6	16.7	26.856
TSKB	36.1	140.1	26.853
STJO	47.6	−52.7	26.675
ONSA	57.4	11.9	26.672
GRAZ	47.1	15.5	26.667
ALGO	46.0	−78.1	26.593
FAIR	65.0	−147.5	26.576
JPLM	34.2	−118.2	26.489
BOR1	52.3	17.1	26.39
GODE	39.0	−76.8	26.223
PIE1	34.3	−108.1	26.223
KOKB	22.1	−159.7	26.103
HERS	50.9	0.3	26.089
METS	60.2	24.4	26.084
MONP	32.9	−116.4	26.067
TIDB	−35.4	149.0	25.999
POTS	52.4	13.1	25.892
WTZR	49.1	12.9	25.796
VILL	40.4	−4.0	25.563
MCM4	−77.8	166.7	25.454
NANO	49.3	−124.1	25.377
AUCK	−36.6	174.8	25.166
HOB2	−42.8	147.4	25.032
TABL	34.4	−117.7	25.002
PERT	−31.8	115.9	24.846
COCO	−12.2	96.8	24.81
KOUR	5.3	−52.8	24.783
AREQ	−16.5	−71.5	24.747
ALIC	−23.7	133.9	24.542
MEDI	44.5	11.6	24.517
SANT	−33.2	−70.7	23.951
LPGS	−34.9	−57.9	23.855
BRAZ	−15.9	−47.9	23.094
HRAO	−25.9	27.7	22.275
HARB	−25.9	27.7	19.721

least median of squares (Rousseeuw 1984). The combination becomes both a separate time series for analysis and a testbed for quality control of each one of the Analysis Centers. A clarification may be needed here. We are using the term combination even though, among the geodetic community, this term mostly refers to an operation at the level of normal equations, which is not the case.

Now, the trends can be computed. For that purpose, three methods will be used: weighted least squares, robust estimation, and non-parametric estimation. The result shown in Sect. 5, only weighted least squares was used.

At this stage, there will be trends originated from each one of the six Analysis Centers, one derived from the combination, one derived from the IGS final product and one derived from ERA-5.

The final analysis of the trends will involve testing their statistical significance. Analysis in frequency domain will also be performed to understand, for a given site, how their frequency bands differ, what the largest discrepancies between trends are and how do they differ with those from the combined solution and ERA-5, here taken as a reference.

6   Results: Station ALBH

As stated before, the results shown in this paper are for station ALBH.

Figure 4 portrays the homogenized ZTD times series of station ALBH, originally provided by GFZ. Black dots indicate homogenized ZTD in their original sampling rate, whereas red dots represent their corresponding daily mean values. Figure 5 displays the homogenized daily mean ZTD time series of all six Analysis Centers and their combination. Colours are indicated in the label.

A careful look at Fig. 5 indicates that there are data gaps in the original time series, which are reflected in the final homogenized daily means. The importance of this fact will be made clear in the sequence.

Table 2 summarizes trends as derived from the homogenized daily mean ZTD time series from each of the six Analysis Centers, and that of the combination. The table shows the trends (mm/decade), the annual amplitudes (mm), the annual phases (degrees), as well as the number of points involved in each solution. The last column indicates that the number of points are different, as the data collected at this particular station ended up being used differently by each Analysis Center. This difference may be the explanation of the large variation seen among the solutions based on different Analysis Centers. The inter-Analysis Center scatter is 1.25 mm/decade for the trends, 0.73 mm for the annual amplitudes and 1.99° for the annual phase.

Table 3 is like Table 2 with a major difference. The trends were computed only using the common epochs between all Analysis Centers, which caused the ZTD time series from ESA and JPL to be disregarded. The inter-Analysis Center scatter decreased to 0.47 mm/decade for the trends, to 0.11 mm for the annual amplitudes, and to 0.29° for the annual phase.

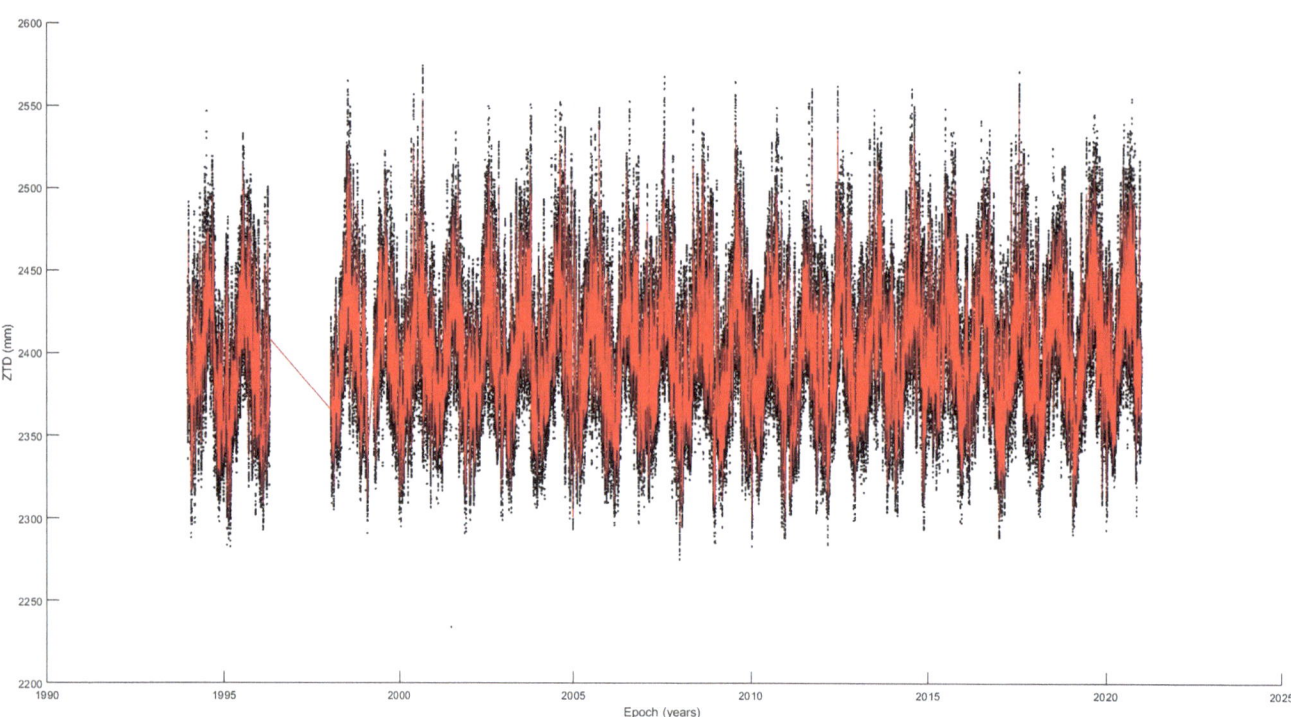

Fig. 4 Homogenized GFZ ZTD times series of station ALBH, original rate (black dots) and their corresponding daily mean values (red dots)

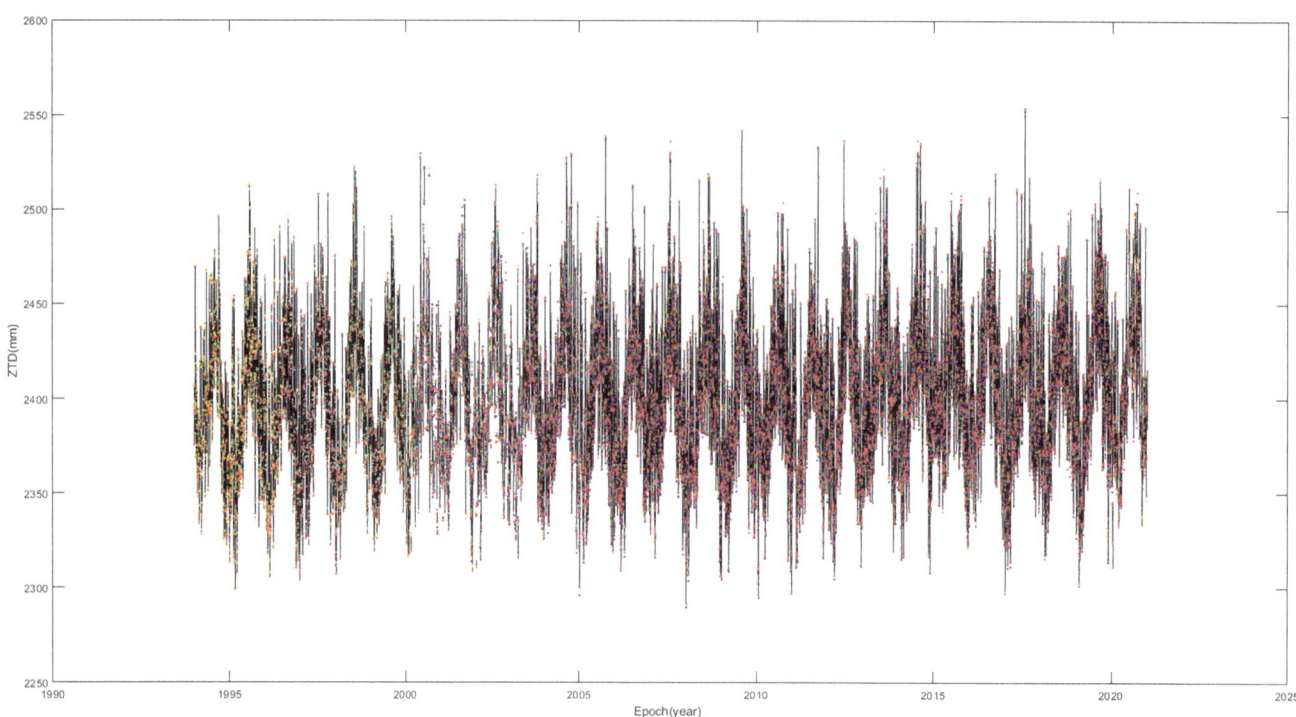

Fig. 5 Homogenized daily mean ZTD times series of all six Analysis Centers and their combination. Colours: combination (black continuous line), COD (navy blue dot), ESA (sky blue dot), GFZ (green dot), GRC (pink dot), JPL (yellow dot) and TUG (red dot)

Table 2 Trend, amplitude, and phase of original ZTD time series

	Trend (mm/decade)	Amplitude (mm)	Phase (°)	Number of points
Combination	3.28	31.99	54.82	9,849
COD	3.06	31.43	54.15	7,170
ESA	6.43	33.39	55.14	3,270
GFZ	2.95	31.78	54.46	9,061
GRG	4.03	31.68	53.25	7,478
JPL	5.05	32.91	59.42	1,337
TUG	4.10	31.80	54.62	9,837

Table 3 Trend, amplitude, and phase of the synchronized ZTD time series

	Trend (mm/decade)	Amplitude (mm)	Phase (°)	Number of points
Combination	3.18	31.44	54.03	7,079
COD	3.19	31.48	54.09	7,079
GFZ	3.00	31.35	54.06	7,079
GRG	3.63	31.58	53.40	7,079
TUG	4.16	31.30	53.80	7,079

A simple look at the statistics shows us that the trend using TUG is slightly away from the mean at 1-sigma, whereas amplitude and phase from GRG are negligibly above the mean at 1-sigma. The reason for that was not established, perhaps some kind of jump that was not detected during the homogenization or such a difference could indicate that those parameters should not be used. Such an analysis lies within the discussion on establishing metrics to determine if a trend can be trusted or not. Further analysis will include the testing of the significance level of the parameters.

7 Lessons Learned

A few statements summarize the lessons learned in this study. The quality of the combination depends on processed data. Combination seems to bring benefits and is a tool for quality control particularly if gappy data are involved in the combination. It would be interesting to understand why some Analysis Centers did not process all data available, if similar happens for other stations too. The process is painstaking, but the effort is being continued and expanded. The overall goal is to have a final report presented during the XXVIII General Assembly of the International Union of Geodesy and Geophysics.

Acknowledgements The first author acknowledges financial support provided by National Sciences and Engineering Research Council of Canada (NSERC). KB is funded by the Deutsche Forschungsgemeinschaft (DFG, German Research Foundation)—Project-ID 434617780—SFB 1464 (TerraQ).

Ethics Approval and Consent to Participate Not applicable.

Consent for Publication Not applicable.

Competing Interests Not applicable.

Authors' Contributions MS leads the project, responsible for the manuscript, and corresponding author. JR did a portion of the data processing (pre-processing and combinations). KB did a portion of the data processing (ERA-5 and trends). AK did the homogenization. RP participated in discussion. All others contributed with the manuscript.

References

Alshawaf F, Zus F, Balidakis K, Deng Z, Hoseini M, Dick G, Wickert J (2018) On the statistical significance of climatic trends estimated from GPS tropospheric time series. J Geophys Res Atmos 123:10,967–10,990. https://doi.org/10.1029/2018JD028703

Bennitt G, Jupp A (2012) Operational assimilation of GPS zenith total delay observations into the UK Met Office numerical weather prediction models. Monthly Weather Rev 140(8):2706–2719. https://doi.org/10.1175/MWR-D-11-00156.1

Bevis M, Businger S, Herring TA, Rocken C, Anthes RA, Ware RH (1992) GPS meteorology, remote sensing of atmospheric water vapour using the global positioning system. J Geophys Res 90(D14):15,787–15,801

Bock O, Pacione R, Ahmed F, Araszkiewicz A, Baldysz Z, Balidakis K, Barroso C, Bastin S, Beirle S, Berckmans J, Böhm J, Bogusz J, Bos M, Brockmann E, Cadeddu M, Chimani B, Douša J, Elgered G, Eliaš M, Fernandes R, Figurski M, Fionda E, Gruszczynska M, Guerova G, Guijarro J, Hackman C, Heinkelmann R, Jones J, Kazanci SZ, Klos A, Landskron D, Martins JP, Mattioli V, Mircheva B, Nahmani S, Nilsson RT, Ning T, Nykiel G, Parracho A, Pottiaux E, Ramos A, Rebischung P, Sá A, Dorigo W, Schuh H, Stankunavicius G, Stepniak K, Valentim H, Van Malderen R, Viterbo P, Willis P, Xaver A (2018) Use of GNSS tropospheric products for climate monitoring (working group 3). In: Jones J, Guerova G, Douša J, Dick G, de Haan S, Pottiaux E, Bock O, Pacione R, van Malderen R (eds) Advanced GNSS tropospheric products for monitoring severe weather events and climate. Springer International Publishing, Cham, pp 267–402. https://doi.org/10.1007/978-3-030-13901-8_5

ECMWF (2019) ERA5 data documentation. European Centre for Medium-Range Weather Forecasts. https://confluence.ecmwf.int/display/CKB/ERA5+data+documentation

Johnston G, Riddell A, Hausler G (2017) The international GNSS service. In: Teunissen PJG, Montenbruck O (eds) Springer handbook of global navigation satellite systems, 1st edn. Springer International Publishing, Cham, pp 967–982. https://doi.org/10.1007/978-3-319-42928-1

Klos A, Bogusz J, Pacione R, Humphrey V, Dobslaw H (2022) Investigating temporal and spatial patterns in the stochastic component of ZTD time series over Europe. GPS Solutions 27. https://doi.org/10.1007/s10291-022-01351-y

Mascitelli A, Federico S, Torcasio RC, Dietrich S (2021) Assimilation of GPS zenith total delay estimates in RAMS NWP model: impact studies over Central Italy. Adv Space Res 68(12):4783–4793. https://doi.org/10.1016/j.asr.2020.08.031

Rousseeuw PJ (1984) Least median of squares regression. J Am Stat Assoc:871–880. https://doi.org/10.1080/01621459.1984.10477105

Van Malderen R, Pottiaux E, Klos A, Domonkos P, Elias M, Ning T, Bock O, Guijarro J, Alshawaf F, Hoseini M, Quarello A, Lebarbier E, Chimani B, Tornatore V, Zengin Kazanci S, Bogusz J (2020) Homogenizing GPS integrated water vapor time series: benchmarking break detection methods on synthetic datasets. Earth Space Sci 7(11). https://doi.org/10.1029/2020EA001121

WMO (2007) The role of climatological normals in a changing climate. WCDMP-No. 61, WMO-TD/No. 1377. World Meteorological Organization

WMO (2011) Guide to climatological practices, WMO-No. 100. World Meteorological Organization, Geneva

Impact of Coordinate- and Tropospheric Ties on the Rigorous Combination of GNSS and VLBI

Iván Darío Herrera-Pinzón and Markus Rothacher

Abstract

In this work, we study the impact of the use of site coordinate and tropospheric ties between VLBI telescopes and GNSS antennas at co-location sites during the CONT17 campaign. We perform the rigorous estimation of all parameter types common to these two techniques: station coordinates, troposphere zenith delays and gradients, and the full set of Earth Orientation Parameters (EOPs) and their rates, including their full variance-covariance information. The core element of our processing scheme is the combination of the techniques via coordinate and tropospheric ties, the later being essential especially for the height estimates. By using and evaluating different weighting schemes, to obtain a unique set of consistent parameters, we analyse coordinate repeatabilities and the behaviour of the EOPs, to discuss the impact of the accuracy and weighting of the coordinate and troposphere ties on the estimation of geodetic parameters. Our work shows that the combined solution with coordinate and troposphere ties generally improves the precision of all the estimated geodetic parameters. In particular, the repeatabilities of the height component, the polar motion estimates, and the LOD, show improvements up to 19%, 35% and 48%, respectively, with respect to the single-technique solutions. These results provide enough evidence of the benefits of our approach.

Keywords

GNSS · Local and tropospheric ties · Rigorous combination · VLBI

1 Combination of Space Geodetic Techniques

In the current realisation of the International Terrestrial Reference Frame (ITRF), Earth Orientation Parameters (EOPs) are heterogeneously determined. Polar motion (x-pole and y-pole) is estimated based on the combination of the four space geodetic techniques, whereas their rates are only

based on two techniques, namely Global Navigation Satellite Systems (GNSS) and Very Long Baseline Interferometry (VLBI). Moreover, the Earth's rotation angle (UT1-UTC) and Length of Day (LOD) are taken solely from the VLBI solution (Altamimi et al. 2016). In addition, the combination of troposphere parameters from VLBI, DORIS and GNSS through the use of tropospheric ties at fundamental sites is not implemented in ITRF's combination strategy. Hence, a rigorous combination of all parameter types common to all techniques, with consistent EOPs and with appropriate inter-technique tropospheric ties, is still missing. A consistent estimation of the TRF, capable of exploiting the advantages of the dense GNSS network with continuous observations and excellent geometry, and the full set of EOP delivered by VLBI, is required to achieve higher precision levels following the requirements given in Rothacher et al. (2009),

I. D. Herrera-Pinzón (✉)
Astronomical Institute, University of Bern (AIUB), Bern, Switzerland
e-mail: Ivan.Herrera@unibe.ch

M. Rothacher
Swiss Federal Institute of Technology in Zurich (ETHZ), Zurich, Switzerland
e-mail: Markus.Rothacher@ethz.ch

© The Author(s) 2023
J. T. Freymueller, L. Sánchez (eds.), *Gravity, Positioning and Reference Frames*,
International Association of Geodesy Symposia 156, https://doi.org/10.1007/1345_2023_195

and it is a pre-requisite for the full exploitation of dedicated co-location satellite mission concepts, such as Delva et al. (2023). A complete definition of the standards, models and parametrisation required for the consistent processing of the different space geodetic techniques is presented by Rothacher et al. (2010), within the scope of the GGOS Germany initiative (GGOS-D). This work discusses the important aspects of a rigorous combination of space geodetic techniques, and emphasises the need for the computation of consistent time series of the parameters relevant to the different techniques, extending the parameter space to link geometry, Earth rotation, and gravity field. In their comprehensive work, Coulot et al. (2007) carried out an early attempt of combining GPS, VLBI, SLR and DORIS data on the observation level. With data covering one year (2002), their work strove to perform the combination by estimating parameters simultaneously, while making use of all their correlation information. Thaller (2008) performed a combination of VLBI, GPS, and SLR normal equations, during the CONT02 campaign, in order to estimate station coordinates, EOPs, and troposphere parameters. Her approach aimed at the homogenisation of the normal equations, through the use of identical a priori models in the estimation of the parameters common to the three techniques. Her work performed the combination at the normal equation level, with all common parameter types included, where the improvement of the combined solution w.r.t. the individual technique solution is evident. In particular, she accomplished a successful estimation of UT1-UTC and LOD, and the stabilisation of the determination of the height component of the coordinates thanks to the common estimation of troposphere zenith delays and gradients. More recently Diamantidis et al. (2021) performed a combination at the observation level of VLBI and GNSS data during the CONT17 campaign, using a unified piece of software based on a batch least-squares estimator. Their work reports an improvement in the coordinate repeatabilities, polar motion, and UT1-UTC of 25%, 20% and 30%, respectively, with respect to the single technique solutions. In a similar fashion, Wang et al. (2022) performs the integrated processing of VLBI and GNSS data, to achieve a combination at the observation level. The main characteristic of their approach was the use of the tropospheric ties among VLBI and GNSS co-located stations, where residual zenith wet delays (ZWD) and gradients for VLBI and GNSS were estimated. As their work used different tropospheric tie setups, the improvement of the coordinate repeatabilites range between 12% and 28%, while for EOPs it goes from 2% up to 18%.

2 Dataset and Processing Strategy

The test scenario to validate our strategy was the data of the Continuous VLBI Campaign 2017 (CONT17). CONT17 was a campaign of continuous VLBI sessions, carried out between November 28, 2017, and December 12, 2017. It was composed of three independent networks observed: two legacy S/X networks with 14 stations each, and one VGOS broadband network consisting of six stations (Behrend et al. 2020). For the scope of our work, we only used the two legacy networks. The geodetic VLBI data of this campaign were extracted from the corresponding NGS cards. Since we were only using the legacy networks, the processing of the data was performed using the S/X part of the source catalogue of the 3rd realization of the International Celestial Reference Frame (ICRF3) of Charlot et al. (2020). To complement the VLBI observations, we selected about 180 GNSS stations of the International GNSS Service (IGS) network (Dow et al. 2009) covering the same time interval, with several stations co-located with the VLBI telescopes (in most of the cases). The integrated processing of the different techniques is done at the observation level, which provides the most rigorous and consistent solution, especially, when all the possible ties are considered. To guarantee the consistency, it is best performed with a single piece of software capable of processing all the techniques with state-of-the-art models and identical parametrisation. To handle the processing of the VLBI and GNSS data at the observation level, we used a modified version of the Bernese GNSS Software v5.2 (Dach et al. 2015), capable of handling VLBI data. This so-called Bernese v5.2 – VLBI Version, inherits all the GNSS & SLR capabilities of the original version: Pre-processing, outlier detection, residual screening, time-series analysis, daily and session processing, generation of normal equations, and more. The main advantages of this development are: (1) the use of an identical parametrisation for all the techniques. (e.g. piece-wise linear estimates, offset-drift estimates, interpolation methods, …), (2) the use of identical models for all techniques, where applicable (e.g station motion, tropospheric refraction, loading, troposphere), with identical handling of parameter constraints, (3) appropriate datum definition such as No-Net-Rotation (NNR), No-Net-Translation (NNT), No-Net-Scale, fixed coordinates, and (4) the implementation of coordinate and tropospheric ties.

Table 1 shows a summary of the modelling and a-priori information used for the rigorous combination of VLBI and GNSS data. For the combination of the data, we estimated

Table 1 Modelling and a-priori information used for the rigorous combination of VLBI and GNSS data

Modelling and a-priori information	
Troposphere	Dry: ECMWF-based mapped w. VMF
	Wet: Piecewise linear
Source catalogue	ICRF3
Observations	GNSS: RINEX
	VLBI: NGS cards
Processing	GNSS: Double diff. + Ambiguity Fixing
	VLBI: Baselines
Datum definition	NNT-NNR
Earth rotation	Piecewise linear functions
Receiver clock	VLBI: Piecewise linear functions
Antenna	VLBI: Axis offset
	GNSS: phase centre variations
Weighting Scheme	Based on repeatabilities

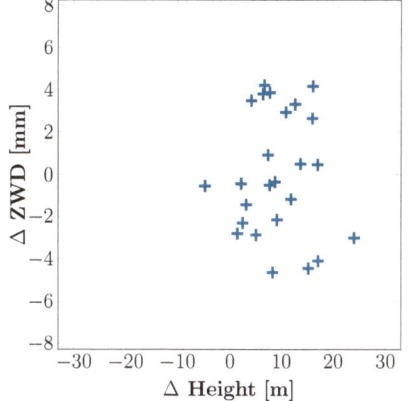

Fig. 1 Summary of the mean zenith wet delay differences for co-located sites, w.r.t. the height difference in the baseline

most of the common parameters with daily resolution: daily station coordinates using the NNR–NNT condition, daily EOPs: polar motion, UT1-UTC, LOD, and celestial pole offsets, and their corresponding rates of change. Zenith tropospheric delays were estimated with 1-h resolution and tropospheric gradients every 24 h. We estimated VLBI clock offsets piece-wise linearly with intervals of 3 h. Finally, for the 15-day rigorous combination, we additionally used the available terrestrial ties and our approach for tropospheric ties.

3 Realisation of Tropospheric Ties

For the modelling of the troposphere, we used as a-priori values for the zenith hydrostatic delays and mapping function the data of the Vienna Mapping Function 1 (VMF1) (Bohm et al. 2006). The use of this type of modelling ensures that the zenith total delay (ZTD) difference between GNSS and VLBI at co-located stations, caused by the height difference, are modelled in advance. The residual wet delays were then estimated as one-hourly piece-wise-linear functions and the tropospheric gradients with daily resolution. In particular, for the baseline FD-VLBA–MDO1 the modelled mean ΔZHD has a significantly large value of 91.7 mm, mostly due to the large height difference between the two stations (ca. 398 m). The statistics associated with the estimated ΔZWD also have an inferior performance: 7.4 mm for the mean, 4.6 mm of standard deviation, and an RMS of 8.7 mm. This clearly shows that the height difference at the co-location site Fort Davis is too large to apply a tropospheric tie and the ZWD parameters for MDO1 and FD-VLBI cannot be stacked. At the remaining co-location sites, we observed that the ΔZWDs were not correlated with the height difference, and

that the ZWD mean values vary within ±5 mm (excluding FD-VLBI–MDO1). These mean differences are shown in Fig. 1. We define the tropospheric tie as the difference in the tropospheric delay between the reference points of the VLBI and the GNSS antennas. Since the a-priori values of these delays are based on state-of-the-art global numerical weather prediction models, the difference between the delays at two stations caused by the height difference is modelled in advance (Wang et al. 2022) and only the delays caused by the residual troposphere should be considered. Moreover, the mean differences shown in Fig. 1 are not taken into account in the tropospheric ties, but are interpreted as resulting from the estimation uncertainty and from small systematic-effects that are not due to the troposphere or the troposphere delays modelling.

4 Optimal Weighting

An important aspect of the combination is the weighting of each technique, as the quality of the individual techniques varies considerably. The large contrast in the formal errors of each solution supports the need for an adequate inter-technique weighting. Our approach follows the idea of Thaller (2008), using coordinate repeatabilities as the base of the weights, since they are directly part of the terrestrial reference frame. First, the quadratic mean repeatability of the station coordinates for all co-located stations over the 15 days of the CONT17 campaign was calculated, as an indicator of the quality of the observations (and the solution):

$$r^2 = \frac{r_e^2 + r_n^2 + r_u^2}{3}$$

Table 2 Results of the calculation of the optimal inter-technique weight for the data of the CONT17 campaign. Repeatabilities are given for east, north and up components, respectively, in millimetres

Indicator			
RMS Rep. VLBI	3.54	3.11	8.07
RMS Rep. GNSS	2.57	2.71	6.16
r_{VLBI}^2		29.10	
r_{GNSS}^2		17.27	
$w_{rep_{ij}}$		1.69	
$\overline{N}$ VLBI		2238.22	
$\overline{N}$ GNSS		1040.80	
Weight factor		0.276	

With this, a relative weighting between techniques i and j was computed:

$$w_{rep_{ij}} = \frac{r_i^2}{r_j^2}$$

Then, the sum of the main-diagonal elements of the normal equation matrix was calculated:

$$\overline{N} = \frac{1}{n_{crd}} \sum_{z=1}^{n_{crd}} N_{zz}$$

where the parameter n_{crd} refers to the number of diagonal elements of the normal equation matrix. Since the weight is based on repeatabilities, only coordinate elements were considered. Moreover, only the coordinates of the co-location sites were used. Finally, the $\overline{N}$ values of each technique were combined with the weight of the corresponding parameter, to obtain the weighting of technique j with respect to technique i:

$$w_{ij} = \frac{\overline{N}_i}{\overline{N}_j} \cdot w_{rep_{ij}}$$

Table 2 shows the results of the calculation of the optimal inter-technique weight. For the data of the CONT17 campaign, an optimal weight for the VLBI NEQs of 0.276 was determined.

5   Validation of the Optimal Weighting

To test the adequacy of the weight determined in Sect. 4, we studied the performance of the repeatabilities of the combined solution, for typical cases of inter-technique weights, taking as reference the GNSS solution and using the parametrisation of Sect. 2. This is, in all the cases the GNSS solution had a weight of 1, while we vary the weight of the VLBI solution. A large number of cases was investigated, but four specific cases give the essence of the behaviour. These are: (1) 100^{-2}, meaning that the GNSS observations had a considerably larger contribution to the final solution. (2) 0.276, the "optimal weight" of Sect. 4. (3) 1, meaning that both techniques were equally weighted. (4) 100^2, meaning that the VLBI observations had a considerably larger contribution to the final solution. Figure 2 shows an example of the repeatabilities for two co-location sites, Pietown and Brewster (USA). For these two particular cases, the repeatabilities of the solution with the optimal weight shows a marginally better performance, especially when it comes to the height component when compared to the solution with equal weights. From these two examples, it is also noticeable that the solutions with larger weights for either VLBI or GNSS underperform when compared to the solution with optimal weights. Moreover, Fig. 3 displays the RMS of the repeatabilities for the combined solutions over the 15 days of the CONT17 campaign, when using different inter-technique weights, using exclusively the stations at co-location sites. While the repeatabilities of the horizontal components remain almost unchanged, there is an improvement in the height component when analysing all stations together (top plot of Fig. 3). The differences are more evident when looking at the stations separated by technique, especially for the GNSS case, where the height component of the solution with the optimal weight outperforms all the other solutions by more than 10%. Since the optimal inter-technique weight was based on the coordinate repeatabilites, it is fair to assume that its influence is not so evident in the remaining parameters.

Fig. 2 Repeatabilities of the combined solution, for different inter-technique weights. The first three values (blue, red and yellow) are the repeatabilities for the GNSS station, while the remaining three (purple, green and cyan) correspond to the VLBI station

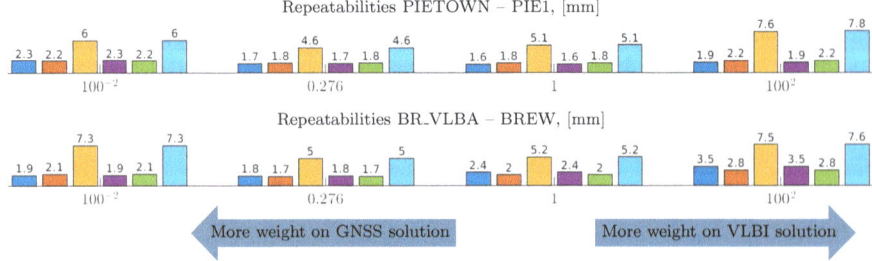

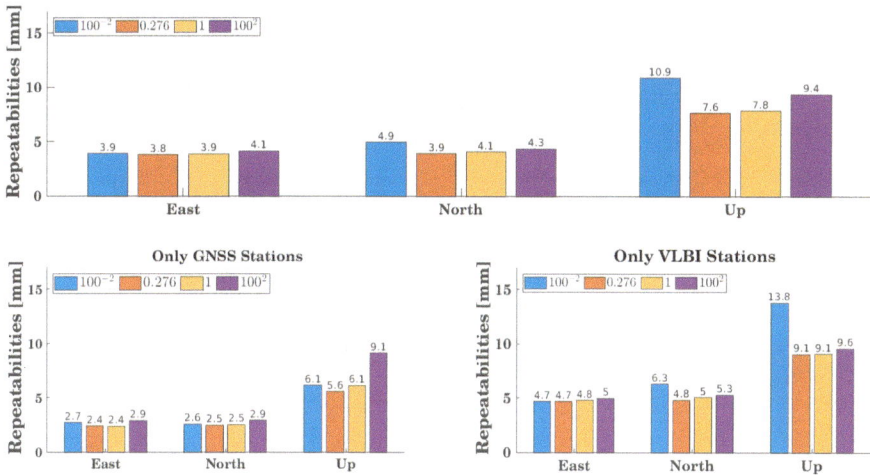

Fig. 3 RMS of the repeatabilities of the combined solution over the 15 days of the CONT17 campaign, for different inter-technique weights. The RMS value was calculated using all the co-located stations

6 Realisation of the Coordinate Ties

A central element in the combination of space geodetic techniques is the use of coordinate ties (Sarti et al. 2013), and in particular, the quality with which they have been determined. To realise the coordinate ties, we used the ITRF2014 coordinates of the GNSS stations and add the coordinate ties to get VLBI coordinates. Then, we applied relative constraints to constrain the vector between the co-located VLBI and GNSS stations with a certain weight. This ensures that the coordinates were consistent with the coordinate tie values. To investigate their quality, we used as relative constraints the formal errors of the coordinates contained in the SINEX files of the coordinate ties of the ITRF2014 solution, from the IERS website.[1] These formal errors (σ_{snx}) were the starting point for the remaining test solutions. We calculated combined solutions with relative constraints of $10^1\sigma_{snx}$, $10^{-1}\sigma_{snx}$ and $10^{-2}\sigma_{snx}$, and analysed the coordinate repeatabilities. Figure 4 shows an example of these repeatabilities for the sites Brewster (USA) and Fortaleza (Brasil). It is expected that a strong constraint on the coordinate tie causes the repeatabilities of the two co-located stations to converge to the same value. We observed that the quality of the coordinate ties varies among the co-location sites, and that different co-location sites have different responses to the relative constraint used. The two co-location sites shown in Fig. 4 represent this behaviour. For the baseline BR_VLBA–BREW, the original relative constraints (σ_{snx}) end up in different repeatabilities for the two co-located sites, especially for the up component. The same is true when using a softer relative constraint. However, when using stronger versions of σ_{snx}, the repeatabilities of the two stations converge to the same (low) values. In contrast, the co-location baseline FORTLEZA–BRFT shows larger

differences in the repeatabilities of the vertical component when using stronger values for σ_{snx} and does not converge to the same values for both co-location sites, indicating strong inconsistencies between the two techniques. In this case, a weaker constraint of the coordinate ties delivers the best results for this co-location site. Based on this analysis, we selected the optimal set of coordinate tie constraints, for each baseline at the co-location sites, so that it minimised the repeatabilities of the two stations, while trying to get them to converge to the same value. Finally, we calculated the RMS of the coordinate repeatabilities when using these appropriate constraints for the coordinate ties, and display them in Fig. 5.

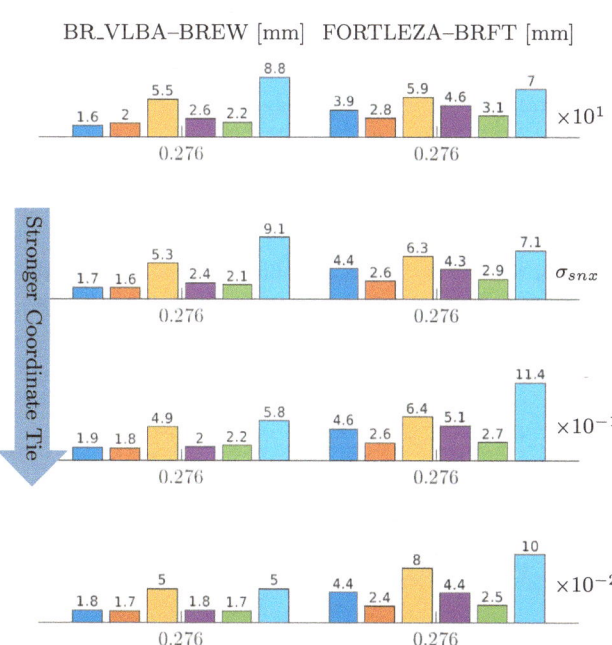

Fig. 4 Coordinate repeatabilities of the combined solution, regarding the type of constraint used for the coordinate tie. The first three values (blue, red and yellow) are the repeatabilities for the GNSS station, while the remaining three (purple, green and cyan) correspond to the VLBI station

---

[1] https://itrf.ign.fr/en/local-ties.

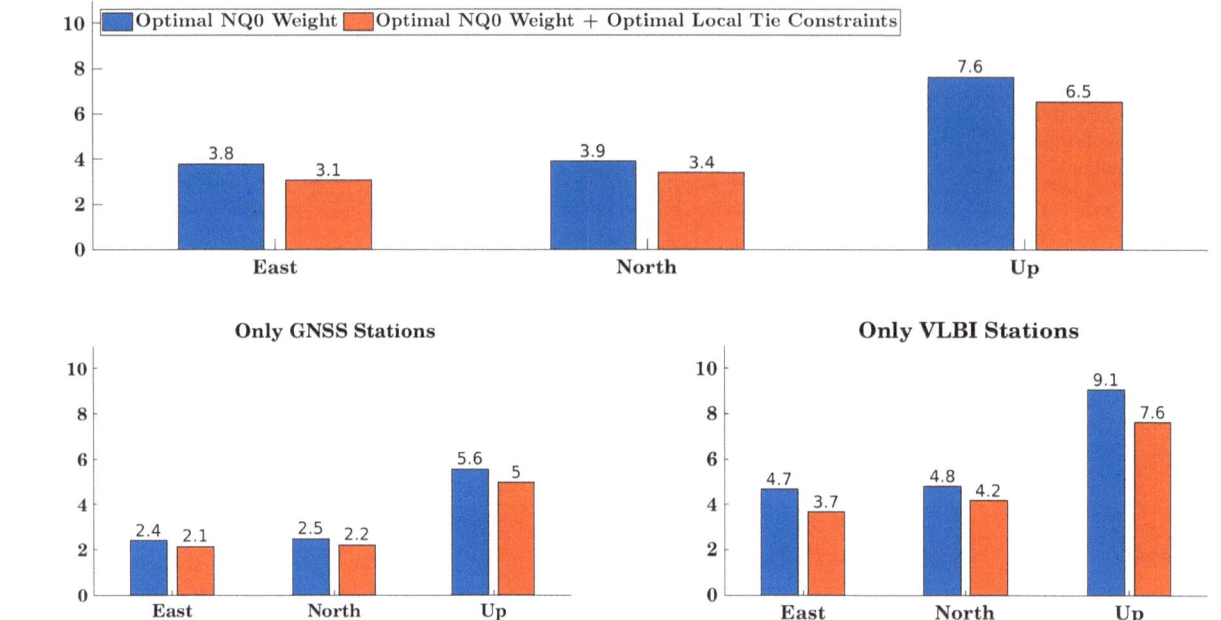

Fig. 5 RMS of the repeatabilities of the combined solution over the 15 days of the CONT17 campaign using appropriate constraints for coordinate ties. All values in mm

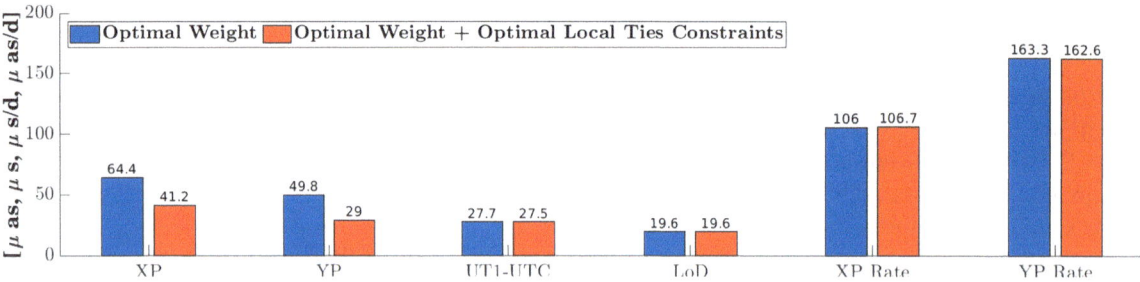

Fig. 6 RMS of differences of the daily estimated EOP, in the combined solutions, with respect to the IGS solution. Notice the different units (left-hand side of the plot) for each type of parameter

The benefits of the solution with optimal relative constraints are evident. The repeatabilities improve by 18%, 13%, and 14%, for the east, north, and height components, respectively (top plot of Fig. 5). When looking only at the GNSS stations, the improvements are 12% for the horizontal, and 11% for the vertical component. The largest improvement can be seen in the repeatabilities of the VLBI stations, with 21% 12%, and 17%, for the east, north and height component, respectively (bottom plots of Fig. 5).

7 Differences of EOPs to IGS Solution

To assess the improvement in the EOPs we used the IGS final solutions[2] as reference for the comparison, and the two 15-days rigorously combined solutions of Sects. 5 and 6 were

analysed. The RMS of the differences between the daily EOP estimates and the IGS solution are displayed in Fig. 6. Both solutions agree with the IGS solution at approximately the same level for the LOD and polar motion rate parameters. However, there is a large improvement in both polar motion components: 36% and 42% for the X and Y components, respectively. It should be mentioned that it is difficult to find a solution that can be used as ground truth for a comparison, as the rigorously combined solution is expected to be better than any other solution.

8 Rigorous and Single Technique Solutions

The final step in the study of the rigorous combination is the comparison of the relevant parameters, to the single-technique solutions. Moreover, as an additional reference

[2]https://www.igs.org/products.

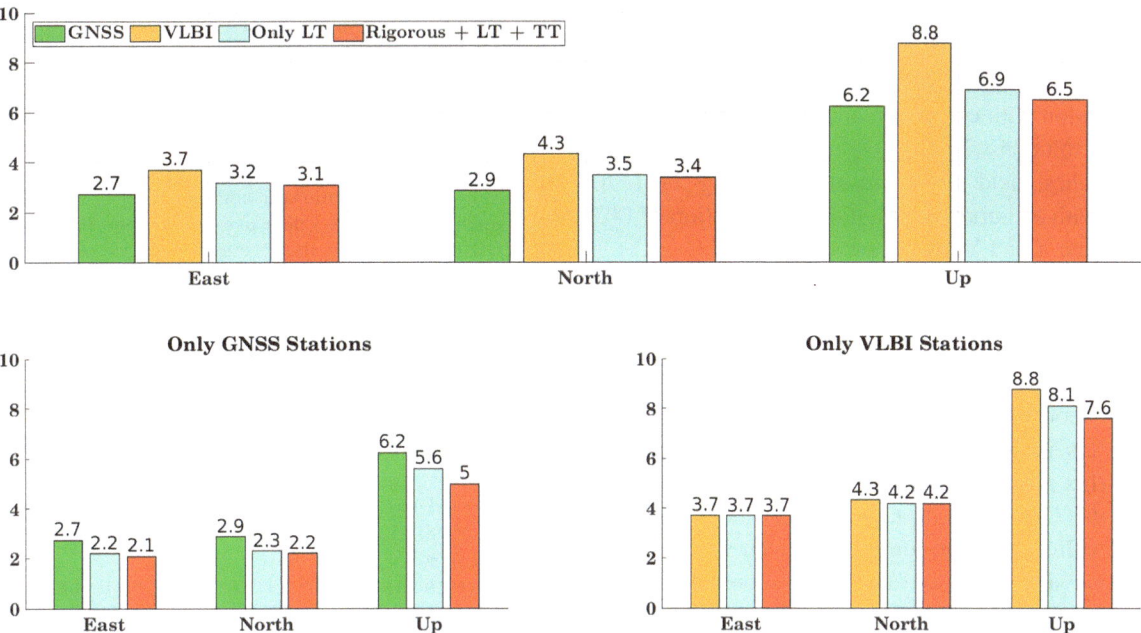

Fig. 7 RMS of coordinate repeatabilities [mm] for the individual technique solutions, and the combined solutions

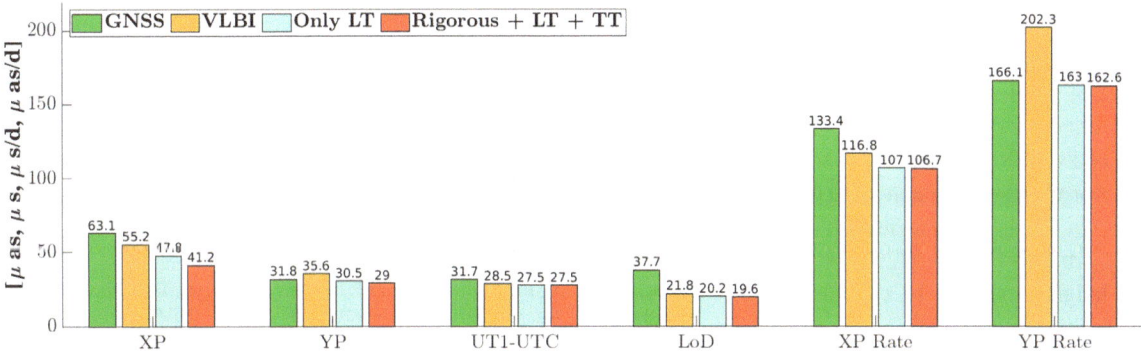

Fig. 8 RMS of daily EOP differences to IGS for the individual technique solutions, and the combined solutions. Notice the different units on the left side for each parameter

for comparison, we included a rigorous combination, where only coordinate ties were used. We start with the analysis of the RMS for the coordinate repeatabilities. While the combined results of both techniques may show a decrease in the performance of the rigorous solution with respect to the GNSS solution (top plot of Fig. 7) when separating the repeatabilities per technique, the benefits of the combined solution are more evident (bottom plot of Fig. 7). The improvement in the repeatabilities of the GNSS stations in the rigorous solution regarding the GNSS-only solution are 22%, 24%, and 19%, for east, north and height, respectively. Similarly, the improvement regarding the VLBI-only solution amounts to 2% and 14% for the north and height component, respectively. We also observe an improvement in the coordinate repeatabilities when comparing the rigorous solution with coordinate and tropospheric ties with the rigorous solution with only coordinate ties, as expected mainly in

height, with the height component of the former improving the performance by 11% (only GNSS stations), 7% (only VLBI stations), and 6% (all stations included). Additionally, the RMS of the difference of the EOPs regarding the IGS final solution is investigated, and displayed in Fig. 8. Once again, the rigorous solution outperforms the single-technique solutions in the polar motion estimates, with an improvement of 35% and 9% regarding the GNSS-only solution, for the X and Y components, respectively, and 25% and 19% regarding the VLBI-only solution, for the X and Y components, respectively. The three solutions agree with the IGS solution at approximately the same level for the UT1-UTC, with the rigorous solution helping to improve the results in the LOD estimate: 48% and 10%, compared to the GNSS-only and VLBI-only solutions, respectively. Polar motion rates show a favourable tendency towards the rigorous solution: 20% and 2% for the rate of the X and Y component, respectively, compared to the GNSS-only solution, and 9% and 20% for the rate of the X and Y component, respectively, regarding the VLBI-only solution. The comparison of the rigorous solution with coordinate and tropospheric ties with the rigorous solution with only coordinate ties showed that both approaches yield similar results regarding the LOD estimation, with an improvement of the polar motion of 14% and 5%, for the X and Y components, respectively.

9   Summary and Outlook

A rigorously combined solution for the estimation of geodetic parameters including GNSS and VLBI data has been achieved. This solution, based on the data of the CONT17 campaign plus GNSS/IGS data, profits from the use of coordinate ties with appropriate constraints, and troposphere ties at co-location sites with carefully chosen constraint levels, as well as from a tailored inter-technique weighting scheme based on the repeatabilities of the station coordinates. The combined solution was processed in a single state-of-the-art software, Bernese v5.2 – VLBI Version, where not only the a priori modelling and the parametrisation for both techniques was exactly the same, but also the full variance-covariance information of all the estimates, and the constraints for all the parameters were used throughout the estimation process. The combined solution with coordinate and troposphere ties generally improves the precision of all the estimated geodetic parameters. In particular, the repeatabilities of the station coordinates are improved by 22%, 24%, and 19%, for east, north, and height, respectively, compared to the GNSS-only solution, and by 2% and 14% for the north and height component, respectively, compared to the VLBI-only solution. Additionally, the EOPs estimates are also improved, with the rigorous solution outperforming the single-technique solutions in the polar motion estimates, by 35% and 9% compared to the GNSS-only solution, for the X and Y components, respectively, and 25% and 19% compared to the VLBI-only solution, for the X and Y components, respectively. The rigorous combination contributes to the stabilisation of the UT1-UTC, with the improvement of the LOD, showing a gain of 48% and 10%, compared to the GNSS-only and VLBI-only solutions, respectively. While there is an improvement when using tropospheric ties, further studies are required to improve the agreement among the VLBI and GNSS tropospheric estimates, which is currently at the level of 1–5 mm. Future activities will include an approach using variance component estimation for the weighting, the study of additional parameters in the combination, additional studies on the combination of intensive VLBI sessions with

GNSS for the estimation of UT1-UTC, and the rigorous triple combination with SLR observations.

Acknowledgements This work has been developed within the project "Co-location of Space Geodetic Techniques on Ground and in Space" in the frame of the DFG funded research unit on reference systems, and founded by the Swiss National Foundation (SNSF, 200021E-160421). Additionally, the authors would like to thank the IERS, CODE, IVS and CDDIS for providing the necessary reference coordinates, orbital products, VLBI data, and GNSS data required for the realisation of this work.

References

Altamimi Z, Rebischung P, Métivier L, Collilieux X (2016) ITRF2014: A new release of the international terrestrial reference frame modeling nonlinear station motions. J Geophys Res Solid Earth 121(8):6109–6131. ISSN:2169–9356. https://doi.org/10.1002/2016JB013098

Behrend D, Thomas C, Gipson J, Himwich E, Le Bail K (2020) On the organization of CONT17. J Geodesy 94(10):100. ISSN:1432–1394. https://doi.org/10.1007/s00190-020-01436-x

Böhm J, Werlb B, Schuh H (2006) Troposphere mapping functions for GPS and very long baseline interferometry from European Centre for Medium-Range Weather Forecasts operational analysis data. Geophys Res (111). https://doi.org/10.1029/2005JB003629.B02406. https://doi.org/10.1029/2005JB003629

Charlot P, Jacobs CS, Gordon D, Lambert S, de Witt A, Böhm J, Fey AL, Heinkelmann R, Skurikhina E, Titov O (2020) The third realization of the international celestial reference frame by very long baseline interferometry. Astron Astrophys 644. https://doi.org/10.1051/0004-6361/202038368

Coulot D, Berio P, Biancale R, Loyer S, Soudarin L, Gontier A-M (2007) Toward a direct combination of space-geodetic techniques at the measurement level: Methodology and main issues. J Geophys Res Solid Earth 112. https://doi.org/10.1029/2006JB004336

Dach R, Lutz S, Walser P, Fridez P (eds) (2015) Bernese GNSS software version 5.2. Astronomical Institute, Bern University. https://doi.org/10.7892/boris.72297

Delva P, Altamimi Z, Blazquez A, et al. (2023) GENESIS: colocation of geodetic techniques in space. Earth Planets Space 75(5). https://doi.org/10.1186/s40623-022-01752-w

Diamantidis P-K, Kłopotek G, Haas R (2021) VLBI and GPS inter- and intra-technique combinations on the observation level for evaluation of TRF and EOP. Earth Planets Space 73(1):68. ISSN 1880-5981. https://doi.org/10.1186/s40623-021-01389-1

Dow J, Neilan R, Rizos C (2009) The international GNSS service in a changing landscape of Global Navigation Satellite Systems. J Geodesy 83(3):191–198. https://doi.org/10.1007/s00190-008-0300-3

Rothacher M, Beutler G, Behrend D, Donnellan A, Hinderer J, Ma C, Noll C, Oberst J, Pearlman M, Plag H-P, Richter B, Schöne T, Tavernier G, Woodworth PL (2009) The future Global Geodetic Observing System. Springer, Berlin, Heidelberg, pp 237–272. ISBN:978-3-642-02687-4. https://doi.org/10.1007/978-3-642-02687-49

Rothacher M, Drewes H, Nothnagel A, Richter B (2010) Integration of space geodetic techniques as the basis for a Global Geodetic-Geophysical Observing System (GGOSD): an overview. Springer, Berlin, Heidelberg, pp 529–537. ISBN:978-3-642-10228-8. https://doi.org/10.1007/978-3-642-10228-843

Sarti P, Abbondanza C, Altamimi Z (2013) Local ties and co-location sites: some considerations after the release of ITRF2008. Springer, Berlin, Heidelberg, pp 75–80. ISBN:978-3-642-32998-2. https://doi.org/10.1007/978-3-642-32998-213

Thaller D (2008) Inter-technique combination based on homogeneous normal equation systems including station coordinates, earth orientation and troposphere parameters. PhD Thesis. https://doi.org/10.2312/GFZ.b103-08153

Wang J, Ge M, Glaser S, Balidakis K, Heinkelmann R, Schuh H (2022) Improving VLBI analysis by tropospheric ties in GNSS and VLBI integrated processing. J Geodesy 96(4):32. ISSN:1432-1394. https://doi.org/10.1007/s00190-022-01615-y

How Do Atmospheric Tidal Loading Displacements Vary Temporally as well as Across Different Weather Models?

Kyriakos Balidakis, Roman Sulzbach, Henryk Dobslaw, and Robert Dill

Abstract

We assess the impact of varying the mass anomaly sources on the calculation of atmospheric tidal displacement harmonics. Atmospheric mass anomalies are obtained from five state-of-the-art numerical weather models (NWM): DWD's ICON-Global, ECMWF's IFS, JMA's JRA55, ECMWF's ERA5, and NASA's MERRA2. To evaluate how the atmospheric tides' representation in the different models displaces Earth's crust, we calculate mass harmonics based on a fixed time span (2019.0–2022.0). To evaluate how temporally variable atmospheric tide manifestations are, we also applied a square-root-information filter on displacements spanning seven decades of ERA5. In addition, the variable harmonic atmospheric forcing is used to excite harmonic sea-surface variations employing the barotropic model TiME. The results from the analysis of the five numerical weather models as well as the monthly updated states of ERA5 harmonics are compared. We find that inter-model differences are larger than temporal harmonic modulations for all waves beating at frequencies higher than 1 cpd. We have confirmed that significant modulations are not an artefact in NWM but rather a true effect, and accounting for them might become of relevance for space geodesy at some point as soon as observations increase in spatio-temporal density and accuracy. The global RMS of radial displacements is 0.07 mm (SNR of 16.2 dB) for the "epoch" ensemble and 0.10 mm (SNR of 8.9 dB) for the "NWM" ensemble. We find discrepancies as large as 0.28 mm between harmonics from MERRA2 and early ERA5 batches, which we attribute to data sparsity in the in situ data assimilated into the NWM during the earlier years of the atmospheric reanalysis.

Keywords

Atmospheric tides · Inter-model variations · Numerical weather model · Temporal modulation · Tidal loading displacements

1   Introduction

Mass redistribution within the Earth's fluid envelope including the atmosphere, the oceans, and the terrestrial water storage elastically deforms Earth's crust hence inducing

Roman Sulzbach, Henryk Dobslaw, Robert Dill contributed equally to this work.

K. Balidakis (✉) · H. Dobslaw · R. Dill
Earth System Modelling, GFZ German Research Centre for Geosciences, Potsdam, Germany
e-mail: kyriakos.balidakis@gfz-potsdam.de; dobslaw@gfz-potsdam.de; dill@gfz-potsdam.de

R. Sulzbach
Earth System Modelling, GFZ German Research Centre for Geosciences, Potsdam, Germany

Institut für Meteorologie, Freie Universität Berlin, Berlin, Germany
e-mail: sulzbach@gfz-potsdam.de

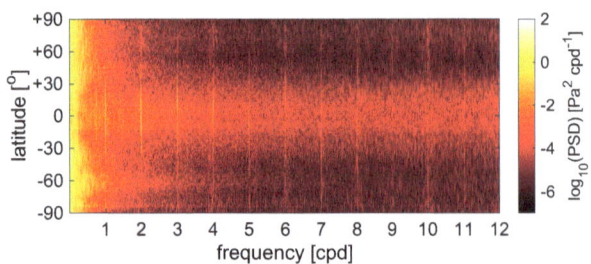

Fig. 1 Power spectral density for ERA5-derived total atmospheric pressure along the −169°W meridian, which crosses the least amount of land according to ERA5. Light pixels indicate latitudes where atmospheric pressure has a strong response at the corresponding frequency

displacements of geodetic markers in excess of 1 cm at sub-daily to seasonal and even inter-annual timescales. This contribution focuses on high-frequency deformation induced by atmospheric tides. Unlike ocean tides that are mostly excited by the gravitational pull of the Moon, atmospheric tides are mostly excited by the Sun, in particular, the periodic absorption of infrared radiation by water vapor in the troposphere and ultraviolet radiation by ozone in the stratosphere, as well as large-scale latent heat release. Inspired by Ray et al. (2021), we have calculated the response of atmospheric pressure anomalies at different frequencies along the −169°W meridian that intersects the least with land (see Fig. 1). We note sharp spectral lines at integer overtones of the solar diurnal wave S_1 as well as its side-bands, and the fact that the highest power spectral density (PSD) values are found in the equatorial belt. The high-frequency waves to which the atmospheric pressure response is the strongest are the S_1 and the S_2. These variations are not artefacts, rather mani-

festations of a well-studied phenomenon called atmospheric tides, which are responsible for high-frequency peak-to-peak pressure anomalies in excess of 500 Pa. Atmospheric tides may be studied by manifestations thereof, which are mainly in atmospheric density and its spatial gradients. For space geodesy, atmospheric tides induce temporal variations in (i) the gravity field (e.g., Boy et al. 2006); (ii) the deformation of Earth's crust due to the loading exerted by the atmospheric mass (e.g., Petrov and Boy 2004); (iii) the coefficients with which we describe how refraction affects signals traversing Earth's electrically neutral atmosphere (e.g., Jin et al. 2008); and (iv) the motion of Earth relative to its spin axis, that is, polar motion and *UT1–UTC* or length-of-day (e.g., Girdiuk 2017), as well as components of Earth's nutation (e.g., Schindelegger et al. 2016). While atmospheric tides are responsible for crustal deformation at the sub-cm-range (see Fig. 2), they should be considered during space geodetic data analysis to mitigate aliasing artefacts. Herein, for the waves that induce the largest mass anomalies in the sub-diurnal frequency band we assess the extent to which the predictions of tidal mass loads and the associated crustal displacements differ depending on the numerical weather model (NWM), as well as how much they differ as a function of time. We retrieve mass anomalies from five state-of-the-art NWMs, namely, ERA5, MERRA2, JRA55, ECMWF's IFS, and ICON.

In this contribution, we assess the extent to which displacements induced by harmonic atmospheric mass variation driven primarily by solar irradiance vary (i) over time, and (ii) between mass variation models. Section 2 describes the atmospheric and oceanic tidal amplitudes from NWM and a

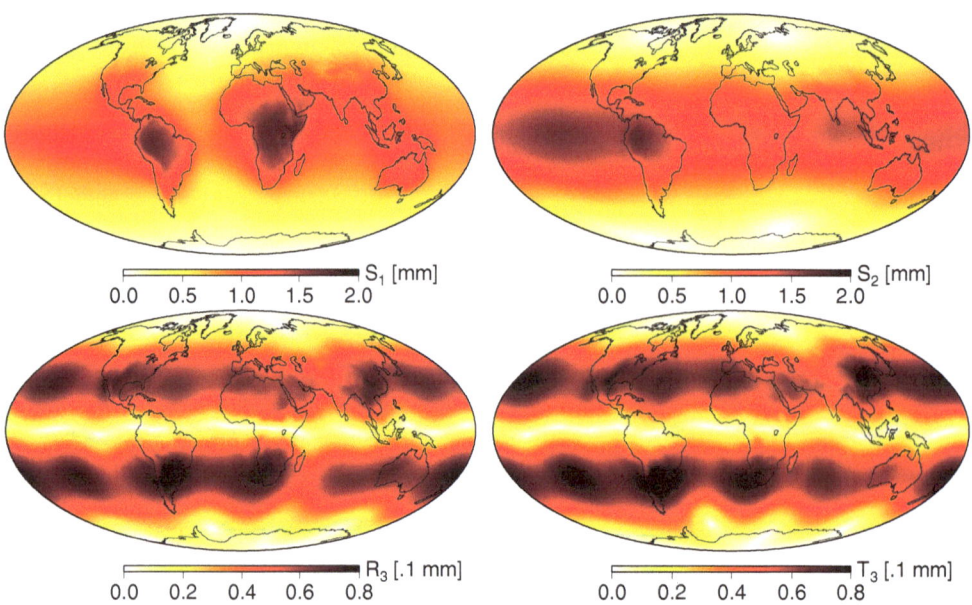

Fig. 2 Radial harmonic atmospheric tidal loading displacement amplitudes employing ERA5 fields spanning the period 1979.0–2022.0, in the center of mass isomorphic reference frame. Shown are the strongest waves in the diurnal (**a**), semi-diurnal (**b**), and ter-diurnal band (**c** and **d**)

barotropic ocean tide model, and presents a relative comparison. Section 3 describes how tidal displacement amplitudes differ between models. Section 4 outlines the estimation of partial tide modulations and discusses the associated estimates from a filter solution. Finally, we summarize our work and draw conclusions in Sect. 5.

2 Atmospheric and Oceanic Tidal Mass Anomalies

To predict loading-induced site displacements, accurate knowledge of the instantaneous mass anomaly is required. In the atmosphere, mass anomalies are inferred from surface pressure anomalies. Ocean mass anomalies are typically obtained by sea-surface heights deduced from the analysis of satellite altimetry observations or by running a model that solves hydrodynamic equations numerically (Dobslaw and Thomas 2005). Since the magnitude of pressure fluctuations depends on the altitude (e.g., for S_1) and the orography of the models, we calculate the pressure at a reference orography employing the three-dimensional atmospheric density which at a given site and epoch is a function of temperature, pressure, specific humidity and geopotential (Dobslaw 2016). In this work, we calculate atmospheric density variations from two operational models, ECMWF's IFS (three-hourly fields on 9 km grids) and DWD's ICON (three-hourly fields on 13 km grids), and three reanalysis models, ECMWF's ERA5 (hourly fields on 31 km grids, Hersbach et al. 2020), NASA's MERRA2 (hourly fields on 50 km grids, Gelaro et al. 2017), and JMA's JRA55 (three-hourly fields on 55 km grids, Kobayashi et al. 2015). Although a considerable fraction of the observations assimilated in these NWM is identical, the underlying data assimilation system as well as the spatio-temporal resolution are largely different ranging from meso-β to meso-γ scale.

Following Balidakis et al. (2022), we have estimated harmonic amplitudes based on several batches of the aforementioned NWM so that we may assess the extent to which atmospheric forcing variations project into harmonic sea-surface heights predicted by the barotropic Tidal Model forced by Ephemerides (TiME, Sulzbach et al. 2021), where self-attraction and loading effects of the ocean mass are rigorously considered. In particular, we focus on the following waves: π_1, P_1, S_1, K_1, ψ_1, M_2, T_2, S_2, R_2, K_2, T_3, S_3, R_3, S_4, S_5, and S_6.

We have estimated atmospheric forcing harmonics and performed TiME simulations (Sulzbach et al. 2021; Balidakis et al. 2022) where we varied the atmospheric forcing by adopting the following scenarios

1. ERA5a: ECMWF's ERA5 (1979.0–1982.0);
2. ERA5b: ECMWF's ERA5 (1989.0–1992.0);
3. ERA5c: ECMWF's ERA5 (1999.0–2002.0);

4. ERA5d: ECMWF's ERA5 (2009.0–2012.0);
5. ERA5e: ECMWF's ERA5 (2019.0–2022.0);
6. ECMWF's IFS (2019.0–2022.0);
7. NASA's MERRA2 (2019.0–2022.0);
8. JMA's JRA55 (2019.0–2022.0); and
9. DWD's ICON (2019.0–2022.0).

The choice of three-year batches is not random. The precision of the harmonic amplitudes increases with the data span. We estimated harmonics for the 16 waves of interest employing variable data spans: from the theoretical minimum of one year of hourly data up to two decades. We found that the increase in precision is quadratic for all diurnal waves as well as for S_2 within the first three years and linear afterwards. For instance, the global RMS between S_1 harmonics estimated based on two decades of ERA5 and only one year of ERA5 is 3.5 Pa, on average. The RMS between the 20-year estimates and three-year estimates is 1.7 Pa, suggesting that the ability to predict pressure anomalies of a three-year estimate is about twice as good compared to an one-year estimate. While we have assessed the temporal tidal variations in atmospheric and ocean bottom pressure from some of the other reanalysis NWM in the framework of (Shihora et al. 2023), here we opt to work with ERA5 driven by its higher reliability (e.g., Ray et al. 2023). Due to the fact that the wind stress contribution to the ocean tide excitation process is considerably smaller in comparison to the contribution of pressure, wind stress harmonics were not used to force these TiME experiments. To evaluate the differences between harmonics estimated employing different data sets, we calculate the RMS which is defined by (e.g., Shihora et al. 2022):

$$\text{RMS}_j(\mathbf{x}) = \sqrt{\frac{\sum_k^K |\zeta_j^k(\mathbf{x}) - \zeta_j^{ref}(\mathbf{x})|^2}{2(K-1)}}, \qquad (1)$$

$$\zeta_j^k(\mathbf{x}) = C_j^k(\mathbf{x}) + i S_j^k(\mathbf{x}),$$

where ζ_j^k is the complex representation of the mass anomaly corresponding to wave j from data set k, which in turn runs over the harmonic amplitude estimates from different data sets up to K. Harmonics with the superscript *ref* stem from the analysis of either longer time series or the combination of the different ensemble members. We have calculated the atmospheric and oceanic mass anomaly RMS upon varying (i) the temporal range, and (ii) the NWM based on which the amplitudes are estimated. For the former, hereinafter "epoch" ensemble, we have employed harmonics from scenarios 1–5, and for the latter, hereinafter "NWM" ensemble, we have employed harmonics from scenarios 5–9. For scenarios 1–5, we choose ERA5 since it features the highest spatio-temporal resolution among the reanalysis NWM within our ensemble. While the assimilation system employed for the production of these data sets does not change, the type, quality, and number of observations ingested within IFS

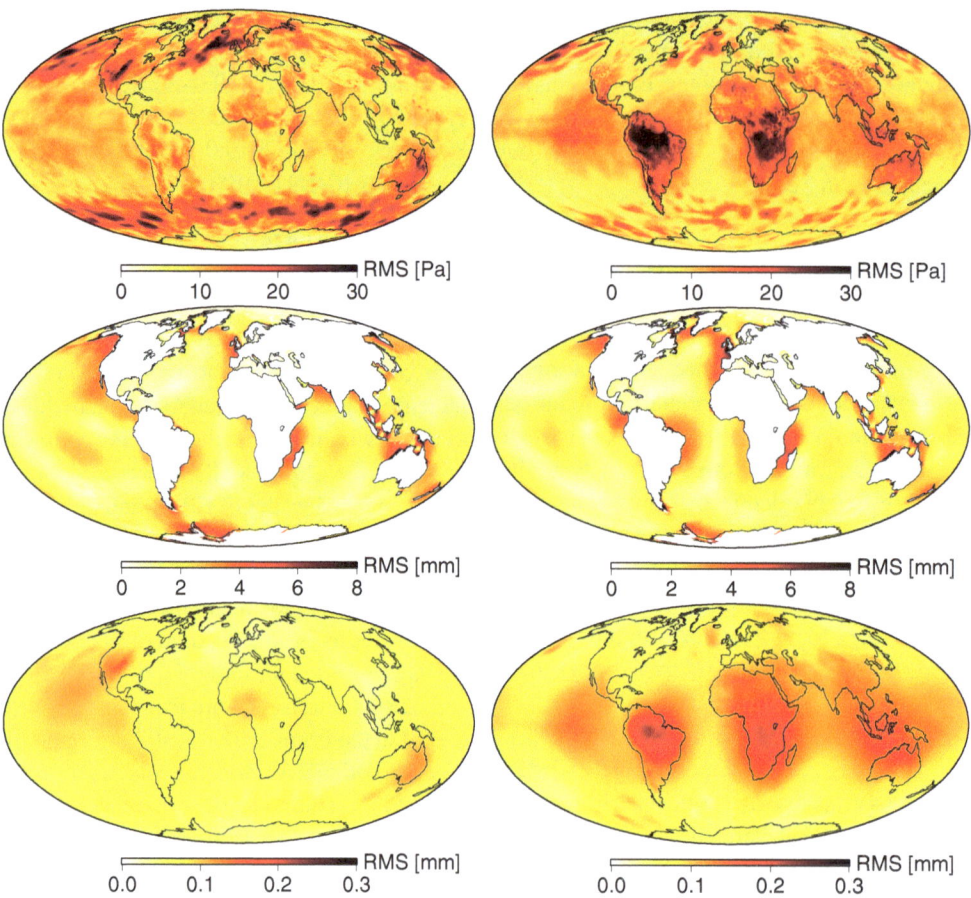

Fig. 3 Atmospheric pressure (1st row) and TiME-derived sea-surface height (2nd row) harmonic RMS following Eq. 1 (all waves), where the data span (left) and the model (right) of the atmospheric forcing have been varied. Atmospheric tidal loading displacement harmonic RMS are also shown in the radial direction (3rd row)

model cycle 41r2 does vary. Together with differences in the input data, climate-related low-frequency changes in parameters such as water-vapor (trends between $-0.10\,\mathrm{kg\,m^{2}a^{-1}}$ and $0.13\,\mathrm{kg\,m^{2}a^{-1}}$ based on ERA5 from 1979 onward) and ozone (trends between $-0.68\,\mathrm{DU\,a^{-1}}$ and $0.41\,\mathrm{DU\,a^{-1}}$ based on ERA5 from 1979 onward) concentration give rise to the differences illustrated in the 1st column of Fig. 3. For the "NWM" ensemble, climate-related variations should not contribute to the differences we observe, rather only discrepancies induced by the assimilation system and the physical formulation of the NWM. The TiME configuration ($\frac{1}{12°}$ mesh, see Balidakis et al. 2022) is identical for all nine scenarios and reflects only differences in the harmonic atmospheric forcing (see 2nd column of Fig. 3). By and large, the assumption we make is that discrepancies within the "epoch" ensemble are due to climate change and availability of observations, whereas discrepancies within the "NWM" ensemble are due to differences in the data assimilation system, spatial resolution, and model physics.

Harmonic atmospheric pressure discrepancies from the "epoch" and "NWM" ensemble are illustrated in Fig. 3. For the "epoch" ensemble, the average RMS is 10.5 Pa. We find large discrepancies in excess of 30 Pa in the Bering

Sea and the North Atlantic Ocean. We also find RMS values over 20 Pa over the North American Great Plains. For the "NWM" ensemble, we observe an average RMS of 11.3 Pa with spatial clusters exceeding 25 Pa over the ocean, similar to the "NWM" ensemble, in the Bering Sea and North Atlantic Ocean as well as the land clusters exceeding 30 Pa in the Andes and the Amazon catchment. The ensemble member responsible for the increased RMS over the equatorial land regions (South America, Central Africa, and Indonesia) is JRA55. Moreover, we observe slightly larger inter-model discrepancies in some regions with steep orographic gradients, which we attribute in large part to the representativeness of the lower-resolution models since the procedure to interpolate pressure given model or pressure level data is very accurate (Dobslaw 2016).

The RMS of the sea-surface height harmonic amplitudes is shown in the 2nd row of Fig. 3. Based on our simulations, we find that varying the NWM (5.2 mm, on average) has a larger impact on the sea-surface height in comparison to varying the period based on which the atmospheric tidal amplitudes were estimated (1.8 mm, on average). We observed deviations in excess of 50 mm at Timor Sea, Bristol Channel, and Hecate Strait. Further, the largest deviations we observe upon varying the time span of the forcing data exceeds 1 cm only at Sea of Okhotsk and the ice shelf at Ross Sea.

The atmospheric forcing harmonic discrepancies do not correlate to the TiME-derived sea-surface height predictions in the spatial domain (see 1st and 2nd row of Fig. 3), since excitations at the frequencies considered here typically lead to hemispheric waves with largest amplitudes at the coasts.

Moreover, we have calculated the pair-wise RMS of the harmonic variations (see Fig. 4). For atmospheric pres-

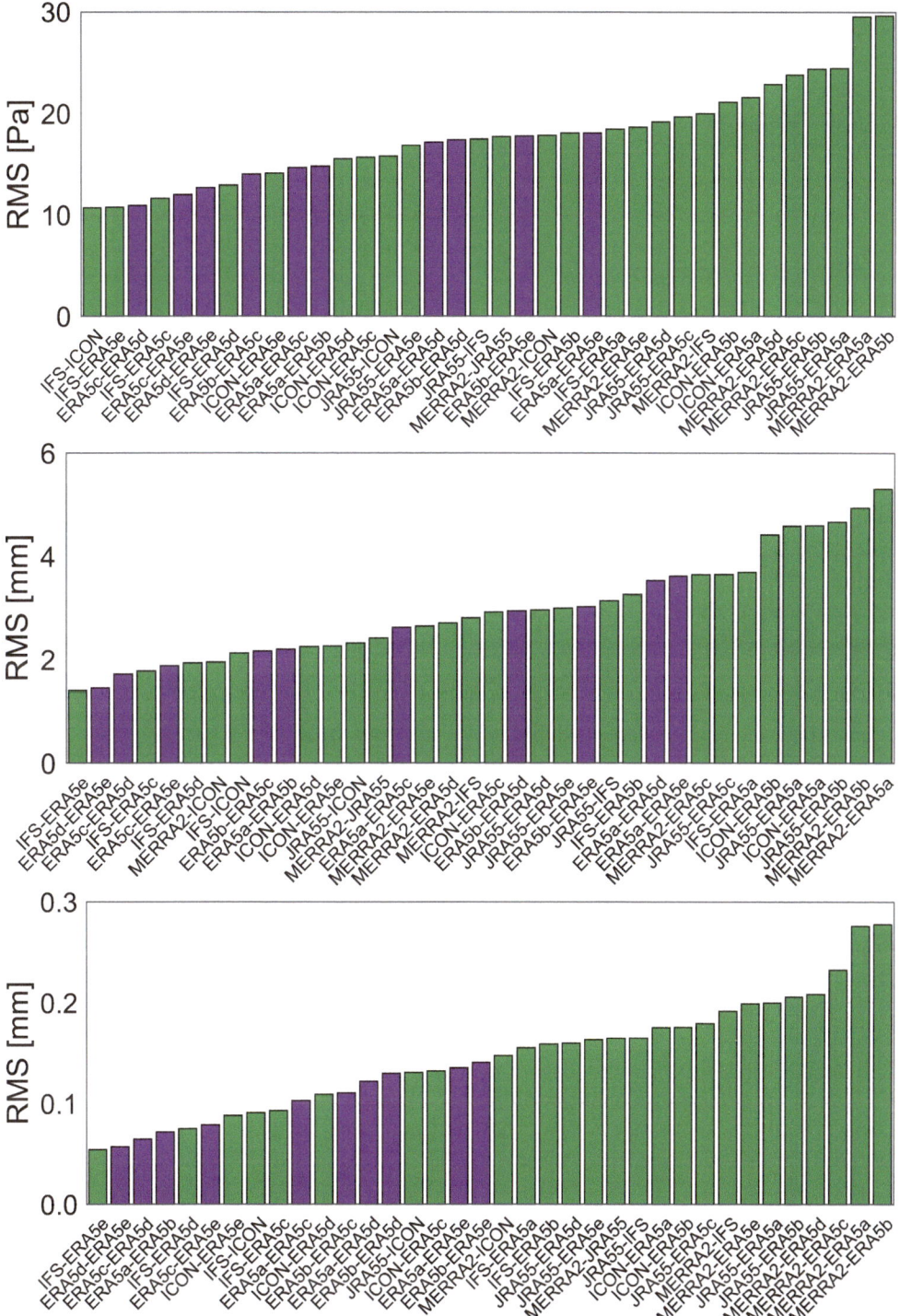

Fig. 4 Sorted harmonic discrepancies' RMS for the pressure anomalies (1st row) and the sea-surface height variations (2nd row), following Eq. 1. RMS for the radial displacements is shown in the 3rd row. In purple shown are the pairs of the "epoch" ensemble and in green shown are the pairs of the "NWM" ensemble

sure, we find the largest discrepancies between early ERA5 data and other non-ECMWF models, namely MERRA2 and JRA55, and the best agreement between IFS, ICON, and recent ERA5 data. For sea-surface heights predicted by TiME, we observe large discrepancies between early ERA5 data and all non-ERA5 experiments. The differences between IFS and late ERA5 data are among the smallest, which justifies the use of current IFS and late ERA5 data in the same analysis; this is certainly due to the fact that the ERA5 system is very close to the operational IFS, also in terms of in situ data assimilation. While the largest discrepancies are found in the "NWM" ensemble, several pairs of the "epoch" ensemble have higher RMS.

3   Tidal Loading Displacements

Approximating the loading mass anomalies as an infinitesimally thin layer, we calculate loading displacement variations by convolving the mass anomaly harmonic amplitudes with load Green functions, following Dill and Dobslaw (2013).

We note that atmospheric loading mass anomalies over the oceans as the atmospheric contribution to ocean tides is not considered, but will be explicitly treated with a global ocean tide model (Sulzbach et al. 2021) for all relevant frequencies, thereby making any assumption about an inverse barometric response of the ocean superfluous. We have calculated displacements induced by the atmospheric harmonic mass loads described in Sect. 2. In Fig. 2 we present the radial harmonic atmospheric tidal loading displacement amplitudes of the strongest atmospheric tidal waves that belong to the diurnal, semi-diurnal and ter-diurnal species, in the center of mass isomorphic frame we calculated employing ERA5 hourly fields spanning the period 1979.0–2022.0. We note that the radial displacements in the ter-diurnal band are about one order of magnitude smaller in comparison to the horizontal displacements in response to the S_1 and S_2 waves, and another three times smaller in comparison to the associated radial displacements. Moreover, we observe that while the spatial pattern for the S_1 and S_2 in-phase and quadrature amplitudes is degree-one and degree-two sectorial harmonic for all coordinate components, for the upper and lower sideband of S_3 it is tesseral of degree/order 4/3 for the radial and eastward component and 5/3 for the northward component.

We find the largest RMS in the radial coordinate component, since the radial Green's function assigns a considerably larger weight to mass anomalies in comparison to the tangential Green's function, especially in the near-field.

The largest discrepancies are found between MERRA2 and ERA5a for S_1 (0.16 mm) and S_2 (0.18 mm for MERRA2-ERA5a). The individual RMS for all other waves is relatively high for P_1, K_1, M_2, R_2, K_2, S_3, however, below 0.05 mm for all pairs. We note that in the ter-diurnal band, while the upper sideband of the S_3, R_3, and its lower sideband T_3 feature larger and more spatially coherent mass anomalies than S_3, the corresponding displacements are not as large. The "typical" wavelength of a ter-diurnal wave is shorter than that of a diurnal approximately by a factor of three, and also shorter than a semi-dirunal approximately by a factor of 3.2. Moreover, the Stokes coefficients are reduced by a factor $(2n + 1)^{-1}$, where n denotes the expansion degree. So, if the wavelength is shorter, the weight typically assigned to the coefficients is smaller by a factor of 3/1 and 3/2, in comparison to the diurnal and semi-diurnal band, respectively. Also, relatively small ter-diurnal amplitudes are partly due to the fact that in the CM a positive mass load placed at a spherical distance larger than about 70° will induce an upward displacement and the harmonic amplitudes of R_3 and T_3 feature a tesseral spherical harmonics pattern of degree/order 4/3. For the eastward coordinate component, the inter-model wave-wise discrepancies are a function of the displacement amplitude, that is, we find RMS up to 0.10 mm for S_1 (MERRA2-ERA5a), up to 0.04 mm for S_2 (JRA55-ICON), and up to 0.02 mm for S_3 (MERRA2-ERA5a). For all other waves, the highest global RMS is below 0.03 mm. Similar to the EW component for NS the largest differences are detected in S_1 (up to 0.09 mm for MERRA2-ERA5a), S_2 (up to 0.03 mm for MERRA2-JRA55), and S_3 (up to 0.02 mm for JRA55-IFS).

The highest RMS is found between MERRA2 and early ERA5 batches and the lowest RMS is found between harmonics estimated based on late ERA5 batches, IFS, and ICON data. For the principal waves S_1 and S_2 the RMS between the models that have the best agreement is about six times smaller in comparison to the RMS between the models that have the largest discrepancies. However, the dynamic range for the S_2 harmonic displacement amplitudes in the tangential components is three time smaller (0.5 dB) in comparison to that of S_1. We observe no relation between the dynamic range and the signal aimplitude. Our results indicate that while temporal variations in harmonic amplitudes are non-negligible, differences in the data assimilation system as well as its input employed by the weather models utilized herein give rise to even larger discrepancies; for the "epoch" ensemble, the spatially average RMS over all waves is 0.07/0.03/0.03 mm and for the "NWM" ensemble the RMS is 0.10/0.05/0.03 mm for the radial/eastward/northward coordinate components, respectively.

4   Estimation of Partial Tide Modulations

Changes in the atmospheric tide exciting mechanisms and forcing agents including but not limited to periodic solar irradiance variability (e.g., Schwabe cycle, seasonal, Carrington rotation), variations in the concentration of water vapor and ozone, deep convective activity in the tropics, fluctuations in Earth's ionosphere and magnetic field are responsible for modulations in the harmonic amplitudes thereof. Section 4 describes their quantification given pressure data.

The first step to our investigations is to vary the data span utilized for the estimation of harmonic amplitudes, given a long time series. For example, given ERA5 atmospheric pressure time series at the site that on average has one of the largest S_2 amplitudes over land (equatorial South America), we have estimated a set of tidal harmonics by different segments of the time series; we have varied both the starting point and length of the time series. The amplitude differences associated with the phase deviations feature peak-to-peak variations of 20 Pa. The shorter the time series employed to calculate the tidal harmonics, the larger the temporal harmonic modulations; analyzing data that span shorter than a decade yields amplitude differences that change as much as 20 Pa within 70 years. In essence, the phase estimate can change as much as $10°$ given only a decade of data. Given the S_2 beating frequency this means that the maximum of the S_2 wave will occur 20 min later than a couple of decades ago.

Below, we present the estimation of harmonic modulations $\mathbf{x} = \begin{bmatrix} \mathbf{x_1} \ \mathbf{x_m} \ \mathbf{x_M} \end{bmatrix}$ parameterized as low-degree B-spline functions by introducing normal equations of monthly intervals, given only three data blocks $\mathbf{y_m}$, $m \in [1, M]$, however without any loss of generality. The solution is given by $\mathbf{x} = \mathbf{N}^{-1}\mathbf{u}$, where the normal equation system is constructed as follows

$$\mathbf{N} = \begin{bmatrix} \mathbf{N_1} & & \\ & \mathbf{N_m} & \\ & & \mathbf{N_M} \end{bmatrix}$$

$$+ \begin{bmatrix} \mathbf{P^r_{1\text{-}m}} & -\mathbf{P^r_{1\text{-}m}} & \\ -\mathbf{P^r_{1\text{-}m}} & \mathbf{P^r_{1\text{-}m}} + \mathbf{P^r_{m\text{-}M}} & -\mathbf{P^r_{m-M}} \\ & -\mathbf{P^r_{m\text{-}M}} & \mathbf{P^r_{m\text{-}M}} \end{bmatrix}$$

$$+ \begin{bmatrix} \mathbf{P^a_1} & & \\ & \mathbf{P^a_m} & \\ & & \mathbf{P^a_M} \end{bmatrix} \qquad (2)$$

$$\mathbf{u} = \begin{bmatrix} \mathbf{u_1} \ \mathbf{u_m} \ \mathbf{u_M} \end{bmatrix}^\top + \begin{bmatrix} \mathbf{P^a_1} \ \mathbf{P^a_m} \ \mathbf{P^a_M} \end{bmatrix}^\top \mathbf{x}_{ref},$$

where $\mathbf{N_m} = \mathbf{J_m^\top P_m^o J_m}$ and $\mathbf{u_m} = \mathbf{J_m^\top P_m^o y_m}$, is the normal equation system, $\mathbf{J_m}$ is the design matrix following observation equation 2 from Balidakis et al. (2022), and $\mathbf{P_m^o}$ is the associated observations' weight matrix. $\mathbf{P_m^r}$ is the matrix that

controls the relative constraints, and $\mathbf{P_m^a}$ is the matrix that controls the absolute constraints for data block m.

Stochastic equivalence constraints stabilize the solution. Relative constraints control how consecutive parameters, that is unknowns referring to one quantity (e.g., the in-phase harmonic amplitude of one wave) at different times, vary under the Markov assumption. They read $\hat{x}(t + dt) = \hat{x}(t) + w(t)dt$, where $\hat{x}$ denotes the a posteriori value of an unknown parameter set up to be estimated at epoch t, and w controls the weight of each pseudo-observation (σ_{rc}^{-2}). We note that imposing very tight relative constraints effectively yields harmonic amplitudes with no temporal variability. In this work, the variations of harmonic amplitudes are parameterized as random walk processes. Absolute constraint observation equations, $\hat{x}(t) - x_{ref} = 0 \pm \sigma_{ac}$, where x_{ref} is a reference value herein derived from a least-squares adjustment involving the full-length time series, and σ_{ac}^{-2} controls the weight of the related pseudo-observation, may be applied as well.

However, it is apparent that a large number of waves, a large number of estimation intervals, as well as a large number of data batches will render the above procedure impractical. For instance, selecting 16 waves and monthly estimation intervals for 70 years will flood the parameter space with more than 26,000 elements. To this end, we have resorted to sequential methods where the dimension of the parameter space depends only upon the wave ensemble. We have adopted the Dyer-McReynolds implementation of the square root information filter (Bierman 1977).

Figure 5 illustrates the results of the S_1 and S_2 phase upon varying the process noise relative to the measurements'

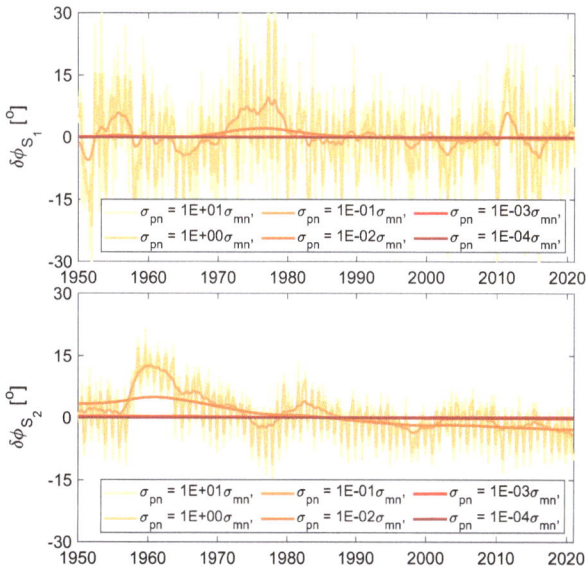

Fig. 5 Temporal S_1 (top) and S_2 (bottom) pressure phase variation estimates from varying the filter process noise (σ_{pn}) in relation to the measurement noise (σ_{nm}) in equatorial South America

noise. As expected, injecting relatively little process noise in the filter yields practically time-invariant harmonics. However, allowing for enough process noise reveals at first inter-annual variations and eventually seasonal modulations at annual frequencies and overtones thereof. In particular, the annual amplitude of the S_2 amplitude and phase modulations for a site in South America is 6.1 Pa and 4.7°, respectively. The S_a modulations for the S_1 wave at that site are 5.4 Pa and 5.8°, respectively. We note that the time-independent amplitude of the S_1 and S_2 is 48.4 Pa and 204.9 Pa, respectively. Furthermore, given the non-negligible deviations from a linear long-term signal evolution, quantifying temporal tidal variations with a first-order polynomial is an oversimplification since varying the interval under consideration will yield considerably different results. We note that temporal variations in harmonic amplitudes may be due to significant changes in the quality, type, and volume of observations assimilated into a NWM. Comparing the evolution of NWM harmonics with the evolution of harmonics from in situ barometers could shed light on whether the former are true or artefacts (Ray 2001; Schindelegger and Ray 2014), what cannot be done solely by studying the NWM-driven harmonics.

We run the procedure described above to the displacements induced by the temporally variable atmospheric pressure harmonics to calculate the associated displacement fields. Our results suggest that given the nominal accuracy of state-of-the-art space geodetic observations (3 ps for VGOS group delays, 5 mm for GNSS P3 observations, and mean-RMS of 20 mm for LAGEOS SLR normal points with 120 s integration interval), the temporal modulations in the tidal atmospheric loading corrections will probably have little impact on the data analysis procedure as well as the products.

5   Conclusions

Atmospheric tides induce atmospheric surface pressure variations that displace the Earth's crust vertically in the sub-cm range. The most prominent periods are 12 and 24 h, but also other relevant lines with neighboring frequencies have been identified. Accounting for such systematic effects in space geodetic data analysis potentially decreases aliasing and facilitates the detection of spurious signals. Herein, we have assessed the extent to which the NWM-predicted tidal atmospheric loading displacements vary temporally as well as across different models. We have utilized mass anomaly fields from five NWM (ERA5, ICON, IFS, JRA55, and MERRA2) from four meteorological agencies (DWD, ECMWF, JMA, and NASA-GMAO) with spatial resolutions ranging between 55 km and 9 km. We have used a square-root-information filter to estimate tidal mass anomaly ampli-

tude modulations, which we have employed to calculate high-frequency tidal loading displacements.

Do atmospheric tidal loading displacement signals vary in time? For the "epoch" ensemble, the RMS for the radial/east/north component ranges between 0.06/0.02/0.02 mm and 0.14/0.06/0.07 mm, respectively. On average, harmonics estimated based on data sets that are close in time show a better agreement. We have also found annual and semi-annual modulations in the mass anomaly fields if we allow in the filter settings for enough process noise relative to the observation noise.

Do models agree in the prediction of tidal atmospheric loading displacements? For the "NWM" ensemble, the RMS for the radial/east/north component ranges between 0.05/0.02/0.02 mm and 0.28/0.10/0.08 mm, respectively. The largest disagreements are found between MERRA2 and early ERA5 data sets. On average the discrepancies between MERRA2 and early ERA5 data sets are 3–12 times larger compared to the RMS between recent ERA5 data and IFS or ICON.

By and large, switching from one NWM to another induces larger tidal harmonic amplitude differences than temporal modulations within a single NWM. However, both inter-model and temporal differences have a global RMS below 0.3 mm for radial displacements, which while it might exceed 10 % of the effect, it is still below the uncertainty of state-of-the-art geometric space geodetic observations. In line with Ray et al. (2023) who did not find systematic errors in ERA5-derived atmospheric tides and in absence of more accurate NWM data, we recommend the application of time-invariant harmonic atmospheric loading displacement corrections derived from multi-year ERA5 data as, e.g., made available recently by Sulzbach et al. (2022).

Acknowledgements KB and RS are funded by the DFG via the Collaborative Research Cluster TerraQ (SFB 1464, Project-ID 434617780). Deutscher Wetterdienst (Offenbach, Germany) and the European Centre for Medium-range Weather Forecasts (Reading, U.K.) are acknowledged for granting access to ICON and ECMWF operational data. Numerical analyses were performed at Deutsches Klimarechenzentrum (DKRZ) in Hamburg (Germany) under Project-ID 0499. Calculations carried out herein were facilitated by employing the CDO software suite (Schulzweida 2022).

References

Balidakis K, Sulzbach R, Shihora L, et al (2022) Atmospheric contributions to global ocean tides for satellite gravimetry. J Adv Model Earth Syst 14(11):e2022MS003,193. https://doi.org/10.1029/2022MS003193, e2022MS003193 2022MS003193

Bierman GJ (1977) Sequential square root data processing. In: Bierman GJ (ed) Factorization methods for discrete sequential estimation, mathematics in science and engineering, vol 128. Elsevier, pp 68–112. https://doi.org/10.1016/S0076-5392(08)60890-5

Boy JP, Ray R, Hinderer J (2006) Diurnal atmospheric tide and induced gravity variations. J Geodyn 41(1):253–258. https://doi.org/10.1016/j.jog.2005.10.010, earth Tides and Geodynamics: Probing the Earth at Sub-Seismic Frequencies

Dill R, Dobslaw H (2013) Numerical simulations of global-scale high-resolution hydrological crustal deformations. J Geophys Res Solid Earth 118(9):5008–5017. https://doi.org/10.1002/jgrb.50353

Dobslaw H (2016) Homogenizing surface pressure time-series from operational numerical weather prediction models for geodetic applications. J Geod Sci 6(1). https://doi.org/10.1515/jogs-2016-0004

Dobslaw H, Thomas M (2005) Atmospheric induced oceanic tides from ECMWF forecasts. Geophys Res Lett 32(10). https://doi.org/10.1029/2005gl022990

Gelaro R, McCarty W, Suárez MJ, et al (2017) The modern-era retrospective analysis for research and applications, version 2 (MERRA-2). J Clim 30(14):5419–5454. https://doi.org/10.1175/jcli-d-16-0758.1

Girdiuk A (2017) Atmospheric tides in earth rotation observed with VLBI. PhD thesis, Technische Universität Wien https://doi.org/10.34726/HSS.2017.30734

Hersbach H, Bell B, Berrisford P, et al (2020) The ERA5 global reanalysis. Q J R Meteorol Soc 146(730):1999–2049. https://doi.org/doi.org/10.1002/qj.3803

Jin S, Wu Y, Heinkelmann R, et al (2008) Diurnal and semidiurnal atmospheric tides observed by co-located GPS and VLBI measurements. J Atmos Sol-Terr Phys 70(10):1366–1372. https://doi.org/10.1016/j.jastp.2008.04.005

Kobayashi S, Ota Y, Harada Y, et al (2015) The JRA-55 reanalysis: General specifications and basic characteristics. J Meteorol Soc Jpn Ser II 93(1):5–48. https://doi.org/10.2151/jmsj.2015-001

Petrov L, Boy JP (2004) Study of the atmospheric pressure loading signal in very long baseline interferometry observations. J Geophys Res Solid Earth 109(B3). https://doi.org/10.1029/2003JB002500

Ray RD (2001) Comparisons of global analyses and station observations of the S2 barometric tide. J Atmos Sol-Terr Phys 63(10):1085–1097. https://doi.org/10.1016/s1364-6826(01)00018-9

Ray RD, Boy JP, Arbic BK, et al (2021) The problematic ψ1 ocean tide. Geophys J Int 227(2):1181–1192. https://doi.org/10.1093/gji/ggab263

Ray RD, Boy JP, Erofeeva SY, et al (2023) Terdiurnal radiational tides. J Phys Oceanogr 53(4):1139–1150. https://doi.org/10.1175/jpo-d-22-0175.1

Schindelegger M, Ray RD (2014) Surface pressure tide climatologies deduced from a quality-controlled network of barometric observations. Month Weather Rev 142(12):4872–4889. https://doi.org/10.1175/mwr-d-14-00217.1

Schindelegger M, Einšpigel D, Salstein D, et al (2016) The global S1 tide in earth's nutation. Surv Geophys 37(3):643–680. https://doi.org/10.1007/s10712-016-9365-3

Schulzweida U (2022) CDO user guide. https://doi.org/10.5281/zenodo.7112925

Shihora L, Sulzbach R, Dobslaw H, et al (2022) Self-attraction and loading feedback on ocean dynamics in both shallow water equations and primitive equations. Ocean Model 169:101,914. https://doi.org/10.1016/j.ocemod.2021.101914

Shihora L, Balidakis K, Dill R, et al (2023) Assessing the stability of AOD1B atmosphere–ocean non-tidal background modelling for climate applications of satellite gravity data: long-term trends and 3-hourly tendencies. Geophys J Int 234(2):1063–1072. https://doi.org/10.1093/gji/ggad119. https://arxiv.org/abs/https://academic.oup.com/gji/article-pdf/234/2/1063/49797154/ggad119.pdf

Sulzbach R, Dobslaw H, Thomas M (2021) High-resolution numerical modeling of barotropic global ocean tides for satellite gravimetry. J Geophys Res Oceans 126(5):e2020JC017,097. https://doi.org/10.1029/2020JC017097, e2020JC017097 2020JC017097

Sulzbach R, Balidakis K, Dobslaw H, et al (2022) TiME22: Periodic disturbances of the terrestrial gravity potential induced by oceanic and atmospheric tides. https://doi.org/10.5880/GFZ.1.3.2022.006

Alternative Strategies for the Optimal Combination of GNSS and Classical Geodetic Networks: A Case-Study in Greece

Dimitrios Ampatzidis, Eleni Tzanou, Nikolaos Demirtzoglou, and Georgios S. Vergos

Abstract

The present study discusses two alternative strategies for the optimal combination of different geodetic reference frames in a rigorous way. The methodological variations stem from the (un)availability and types of the 3D network observables. The alternative strategies are tested in Drama region, Northern Greece, where two local networks were established; a 3D one expressed in ITRF2008 (a modern GNSS network established for precise surveying) and a classical one which refers to the official Greek Geodetic Reference System, the Hellenic Geodetic Reference System of 1987. The concept of the proposed strategy is based on the rigorous combination of the different networks at the Normal Equation (NEQ) level. The zenith angles play crucial role for the implementation of the alternative strategies, especially for the correct use of the vertical information. The results of the case study show that the combined solutions provide generally a good level of consistency with the individual networks (GNSS and conventional land surveying).

Keywords

Geodetic network · GNSS · Optimal combination · TRF · Zenith angles

1 Introduction

The advent of GNSS, particularly after 1990, revolutionized the everyday geodetic and surveying workflow worldwide. The time-demanding and often cumbersome terrestrial measurements were replaced by GNSS occupations in static and real-time modes. However, until today, the majority of current high-level geodetic networks, which were established through observations collected during conventional terrestrial surveying campaigns as far as a century ago, are

in 2D. Therefore, the combination of 3D network (mainly now from GNSS, earlier e.g., from Doppler measurements) and classical 2D networks should be materialized. Such combination schemes have been extensively studied (Gargula 2021; Ilie 2016; Kadaj 2016; Peterson 1974; Weiss et al. 2022). In addition, the local ties used to establish the International Terrestrial Reference Frame (ITRF, Altamimi et al. 2023) combine both space and terrestrial observations (Abbondanza et al. 2009; Lösler et al. 2023).

The main scope of the combination is to align the combined network to a unified global reference frame, -e.g., the ITRF, which may be a version of or even a regional one, e.g. ETRS89 or SIRGAS (Kenyeres et al. 2019; Sánchez and Drewes 2020). We may categorize the combination methodologies as follows:

a. Common Adjustment (CA): The GNSS observations are introduced as 3D baselines (dX, dY, dZ) and the 2D/1D conventional surveying observations are treated as they are observed. In most cases, these observations are incor-

D. Ampatzidis (✉) · E. Tzanou
Department of Surveying and Geomatics Engineering, GENIE Lab, International Hellenic University, Serres, Greece
e-mail: dampatzi@teicm.gr

N. Demirtzoglou
Freelancer MSc. Dipl. Surv. Engineer, Drama, Greece

G. S. Vergos
Department of Geodesy and Surveying, GravLab, Aristotle University of Thessaloniki, Thessaloniki, Greece

© The Author(s) 2023
J. T. Freymueller, L. Sánchez (eds.), *Gravity, Positioning and Reference Frames*,
International Association of Geodesy Symposia 156, https://doi.org/10.1007/1345_2023_237

porated into an appropriate Least Squares (LS) adjustment software.

b. Helmert transformation (HLMT): Through common points, one set of stations is transformed to the other's reference frame.

Despite the fact that these two methodologies are dominant throughout the geodetic literature, there are some limitations for each. For the case of CA, the main problem arises when the System Independent Exchange (SINEX, Blewitt et al. 1994) file is used. It is not clear how observations such as baselines, spatial distances or angles can be derived, since the SINEX file format focuses on sets of coordinates or/and velocities and their associated Normal Equation or Covariance Matrix. Additionally, typically, the CA method is realized to a well-designed network with favourable geometry and a substantial number of stations are occupied using both classical and GNSS observations. However, this is not always the case.

On the other hand, the HLMT methodology exhibits limitations when it is applied in cases of poor networks' geometry, e.g., the common stations do not cover the whole area or when there are sparse areas in the network and in cases of small areas, where the correlations of the estimated parameters could become notable. Finally, when 2D information is used, the vertical information is lost. In that case, even though the existing methods are widely and successfully used, they have some pitfalls which potentially can lead to less accurate results. A common pitfall is the lack of the full Covariance matrix (CV matrix) of the estimated coordinates (most of the times only the standard deviations of the points are known) which leads to non-rigorous results.

The present study deals with the description of an alternative strategy in the direction of the optimal combination of modern 3D and classical networks, exploiting existing zenith angle measurements and mathematical models. The proposed strategy is based on the combination of Normal Equations (NEQs) which are properly converted, added, and restored according to existing methodologies. Special treatment is applied to remedy datum-related information. In addition, two different schemes are presented for the combination of a modern 3D GNSS network and a classical one. These two schemes pertain to the use of the vertical part of the classical networks at the combination process. The alternative strategy is implemented in the region of Drama, Greece using two local networks (GNSS and classical).

2 Combination Strategies

The aim of our study is to build rigorous yet easily-applied algorithms to combine 3D (GNSS) networks from a SINEX file format, and 2D classical ones including spatial distances, horizontal and zenith angles (vertical circular reading of the instrument with respect to the plumb line) and/or azimuths (astronomical azimuths reduced to geodetic and grid ones, respectively). The proposed approach focuses on the following aspects:

1. Utilization of both 3D (GNSS) and 2D (classical) information.
2. Explicit definition of the Reference System, including datum specification.
3. Adaption to various scenarios based on observation accuracy and requirements.

The observed zenith angles and the deflections of the vertical play a key role in our approach, as we elaborate below. Zenith angles are first corrected due to refraction and curvature of the earth (as e.g., in Torge and Müller 2012). The usage of the zenith angles is a matter of great importance, as they practically enable the extension of the 2D networks to complete 3D representation. Initially we present the common steps for all different strategies.

2.1 Common Steps of the Alternative Strategies

The alternative strategies are based on particular algorithmic steps. The first three steps are common and described below. Assuming that there is a file in SINEX format containing a Covariance Matrix of the solution:

1. Convert the 3D cartesian coordinates (XYZ) to the topocentric (ENU) system. This holds for both the approximate and estimated coordinates. The 3D network refers to a modern 3D Terrestrial Reference Frame (TRF). The conversion is realized through the following pointwise formula:

$$\mathbf{q}_i = \mathbf{R}\mathbf{x}_i, \qquad (1)$$

where $\mathbf{q}_i = [E_i \ N_i \ U_i]^T$ represents the topocentric coordinates (East, North, Up components),

$$\mathbf{R} = \begin{bmatrix} -\sin\lambda_m & \cos\lambda_m & 0 \\ -\sin\varphi_m\cos\lambda_m & -\sin\varphi_m\sin\lambda_m & \cos\varphi_m \\ \cos\varphi_m\cos\lambda_m & \cos\varphi_m\sin\lambda_m & \sin\varphi_m \end{bmatrix}$$ the

orthogonal conversion matrix, $\mathbf{x}_i = \begin{bmatrix} X_i \ Y_i \ Z_i \end{bmatrix}^T$ the Cartesian coordinates, φ_m, λ_m the average/reference geodetic latitude and longitude, respectively, of the area. We may refer that for our case study (see Sect. 3 ibid.), the average latitude and longitude are estimated for an area not larger than 10×10 km.

2. Transform the Covariance matrix (CV) and the Right-Hand Side (RHS) of the 3D network from geocentric to topocentric system, using error propagation theory:

$$\mathbf{C}_q^{3D} = \mathbf{J}\mathbf{C}_x^{3D}\mathbf{J}^T, \qquad (2a)$$

and

$$\mathbf{u_q^{3D}} = \mathbf{J^T u_x^{3D}}, \qquad (2b)$$

where $\mathbf{C_q^{3D}}$, $\mathbf{C_x^{3D}}$ the full 3D CV matrices of the topocentric and geocentric coordinates, respectively, $\mathbf{u_q^{3D}}$, $\mathbf{u_x^{3D}}$ the full 3D RHS of the topocentric and geocentric coordinates, respectively, $\mathbf{J} = \begin{bmatrix} \mathbf{R} & & \\ & \ddots & \\ & & \mathbf{R} \end{bmatrix}$ the total transition matrix (Jacobian) from geocentric to topocentric coordinates. The RHS for the geocentric coordinates can be computed by the following equation:

$$\mathbf{u_x^{3D}} = \mathbf{C_x^{3D}} \left(\mathbf{x^{est}} - \mathbf{x^{apr}} \right), \qquad (3)$$

where $\mathbf{x^{est}}$, $\mathbf{x^{apr}}$ the vectors of the estimated and approximate geocentric coordinates, respectively (expect to be found in a typical SINEX file). The RHS is the part of the Normal Equation related to the observations of the 3D (GNSS) network.

3. Re-calculate the approximate topocentric coordinates of classical network's stations, through analytical geometry formulation. This could be easily estimated for the horizontal components E and N utilizing the well-known formulas for the traverse solution (latitudes and departures), considering at least two stations as fixed. The approximate up-components can be determined by trigonometric calculations, fixing at least one station. The fixed stations are the common stations of the two networks (3D and classical) whose topocentric coordinates are straightforwardly computed from the geocentric coordinates, as we already mention in Step 1. The new approximate values of the classical network refer now to the modern 3D TRF. Hence, the two different networks refer to a common reference system (datum).

4. Invert the CV matrix of the topocentric coordinates to obtain the associated Normal Equation Matrix (NEQ):

$$\mathbf{N_q^{3D}} = \left(\mathbf{C_q^{3D}} \right)^{-1}. \qquad (4)$$

2.2 First Alternative Strategy: Transition from a 2D to a 3D Network Employing Zenith Angles

Normally, the classical network includes observations of zenith angles referring to the physical surface (plumbline). Through the Deflections of Vertical (DoV), they are reduced to an ellipsoid (Barzaghi et al. 2016). The corrected andreduced zenith angles (with respect to the vertical of

geodetic system) are estimated as follows (Rossikopoulos 1999):

$$z_{ij} = \zeta_{ij} + \underbrace{\frac{k_{ij}}{2R} \left(\rho_{ij} \, sin \, \zeta_{ij} \right)^2}_{refraction \; term \; correction}$$
$$- \underbrace{\frac{1}{2R} \left(\rho_{ij} \, sin \, \zeta_{ij} \right)^2}_{Earth's \; cruvature \; term \; correction} \qquad (5)$$
$$+ \underbrace{sin \, a_{ij} \eta_i + cos \, a_{ij} \xi_i}_{DoV},$$

where z_{ij} is the reduced zenith angle (from station i to station j) with respect to the vertical, ζ_{ij} the observed zenith angle, k_{ij} the refraction term, ρ_{ij} the spatial distance, R the Earth's mean radius for the area, α_{ij} the azimuth and ξ_i, η_i the deflections of the vertical. If the refraction term is unknown, then a mean value of 0.13 can be assumed for Greece (Labrou and Pantazis 2010).

Continuing from 2.1

5. Solve the classical network in 3D (topocentric system), using the original observations which can include: spatial distances horizontal angles, directions and azimuths (geodetic or grid ones). The zenith angles are reduced according to Eq. (5). Now the solution offers a complete 3D dataset. The observation equations of the spatial distance and the zenith angle with respect to the topocentric coordinates, are as follows:

$$\rho_{ij} = \sqrt{\left(\Delta E_{ij}^2 + \Delta N_{ij}^2 + \Delta U_{ij}^2 \right)}, \qquad (6a)$$

and

$$z_{ij} = arctan \frac{\sqrt{\left(\Delta E_{ij}^2 + \Delta N_{ij}^2 \right)}}{\Delta U_{ij}}. \qquad (6b)$$

There are some major issues regarding the reliability of the measured zenith angles: (a) their accuracy is strongly dependent on the atmospheric refraction and (b) the requisite knowledge of the deflections of the vertical (DoV), is many times either not applicable or it is rather problematic. Practically, the first alternative strategy imposes a "three-dimensionalization" of the classical network through the corrected and reduced zenith angles. The solution of the classical network results the NEQ and the RHS, $\left(\mathbf{N_q^{3D}} \right)^{classical}$ and $\left(\mathbf{u_q^{3D}} \right)^{classical}$, respectively.

6. Combine GNSS and classical network in full 3D by NEQ stacking. The combined NEQ and RHS yield:

$$\left(\mathbf{N_q^{3D}} \right)^{combined} = \left(\mathbf{N_q^{3D}} \right)^{GNSS} + \left(\mathbf{N_q^{3D}} \right)^{classical}, \qquad (7a)$$

and

$$\left(\mathbf{u_q^{3D}}\right)^{combined} = \left(\mathbf{u_q^{3D}}\right)^{GNSS} + \left(\mathbf{u_q^{3D}}\right)^{classical}. \quad (7b)$$

7. Define a stable and accurate reference system, by implementing the Controlled Datum Removal (CDR, (Kotsakis and Chatzinikos 2017)). Using CDR, the user selects which of the fundamental datum quantities (origin, scale and orientation) should be externally imposed with minimum constraints. The CDR-related NEQ and RHS now yield:

$$\left(\mathbf{N_q^{3D}}\right)^{CDR} = \left[\begin{array}{cc} \left(\mathbf{N_q^{3D}}\right)^{combined} & \left(\mathbf{N_q^{3D}}\right)^{combined}\mathbf{E}^T \\ \mathbf{E}\left(\mathbf{N_q^{3D}}\right)^{combined} & \mathbf{E}\left(\mathbf{N_q^{3D}}\right)^{combined}\mathbf{E}^T \end{array} \right], \quad (8a)$$

and

$$\left(\mathbf{u_q^{3D}}\right)^{CDR} = \left[\begin{array}{c} \left(\mathbf{u_q^{3D}}\right)^{combined} \\ \mathbf{E}\left(\mathbf{u_q^{3D}}\right)^{combined} \end{array} \right], \quad (8b)$$

where $\mathbf{E}$ is a properly selected transformation matrix (with respect to the Helmert parameters), defining the datum parameters that should be externally defined. The $\left(\mathbf{N_q^{3D}}\right)^{CDR}$ has a rank deficiency which corresponds to the number of the rows of the $\mathbf{E}$ matrix. The solution is then achieved by imposing minimum constraints.

8. Convert the coordinates and CV matrix from the topocentric to a geocentric system, respectively.

2.3 Second Alternative Strategy: Transition from a 2D to a Quasi-3D Network Through the Zenith Angles

This scenario is practically a special case of the first one: In order to mitigate—as much as possible—the effect of the low accuracy of the observed zenith angles, we divide the classical network into two components:

a. Classical 2D solution where the horizontal part (for E, N components) NEQ and RHS ($\mathbf{N_q^{hor}}$, $\mathbf{u_q^{hor}}$) are estimated.
b. The spatial distances and the reduced zenith angles used for geometric height differences observations Δh_{ij} (trigonometric levelling, Rossikopoulos 1999):

$$\Delta U_{ij} = \Delta h_{ij} = \rho_{ij} \cos z_{ij} + s_i - t_j, \quad (9)$$

where s and t are the heights of the instrument and the target, being derived from the record of the classical observations. The height difference observations lead to vertical (Up) NEQs and RHS ($\mathbf{N_q^{vertical}}$, $\mathbf{u_q^{vertical}}$). The geometric height differences (on an ellipsoid) correspond to the Up-

component differences of the topocentric system (Vanicek and Krakiwsky 1986, p. 334).

Continuing from 2.1

5. Stack NEQs and RHS. The stacking (GNSS-derived and classical one) is realized separately for horizontal and vertical NEQs and RHS parts of the contributing networks, forming finally a consistent 3D NEQ and RHS for the topocentric coordinates. The combined NEQ and RHS are formulated as follows:

$$\left(\mathbf{N_q^{3D}}\right)^{combined} = \left[\begin{array}{cc} \mathbf{N_q^{hor}} & \mathbf{N_q^{hor,vertical}} \\ \left(\mathbf{N_q^{hor,vertical}}\right)^T & \mathbf{N_q^{vertical}} \end{array} \right]^{GNSS}$$
$$+ \left[\begin{array}{cc} \mathbf{N_q^{hor}} & \\ & \mathbf{N_q^{vertical}} \end{array} \right]^{classical}, \quad (10a)$$

and

$$\left(\mathbf{u_q^{3D}}\right)^{combined} = \left[\begin{array}{c} \mathbf{u_q^{hor}} \\ \mathbf{u_q^{vertical}} \end{array} \right]^{GNSS} + \left[\begin{array}{c} \mathbf{u_q^{hor}} \\ \mathbf{u_q^{vertical}} \end{array} \right]^{classical}. \quad (10b)$$

6. Apply CDR and solve the NEQ system. The procedure is identical to the 7th step of the first alternative strategy.
7. Identical to Step 8 of the first alternative strategy.

The second alternative strategy brings two sets of different NEQs and RHS which artificially "de-correlate" the horizontal and the vertical parts. This approach serves two primary objectives (a) ensuring the 3D nature of the final network and (b) mitigating the impact of the relatively larger errors of the zenith angle measurements on the 3D result.

We may also underline that, even though our present study is dealing with GNSS networks, the aforementioned strategies can be easily applied for other space techniques (VLBI, SLR and DORIS), since the core of the method is the use of the SINEX format. This can be useful for some applications such as local ties.

3 Case Study

The alternate strategies are implemented over two geodetic networks, located in Drama Prefecture in Greece (Fig. 1). In 1998 Drama's Municipal Enterprise of Water Supply and Sewerage (DEYAD) established a classical 2D geodetic network, occupied with spatial distances (accuracy: 0.5 mm + 5 ppm), directions and zenith angles (measurement accuracy for the angular quantities: 1 mgon). In total, eight stations have been occupied, with two of them being part of the National Triangulation Network (NTN-state's benchmarks). The classical network was aligned to the Hellenic Geodetic Reference System of 1987 (HGRS 1987; Veis 1996).

Fig. 1 The location of the networks. The red polygon defines the study area in Drama (Kallifitos village) (from [Open Street Map], licensed under (CC BY-SA 2.0))

In 2014, a GNSS campaign was conducted for the needs of DEYAD. In total, nine sites were occupied with static GNSS observation (for at least 2 h). Four of them belong to the NTN. The GNSS network was aligned to ITRF2008, epoch 2014.35. The solution of the GNSS solution is expressed in SINEX file format. Figure 2 visualizes the two described networks.

As shown in Fig. 2, only two benchmarks are common between the two networks, both located only in the middle of the study area. In fact, the geometry of the networks does not support either the implementation of the HLMT methodology (only two common sites, not enclosing the area) or CA (the 3D network solution is expressed in SINEX format, providing coordinates and their associated CV matrix. The sites M0, M1 and M4 are now covered with dense canopy (probably was not the case back in 1998) and there is no mutual visibility between the 3D and 2D network sites, thus there is no way to connect them with classical observations. To proceed with the Alternative Strategies application, the DoV values for the area need to be estimated (see Eq. 5, ibid). Since we do not have any information from any local agency regarding the DoVs, we employ those calculated

from the XGM2019e model (Zingerle et al. 2020) complete to degree and order 2190. The CDR was applied for all datum parameters (origin, scale, and orientation). The rank deficiency (after the application of CDR) was compensated by the application of minimum constraints to the set of the four NTN stations. Table 1 shows the results of (a) Each individual 3D and 2D networks solution and (b) after the implementation of the Alternative Strategies.

Next, we proceed with an additional quality check criterion. We compare the results of the combined solution (for both alternative strategies) with the individual ones (as they solved solely). We estimate the following discrepancies, pointwise:

$$\delta_{s_i} = \sqrt{\left(X_i^{GNSS} - X_i^{comb}\right)^2 + \left(Y_i^{GNSS} - Y_i^{comb}\right)^2 + \left(Z_i^{GNSS} - Z_i^{comb}\right)^2} \tag{11}$$

for the GNSS network (points labelled with S according to Fig. 2).

$$\delta_{d_i} = \sqrt{\left(E_i^{classical} - E_i^{comb}\right)^2 + \left(N_i^{classical} - N_i^{comb}\right)^2} \tag{12}$$

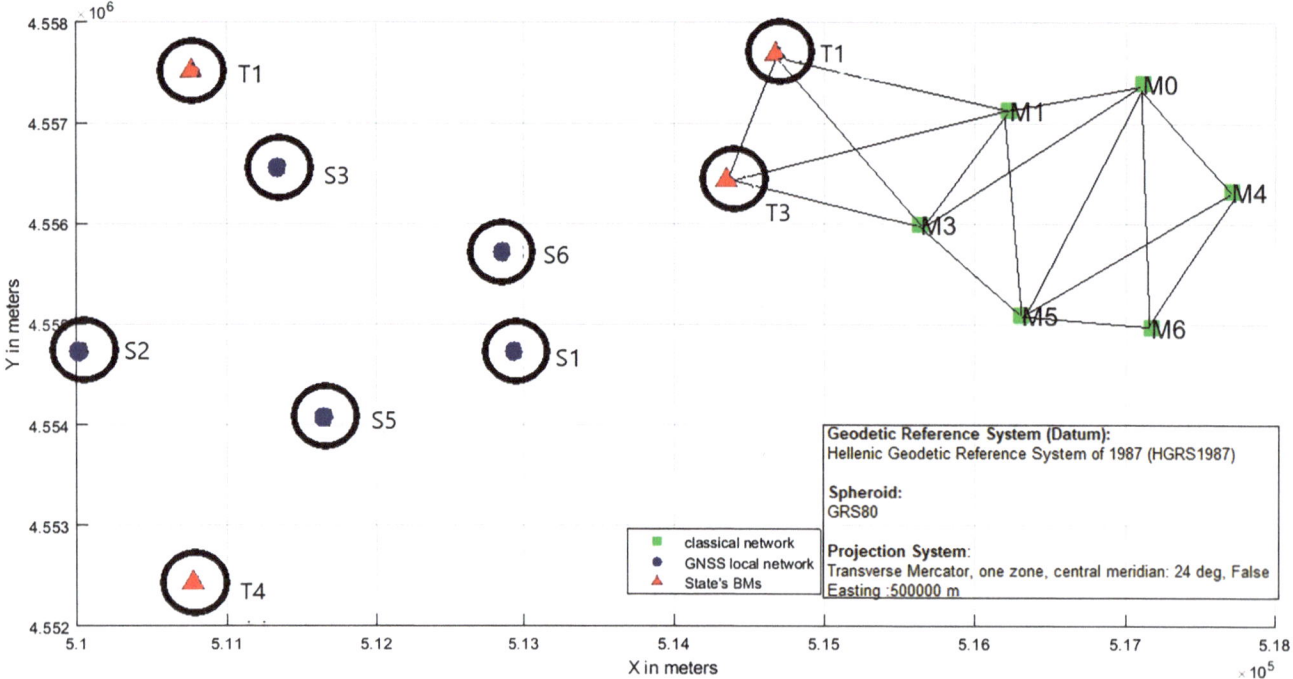

Fig. 2 The classical and the GNSS networks established in Drama. Each symbol corresponds to different type of network (GNSS and classical one). T1 and T3 are the common stations of the two networks

Table 1 The results of the individual solutions and the Alternative Strategies [Units: cm]

Strategy	Mean spherical error[a]	Max. spherical error[b]	Mean horizontal error[c]	Max. horizontal error	Max. vertical error
Individual 2D network	–	–	1.08	1.74 (at M4)	–
Individual 3D network	0.67	0.95 (at S6)	0.32	0.57 (at S6)	0.88 (at S1)
First alternative strategy	1.85	4.45 (at M6)	1.21	2.41 (at M6)	3.71 (at M6)
Second alternative strategy	1.01	2.84 (at M6)	0.85	1.22 (at M4)	2.61 (at M4)

[a] Spherical Error $\sqrt{\dfrac{trace\left(C_q^{3D}\right)}{3n}}$

[b] $max\left(\sqrt{\left(\sigma_{E_i}^2 + \sigma_{N_i}^2 + \sigma_{U_i}^2\right)}\right)$

[c] Horizontal Error $\sqrt{\dfrac{trace\left(C_q^{2D}\right)}{2n}}$

Table 2 Statistics of the discrepancies between the individual GNSS and the combined solutions, respectively [Units: cm]

	First alternative strategy	Second alternative strategy
min (δs_i)	0.2	0.2
max (δs_i)	2.6	1.6
mean (δs_i)	0.9	0.6
std (δs_i)	1.0	0.7

Table 3 Statistic of the discrepancies between the individual classical and the combined solutions, respectively [Units: cm]

	First alternative strategy	Second alternative strategy
min (δd_i)	0.2	0.1
max (δd_i)	1.8	1.1
mean (δd_i)	0.6	0.5
std (δd_i)	0.5	0.4

for the classical network (points labelled with M with according to Fig. 2). For this, test, the classical network was solved with respect to the ITRF2008 (topocentric coordinates, see Sect. 2.1 *ibid.*). Table 2 refers to the GNSS network comparisons, while Table 3 to the classical network, respectively.

The solutions from the two alternative strategies yield some notable findings. First, the second alternative strategy (NEQs are separated into horizontal and vertical part and combined), performs better than the first alternative strat-

egy (1.01 vs 1.85 cm mean spherical error, respectively). It seems that the discarding the zenith angles uncertainty leads to large errors. On the other hand, the first alternative strategy gives worse results as the zenith angles play indeed a significant role, contaminating the achieved accuracy for both the horizontal and vertical components. Even though the accuracy is slightly worse in the second alternative strategy after the combination, compared to the individual 3D network solution (1.01 cm compared to 0.67 cm), the results are still suitable for surveying applications and the existing infrastructure could directly refer to the combined solution aligned to a modern global TRF.

Furthermore, Tables 2 and 3 show that (a) the combined solutions perform at good level of consistency with the individual ones (better than 1 cm for the mean discrepancies for both GNSS and classical networks) and (b) the second alternative strategy provides better results compared to the first one (as mean average and as maximum values, respectively). These findings confirm that for the tested network the alternative strategies can be beneficial towards the alignment of a classical network to an accurate global TRF.

4   Conclusions

The suggested strategies for the optimal combination between 3D and 2D networks can stand as alternative scenarios for the cases were the Common Adjustment (CA) and the Helmert Transformation (HLMT) methodologies encounter challenges. A usual problem is the poor geometry of the combined networks (few common stations, not well-designed observations, gaps). Furthermore, the alternative strategies can be easily applied under the existence of a SINEX format file since there is no need of special treatment of the observations.

The second alternative strategy (separating the horizontal and the vertical part of the NEQ) emerges as superior option compared to the first one. Finally, the alternative strategies can be applied in combination cases involving space geodetic techniques (VLBI, SLR, and DORIS) following the same conceptual manner. This can be useful for, e.g., the co-location sites which contribute to the inter-system ITRF construction.

Acknowledgments The two anonymous reviewers, the Associate Editor M. Craymer and the Editor in Chief J. Freymueller are kindly acknowledged for their comments which led to a significant improvement of the initial manuscript.

References

Abbondanza C, Altamimi Z, Sarti P, Negusini M, Vittuari L (2009) Local effects of redundant terrestrial and GPS-based tie vectors in ITRF-like combinations. J Geod 83:1031. https://doi.org/10.1007/s00190-009-0321-6

Altamimi Z, Rebischung P, Collilieux X, Métivier L, Chanard C (2023) ITRF2020: an augmented reference frame refining the modeling of nonlinear station motions. J Geod 97:47. https://doi.org/10.1007/s00190-023-01738-w

Barzaghi R, Betti B, Biagi L, Pinto L, Visconti M (2016) Estimating the baseline between CERN target and LNGS reference points. J Surv Eng 142:04016012. https://doi.org/10.1061/(ASCE)SU.1943-5428.0000173

Blewitt G, Bock Y, Kouba J (1994) Constraining the IGS polyhedron by distributed processing. In: Workshop proceedings: densification of ITRF through regional GPS networks, held at JPL, Nov 30–Dec 2, pp 21–37

Gargula T (2021) Adjustment of an integrated geodetic network composed of GNSS vectors and classical terrestrial linear pseudo-observations. Appl Sci 11(10):4352. https://doi.org/10.3390/app11104352

Ilie A-S (2016) Adjusting 3D geodetic network using both global navigation satellite systems technology (GNSS) and terrestrial measurements. Environ Eng Manag J 15(6):1223–1235

Kadaj R (2016) The combined geodetic network adjusted on the reference ellipsoid – a comparison of three functional models for GNSS observations. Geod Cartogr 65. https://doi.org/10.1515/geocart-2016-0013

Kenyeres A, Bellet JG, Bruyninx C, Caporali A, de Doncker F, Droscak B, Duret A, Franke P, Georgiev I, Bingley R, Huisman L, Jivall L, Khoda O, Kollo K, Kurt AI, Lahtinen S, Legrand J, Magyar B, Mesmaker D, Morozova K, Nágl J, Özdemir S, Papanikolaou X, Parseliunas P, Stangl G, Ryczywolski M, Tangen OM, Valdes M, Zurutuza J, Weber M (2019) Regional integration of long-term national dense GNSS network solutions. GPS Solut 23:122. https://doi.org/10.1007/s10291-019-0902-7

Kotsakis C, Chatzinikos M (2017) Rank defect analysis and the realization of proper singularity in normal equations of geodetic networks. J Geod 91:627–652. https://doi.org/10.1007/s00190-016-0989-3

Labrou E, Pantazis G (2010) Applied geodesy. Ziti Publications, Thessaloniki. (in Greek)

Lösler M, Eschelmbach C, Mähler S, Guillory J, Truong D, Wallerand J-P (2023) Operator-software impact in local tie networks. Appl Geomat. https://doi.org/10.1007/s12518-022-00477-5

Peterson AE (1974) Merging of the Canadian Triangulation Network with the 1973 Doppler Satellite Data. The Canadian Surveyor 28:487–495

Rossikopoulos D (1999) Surveying networks and computations, 2nd edn. Ziti Publications, Thessaloniki. (in Greek)

Sánchez L, Drewes H (2020) Geodetic monitoring of the variable surface deformation in Latin America. In: International Association of Geodesy Symposia Series, Springer, Berlin, 12 pp. https://doi.org/10.1007/1345_2020_91

Torge W, Müller J (2012) Geodesy. De Gruyter, Berlin, Boston. https://doi.org/10.1515/9783110250008

Vanicek P, Krakiwsky EJ (1986) Geodesy: the concepts, 2nd edn. North-Holland, Amsterdam

Veis G (1996) National report of Greece. Report on the Symp. of the IAG Sub-commission for the European Reference Frame (EUREF), Ankara, 22–25 May 1996. Report, Verlag der Bayerischen Akademie der Wissenschaften, Heft Nr. 57

Weiss G, Lebant S, Gasinec J, Stankova H, Cernota P, Weiss E, Weiss R (2022) Establishment of local geodetic networks based on least-squares adjustments of GNSS baseline vectors. Adv Geod Geoinf 71(1). https://doi.org/10.24425/gac.2022.141168

Zingerle P, Pail R, Gruber T, Oikonomidou X (2020) The combined global gravity field model XGM2019e. J Geod 94:66. https://doi.org/10.1007/s00190-020-01398-0. Springer

Webpages

https://gssc.esa.int/navipedia/index.php/Transformations_between_ECEF_and_ENU_coordinates

A Concept of Precise VLBI/GNSS Ties with Micro-VLBI

Leonid Petrov, Johnathan York, Joe Skeens, Richard Ji-Cathriner, David Munton, and Kyle Herrity

Abstract

We present here a concept of measuring local ties between collocated GNSS and VLBI stations using the microwave technique that effectively transforms a GNSS receiver to an element of a VLBI network. This is achieved by modifying the signal chain that allows to transfer voltage of the GNSS antenna to a digitizer via a coaxial cable. We discuss the application of this technique to local tie measurement. We have performed observations with a GNSS antenna and FD-VLBA radiotelescope and detected a strong interferometric signal from both radiogalaxies and GNSS satellites.

Keywords

GNSS · Local ties · VLBI

1 Introduction

Space geodetic observation with Very Long Baseline Interferometry (VLBI), Global Navigation Satellite System (GNSS), Satellite Laser Ranging (SLR), or Doppler Orbitography and Radiopositioning Integrated by Satellite (DORIS) involves a measurement of travel time of electromagnetic radiation between an emitter and a receiver and/or rates of its change. The position of a ground space geodesy instrument is referred to its own unique reference point. In a similar way, the position of a space-borne emitter or receiver is referred to its own reference point.

Space geodesy techniques have their strengths and weaknesses. VLBI provides a reference to inertial space, SLR provides a reference to the Earth's center of mass, including the solid earth, oceans, cryosphere and atmosphere,

L. Petrov (✉)
NASA GSFC, Greenbelt, MD, USA
e-mail: Leonid.Petrov@nasa.gov

J. York · J. Skeens · R. Ji-Cathriner · D. Munton · K. Herrity
University of Texas at Austin, Austin, TX, USA
e-mail: york@arlut.utexas.edu; jskeens1@utexas.edu;
rcathriner@arlut.utexas.edu; dmunton@arlut.utexas.edu;
kherrity@arlut.utexas.edu

DORIS provides wide spatial coverage, and GNSS is able to sense site deformations with a fine time resolution. It was recognized over 20 years ago that a combination of all space geodesy techniques has the potential to provide the most accurate results by mitigating the weaknesses of each individual technique (see for example, Altamimi et al. 2002). Combination implies that observations necessarily must have something common that ties them together. Ties can be direct in the form of a position vector between either a space-borne or a ground-based reference point that is precisely known, or indirect, for instance in the form of Earth rotation parameters that affect all ground stations.

A number of sites have instruments of more than one technique at distances of 30–500 m. Direct measurement of their positions with respect to each other can establish direct ties. Survey techniques measure angles and distances *between markers*. These measurements can reach accuracy of 1–3 mm (Matsumoto et al. 2022). However, the ties should provide the positions of *technique reference points*. Reference points of microwave techniques, such as VLBI and GNSS, cannot be directly pin-pointed by markers. An offset of a reference point with respect to a marker is inferred. In the case of GNSS ground stations, an offset between a marker on the instrument and its phase center can be calibrated, for instance, in an anechoic chamber. In the case

© The Author(s) 2023
J. T. Freymueller, L. Sánchez (eds.), *Gravity, Positioning and Reference Frames*,
International Association of Geodesy Symposia 156, https://doi.org/10.1007/1345_2023_211

of VLBI radiotelescopes, markers are put on the antenna, and the position of a geometric reference point on a fixed axis that is a projection of a moving axis is derived from processing a cloud of points measured with a total station at different antenna azimuths and elevations. Then an assertion is made that the geometric reference point coincides with the reference point estimated in VLBI data analysis. The validity of that assertion cannot be evaluated.

The positions of microwave reference points provided by data analysis of both the GNSS and VLBI techniques may have biases with respect to the geometric reference points. As long as these biases are permanent and do not depend on any other variable parameters, they can remain unnoticed. For a number of applications, for instance for a study of motions caused by plate tectonics or for mean sea level determination, permanent biases are irrelevant. Sarti et al. (2011) showed that antenna gravity deformation caused a 7 mm offset of the microwave reference point of VLBI station MEDICINA, whose position was determined from analysis of VLBI group delay with respect to a geometric refrence point determined from a local survey. Mismodeling antenna gravity deformation will not affect the least square fit and may not be noticed, but biases in local tie measurements comparable or exceeding the internal accuracy of GNSS and VLBI techniques make them close to useless.

The fundamental problem of tie measurements with local surveys is that the optical technique used by a survey instrument, such as a total station, cannot measure the phase center of a microwave technique. While the accuracy of measurements of a vector between markers can be evaluated from a scatter of residuals, the accuracy of a vector between a marker to a phase center is poorly known. That makes tie vectors determined with local surveys unreliable. A typical discrepancy between VLBI positions determined from analysis of group delays and GNSS positions reduced to the VLBI reference point via tie vectors determined with local surveys is 5–20 mm (Ray and Altamimi 2005). Lack of realistic uncertainties of tie vectors does not allow us to interpret these discrepancies because we do not know whether they are due to systematic errors of space geodetic techniques or due to error in tie vectors. That motivated us to seek for alternative measurements of tie vectors.

2   A Microwave Technique for VLBI/GNSS Tie Measurements

We are leveraging the High Rate Tracking Receiver (HRTR) to serve as both an advanced software-defined GNSS receiving system and a general purpose L-band receiver (York et al. 2012, 2014). It directly digitizes voltage from the receiver in a range of 1 to 2 GHz at a rate of 2 gigasamples per second. To access a larger extent of the signal in the 1–2 GHz range,

we modified a commercial GNSS antenna. Specifically, we removed the internal amplifier and narrowband bandpass filters that are provided, and we have replaced them with an alternate amplification and filtering stage of our own design. The aggregate RF system including modified components has a passband of approximately [1.10, 1.65] GHz. The antenna elements are not altered in this modification. Therefore, phase center offset/variation corrections of the modified antennas are the same as of the original antenna.

The HRTR performs digital downconversion of the input samples and produces up to nine independently chosen frequency bands 40.912 MHz wide. The HRTR also allows us to configure the bit depth of the received signal. We utilized complex encoding for our work, using one bit for the in-phase and one bit for the quadrature voltage. Datastreams with the baseband signal from each band are recorded to a general purpose RAID of magnetic hard drives.

In addition to recording voltage from the receiver, HRTR simultaneously computes conventional GNSS observables on civil GNSS signals in real time and provides an output convertible to RINEX format. We can recompute conventional GNSS observables by processing digital records of the baseband signal if needed.

We noticed that the HRTR has a striking similarity to a radiotelescope that is an element of a VLBI network. Like a radiotelescope, the HRTR digitizes voltage from the antenna and records the data with time stamps from a precise clock. HRTR data are processed after the experiment in a similar way as VLBI. Extending this analogy further, we came to the idea of using a HRTR itself as an element of a VLBI network. A GNSS antenna with an effective diameter of ~ 0.08 m surrounded by 0.38 m wide choke ring is roughly four orders of magnitude less sensitive than a 12–30 m radiotelescope, it operates at a lower frequency, and at a first glance does not look competitive. However, the use of a GNSS antenna as an element of a VLBI network is very promising for local tie measurement. Atmospheric contribution is negligible at short baselines of 30–3000 m. Varenius et al. (2021) demonstrated that baseline length repeatability at a sub-millimeter level has been achieved from processing of phase delays at short baselines. We should stress that the baseline vector between two antennas evaluated from VLBI observations is between the microwave reference points of the radiotelescopes. Therefore, by processing GNSS/radiotelescope data, we can eliminate the weakest link in the measurement chain of a tie vector with the use of conventional local surveys: the offset between a marker and a microwave reference point.

The measurement concept is presented in Fig. 1. The voltage of the emission received by the GNSS antenna in a range of 1.0 to 2.0 GHz is transferred via a coaxial cable to an analog-to-digital converter and recorded. The digital records are re-sampled and re-coded to VDIF format that is

Fig. 1 Measurement concept of VLBI/GNSS ties with a microwave technique

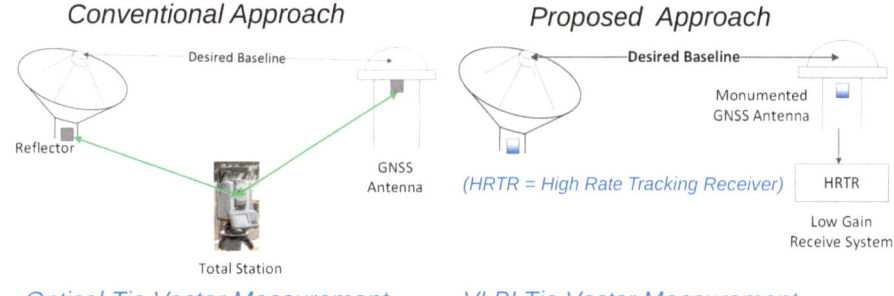

Conventional Approach   Proposed Approach

Optical Tie Vector Measurement   VLBI Tie Vector Measurement

Determine the baseline between electrical phase centers directly

commonly accepted in radio astronomy. Emission received by a VLBI antenna is processed the same way, but with different hardware. The output of both the GNSS and VLBI antennas is recorded in the same format. Further processing is performed exactly the same way as processing of any other VLBI data: the data are correlated, the fringe fitting procedure finds group delays and phase delay rates that maximize the fringe amplitude coherently averaged over time and frequency, and finally, group and phase delays are used for determination of the baseline vector. Therefore, we expect that a position vector of a GNSS antenna/radiotelescope baseline will be determined with the same sub-millimeter accuracy as position vectors of other short baselines.

3   Early Results

Implementation of VLBI with a GNSS antenna requires overcoming a number of difficulties (Skeens et al. 2023). It is essential that the HRTR does not perform any analog signal transformation. It simply digitizes signal as is, performs digital filtering into several bands, and writes the digital signal. This early digitization approach shifts the burden of signal processing to programming. This facilitates the tuning of the processing pipeline since digital recordings can be reprocessed as many times as needed.

We performed three 3 hr observing sessions in 2022 between two transportable HRTRs and a 25 m radiotelescope FD-VLBA. That radiotelescope is a part of the Very Long Baseline Array dedicated for VLBI and has been operating since 1991. It is equipped with an H-maser clock. The antenna has a number of very sensitive receivers, including the one that operates at L-band. We put the first HRTR within 90 m of the FD-VLBA. That HRTR was stabilized by a Rubidium clock. We put the second HRTR within 9000 m of FD-VLBA near the NASA VLBI station MACGO12M. This HRTR was stabilized by the H maser clock used by MACGO12M. Since MACGO12M does not have the technical capability to observe below 2 GHz, we performed observations at only the two HRTRs and FD-VLBA.

The observing schedule included observations of seven of the brightest radiogalaxies and a number of GNSS satellites. We have detected all but one radiogalaxy at the short baseline with FD-VLBA and some sources at the 9 km long baseline. As expected, no detection was found between the two HRTR stations. Figure 2 shows fringe phases and normalized fringe amplitudes of radiogalaxy Cyg-A located at a distance of $7.2 \cdot 10^{21}$ m. This goes well beyond (fourteen orders of magnitude!!) the intended use of the GNSS equipment. The interferometric fringes of radiogalaxies were stable over time, and integration could be extended up to 20 min without a noticeable degradation of fringe amplitude.

Figure 3 shows fringe plots of a GPS satellite over 10 s integration time. We processed GPS signal as a random noise (Skeens et al. 2023). The fringe amplitude has a peak at the carrier frequency of 1575.42 MHz, emission near 1 MHz of the peak due to the C/A signal, and a broad emission due the binary offset carrier modulation of the M-code that has a detectable power within ±15 MHz of the carrier. This allows us to compute group delay over the total bandwidth of ∼ 30 MHz with a precision of 60–90 ps over 10 s. This should be sufficient for resolving phase delay ambiguities with spacings of 635 ps and then use it for data anlysis. Interferometric responses have been detected at the 9 km long baselines as well. At this stage of the project we did not yet attempt to perform geodetic analysis.

4   Discussion

A vector tie can be determined from both observations of natural sources, such as radiogalaxies, and from observations of navigational satellites. The broadband GPS signal due to the modulation of the M-code has flux density around −200 dBW/m²/Hz (Thoelert 2019) within 30 MHz, i.e. ∼ 1 MJy, while there are only 10 natural radio sources brighter that 0.02 MJy at 1.5 GHz, i.e. a factor of 50 fainter. The scarcity of very strong radiogalaxies makes preparation of a VLBI schedule optimized for geodesy difficult.

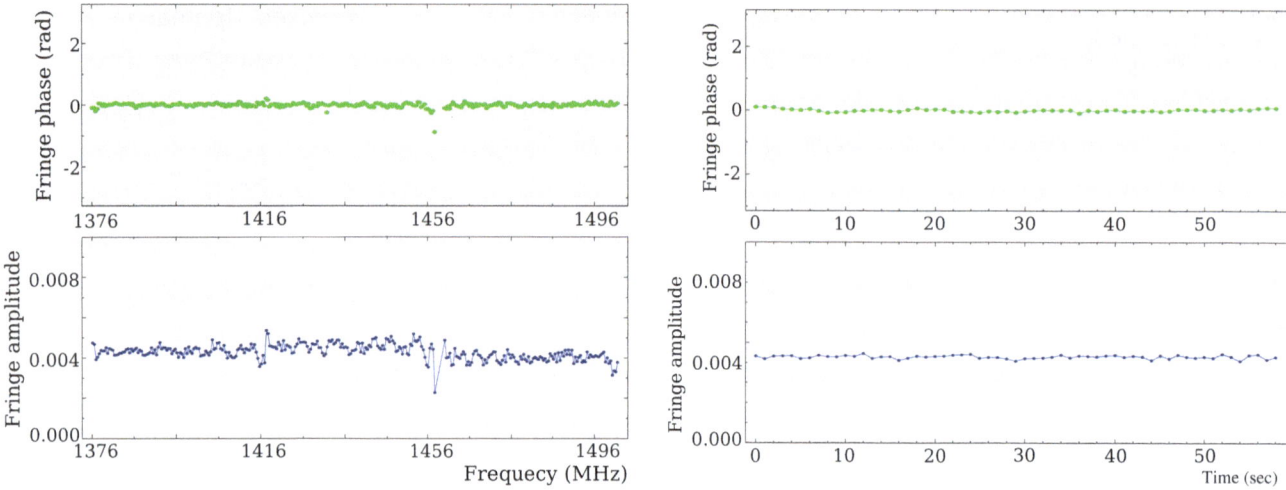

Fig. 2 Fringe phase (upper) and normalized fringe amplitude of Cyg-A at a 90 m long baseline HRTR/FD-VLBA. Signal to noise ratio (SNR) 387 was achieved for 60 s of integration time. Left plot shows fringe phase and amplitude versus frequency and right plot shows fringe phase and amplitude versus time

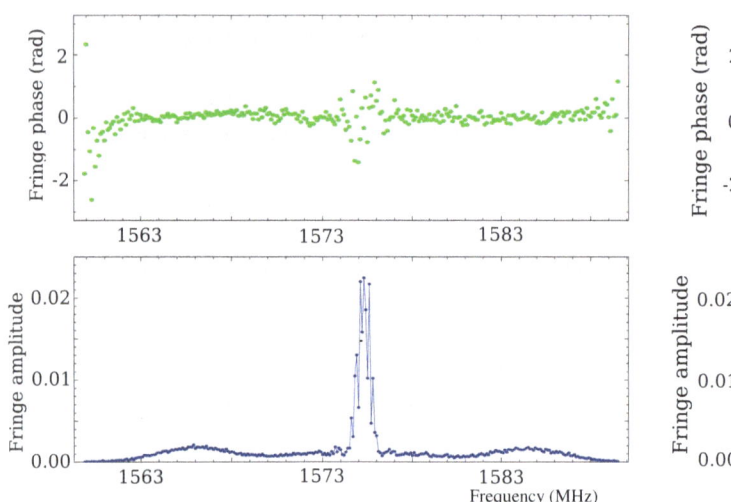
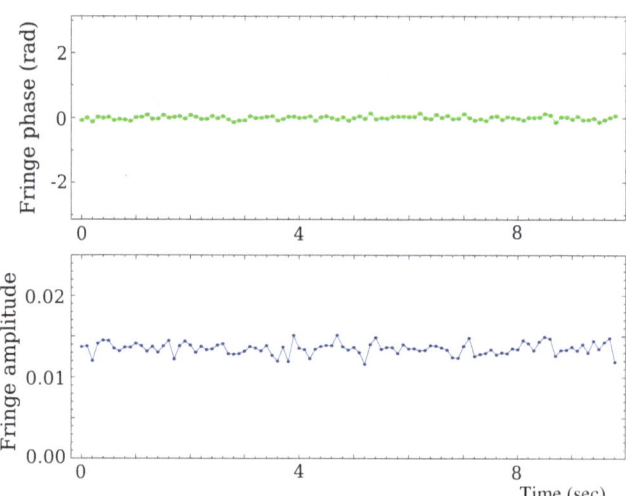

Fig. 3 Fringe phase (upper) and normalized fringe amplitude of a GPS satellite at a 90 m long baseline HRTR/FD-VLBA. SNR 138 was achieved for 10 s of integration time. Left plot shows fringe phase and amplitude versus frequency and right plot shows fringe phase and amplitude versus time

Detection of a radio source requires observations with a sensitive VLBI antenna. FD-VLBA is dedicated for astronomy and its L-band receiver has a system equivalent flux density (SEFD) of 289 Jy. Compare with SEFD 2500 Jy at the 2–3 GHz band at 12 m geodetic VLBI antennas. Since sensitivity of an interferometer is proportional to the square root of the product of SEFD of individual antennas, VLBI observations of a GPS satellite between a HRTR and a radiotelescope requires $50^2 = 2500$ times less sensitive radiotelescope than observations of radiogalaxies. Figures 2, 3 seem to contradict that statement. It turned out FD-VLBA receiver worked in a saturated regime when observing a megajansky source, and fringe amplitude was strongly underestimated. Applying an additional attenuation is required to mitigate the problem.

But in general, it is much easier to reduce the sensitivity than to increase it.

The primary observable that will be used for determination of the tie vector will be phase delay. Phase delay ambiguities can be resolved if group delay can be determined with an accuracy 1/4 of the phase ambiguity spacing or better and short-term systematic differences between phase and group delays have a scatter not exceeding that number. Since precision of group delay is reciprocal to the bandwidth of a detected signal, observations of navigational satellites with a relatively broad spectrum, such as GPS and Galileo, is preferable.

Determination of a tie vector between a HRTR and a geodetic VLBI antenna is interesting, but not exactly what is needed. We need to determine a tie vector between a VLBI antenna and a permanent GNSS antenna. There are two possible solutions. First, one can make HRTR a permanent GNSS site. It is expected that in 2023–2024, permanent HRTRs will be installed within a hundred meters of each of the ten VLBA antennas. Second, one can install a HRTR on a temporary monument and determine a tie vector HRTR/radiotelescope using VLBI and subsequently a tie vector between the HRTR and the existing GNSS antenna by processing double-differenced GNSS carrier phase measurements. Processing differential GNSS data at baselines of ~ 100 m long provides a millimeter level accuracy owing to cancellation of the atmospheric contribution. Then combining the VLBI/HRTR and HRTR/GNSS tie vectors, we get a VLBI/GNSS tie.

Radiotelescopes used for radio astronomy often have receivers in 1.2–1.8 GHz ranges, although very few instruments have the ability to simultaneously record within a band of 1.15–1.65 GHz that covers navigation signals L1, L2, and L5. Radiotelescopes dedicated for geodesy usually cannot receive emission below 2 GHz because of strong high-pass filters installed to mitigate the impact of radio interference. Some new generation broad-band VLBI Geodetic Observing System (VGOS) radiotelescopes have a low cutoff frequency as high as 3.0 GHz because broadband receivers are much more susceptible to radio interference. A solution is to equip existing geodetic radio telescopes with auxiliary receivers operating in the 1–2 GHz range dedicated to VLBI observations of navigational satellites. Since the GNSS signal is so strong, such receivers do not require cooling. The navigational receiver can be installed alongside the main receiver, and the signal can be directed to it either with a deployable mirror or with a dichroic plate.

5   Summary

We propose a novel concept of GNSS/VLBI tie measurements based on a microwave technique. We essentially transform a GNSS antenna into an element of the VLBI network. This method will allow us to estimate the tie vector between the VLBI and GNSS reference points directly using the microwave technique without the need to determine the position offsets of the microwave reference points with respect to markers accessible to local surveys.

We expect that the application of this method will have a profound impact because we expect this method will be bias-free. As a result, vector tie repeatability could be used

as a measure of the accuracy of tie vector determination. Knowing errors of tie vectors will enable us to close the budget of the differences of the VLBI reference points reduced to the GNSS reference points and make an inference about whether these differences are statistically significant or not.

We ran three 3 h long observing sessions between the FD-VLBA radiotelescope and a high rate GNSS receiver co-located within 90 m. We have detected fringes from both natural extragalactic radio sources and GPS satellites with the SNR well above 100. Thus, we have demonstrated that technical problems related to GNSS/radiotelescope VLBI can be solved. Future work will be focused on the determination of vector ties using this technique.

References

Altamimi Z, Boucher C, Sillard P (2002) New trends for the realization of the international terrestrial reference system. Adv Space Res 30(2):175–184. https://doi.org/10.1016/S0273-1177(02)00282-X

Matsumoto S, Ueshiba H, Nakakuki T, Takagi Y, Hayashi K, Yutsudo T, Mori K, Sato Y, Kobayashi T (2022) An effective approach for accurate estimation of VLBI-GNSS local-tie vectors. Earth Planets Space 74(147). https://doi.org/10.1186/s40623-022-01703-5

Ray J, Altamimi Z (2005) Evaluation of co-location ties relating the VLBI and GPS reference frames. J Geodesy 79(4–5):189–195. https://doi.org/10.1007/s00190-005-0456-z

Sarti P, Abbondanza C, Petrov L, Negusini M (2011) Height bias and scale effect induced by antenna gravitational deformations in geodetic VLBI data analysis. J Geodesy 85(1):1–8. https://doi.org/10.1007/s00190-010-0410 6

Skeens J, York J, Petrov L, Munton D, Herrity K, Ji-Cathriner R, Bettadpur S, Gaussiran T (2023) First observations with a GNSS antenna to radio telescope interferometer. https://doi.org/10.48550/arXiv.2304.11016. arXiv:2304.11016

Thoelert S (2019) Latest GNSS signal in space developments – GPS, QZSS & the new Beidou 3 under examination. E3S Web Conf 94:03016. https://doi.org/10.1051/e3sconf/20199403016

Varenius E, Haas R, Nilsson T (2021) Short-baseline interferometry local-tie experiments at the Onsala Space Observatory. J Geodesy 95(5):54. https://doi.org/10.1007/s00190-021-01509-5

York J, Joplin A, Bratton M, Munton D (2014) A detailed analysis of GPS live-sky signals without a dish. Navigation 61(4):311–322. https://doi.org/10.1002/navi.69

York J, Little J, Nelsen S, Caldwell O, David M (2012) Novel software defined GNSS receiver for performing detailed signal analysis. In: Proceedings of the 2012 International Technical Meeting of The Institute of Navigation. AAS/Division for Planetary Sciences Meeting Abstracts, pp 1910–1921

Status of the SIRGAS Reference Frame: Recent Developments and New Challenges

Sonia M. Alves Costa, Laura Sánchez, Diego Piñón, José A. Tarrio Mosquera, Gabriel Guimarães, Demián D. Gómez, Hermann Drewes, María V. Mackern Oberti, Ezequiel D. Antokoletz, Ana C. O. C. de Matos, Denizar Blitzkow, Alberto da Silva, Jesarella Inzunza, Draco España, Oscar Rodríguez, Sergio Rozas-Bornes, Hernan Guagni, Guido González, Oscar Paucar-Llaja, José M. Pampillón, and Álvaro Alvarez-Calderón

Abstract

In accordance with recent developments of the International Association of Geodesy (IAG) and the policies promoted by the Subcommittee on Geodesy of the United Nations Committee of Experts on Global Geospatial Information Management (UN-GGIM), a main goal of the Geodetic Reference System for the Americas (SIRGAS) is the procurement of an integrated regional reference frame. This frame should support the precise determination of geocentric coordinates and also provide a unified physical reference frame for gravimetry, physical heights, and a geoid. The geometric reference frame is determined by a network of about 500 continuously operating GNSS stations, which are routinely processed by ten analysis centers. The GNSS solutions from the analysis centers are used to generate weekly station positions aligned to the International Terrestrial Reference Frame (ITRF) and multi-year (cumulative) reference frame solutions. This processing is also the basis for the generation of precise tropospheric zenith path delays with an hourly sampling rate over the Americas. The reference frame for the determination of physical heights is a regional

S. M. Alves Costa · A. da Silva ·
Brazilian Institute of Geography and Statistics (IBGE), Rio de Janeiro, Brazil

L. Sánchez · H. Drewes
Deutsches Geodätisches Forschungsinstitut, Technische Universität München (DGFI-TUM), Munich, Germany

D. Piñón · H. Guagni
Instituto Geografico Nacional, Buenos Aires, Argentina

J. A. Tarrio Mosquera · J. Inzunza
Centro de Procesamiento y Análisis Geodésico USC
Universidad Santiago de Chile (USACH), Santiago de Chile, Chile

G. Guimarães
Federal University of Uberlandia (UFU), Uberlândia, Brazil

D. D. Gómez
Division of Geodetic Science, School of Earth Sciences, Ohio State University (OSU), Columbus, OH, USA

M. V. Mackern Oberti (✉)
Facultad de Ingeniería, Universidad Nacional de Cuyo, CONICET, Universidad Juan Agustín Maza, Mendoza, Argentina
e-mail: vmackern@mendoza-conicet.gob.ar

E. D. Antokoletz
Facultad de Ciencias Astronómicas y Geofísicas, Universidad Nacional de La Plata (UNLP), La Plata, Argentina

Federal Agency for Cartography and Geodesy (BKG), Leipzig, Germany

A. C. O. C. de Matos · D. Blitzkow
Laboratório de Topografia e Geodesia, Programa de Pós-graduação em Engenharia de Transportes, Escola Politécnica da Universidade de São Paulo (EPUSP), São Paulo, Brazil

D. España
Instituto Geográfico Militar, Quito, Ecuador

O. Rodríguez
Instituto Geográfico Agustín Codazzi (IGAC), Bogotá, Colombia

S. Rozas-Bornes
Instituto Geográfico Militar, Santiago de Chile, Chile

G. González
Instituto Nacional de Estadística y Geografía (INEGI), Aguascalientes, Mexico

O. Paucar-Llaja
Instituto Geográfico Nacional, Lima, Peru

J. M. Pampillón
Instituto Geográfico Militar, Montevideo, Uruguay

Á. Alvarez-Calderón
Instituto Geográfico Nacional, Registro Nacional, San José, Costa Rica

© The Author(s) 2023
J. T. Freymueller, L. Sánchez (eds.), *Gravity, Positioning and Reference Frames*,
International Association of Geodesy Symposia 156, https://doi.org/10.1007/1345_2023_227

densification of the International Height Reference Frame (IHRF). Current efforts focus on the estimation and evaluation of potential values obtained from high resolution gravity field modelling, an activity tightly coupled with geoid determination. The gravity reference frame aims to be a regional densification of the International Terrestrial Gravity Reference Frame (ITGRF). Thus, SIRGAS activities are focused on evaluating the quality of existing absolute gravity stations and to identify regional gaps where additional absolute gravity stations are needed. Another main goal of SIRGAS is to promote the use of its geodetic reference frame at the national level and to support capacity building activities in the region. This paper summarizes key milestones in the establishment and maintenance of the SIRGAS reference frame and discusses current efforts and future challenges.

Keywords

GNSS reference networks · IHRF regional densification · ITGRF regional densification · ITRF regional densification · Regional reference frames · SIRGAS

1   Introduction

The Geodetic Reference System for the Americas (*Sistema de Referencia Geodésico para las Américas*, SIRGAS) was established in 1993 at an international conference in Asunción, Paraguay, organized by the International Association of Geodesy (IAG), the Pan-American Institute for Geography and History (PAIGH), the Deutsches Geodätisches Forschungsinstitut (DGFI), and the U.S. Defense Mapping Agency (DMA) (see e.g. Drewes 2022). During this meeting, participants defined the main goal of SIRGAS: the unification of the South American Datum using the Global Positioning System (GPS). Two Working Groups (WG) were formed to achieve this goal, "Reference Frame" (WGI) and "Geodetic Datum" (WGII, now called "SIRGAS at National level"). Their charge was to define, realize, and maintain a geocentric reference system and to support its integration to the national densifications.

The first frame realization, SIRGAS95, included stations only in South America (SIRGAS 1997). The second one, SIRGAS2000, included stations in countries in all the Americas (Drewes et al. 2005). Some years later, SIRGAS implemented a reference frame using only continuously operating GNSS (Global Navigation Satellite Systems) stations (see e.g. Brunini et al. 2012). In 1997 SIRGAS created the "Vertical Datum" WGIII for the determination of a vertical reference frame for South America that aimed to connect the existing levelling networks (Drewes 2022). This WGIII is currently dedicated to establish regional densifications of the International Height Reference Frame (IHRF) and the International Terrestrial Gravity Reference Frame (ITGRF) as to provide consistency for gravimetry, physical heights, and geoid.

Over the years, the SIRGAS WGI provided weekly products such as, coordinates, hourly zenith path delays (ZPD), and long-term products such as velocity models (VEMOS) and multi-year solutions.

Another important and strategic task carried out by SIRGAS is knowledge transfer and capacity building (see https://sirgas.ipgh.org/eventos-sirgas/cursos/). This paper summarizes activities carried out by the SIRGAS WGs, including efforts from the IGS Regional Network Associate Analysis Centre for SIRGAS (IGS RNAAC SIRGAS). Also, the SIRGAS executive committee recently joined the Regional Committee (UN-GGIM:Americas) of the Geodetic Reference Frame for the Americas WG. We discuss the activities and new responsibilities of SIRGAS within this WG.

2   Main SIRGAS Objectives and International Networking

SIRGAS mainly interacts with four international bodies: IAG, which provides guidance for the scientific and technical SIRGAS activities; the International GNSS Service (IGS), which provides support for the proper analysis of the SIRGAS reference frame; PAIGH, which provides a direct link to the national agencies responsible for the geodetic reference frames; and the chapter Americas of the United Nations Committee of Experts on Global Geospatial Information Management (UN-GGIM: Americas), which provides a policy framework for geodetic capacity building at the regional level. Based on this networking, the main objectives of SIRGAS are:

- To establish and maintain a continental geocentric reference frame that is a regional densification of the International Terrestrial Reference Frame (ITRF);
- To define and maintain a unified vertical reference system by means of physical and geometric heights consistent with IHRF;

- To develop and maintain updated a gravimetric geoid model of continental coverage; and
- To establish and maintain a continental absolute gravity reference network consistent with the ITGRF.

These goals are faced by the WGI and WGIII, whose chairs are also responsible for the IAG Sub-Commissions 1.3.b (Regional Reference Frames – South and Central America) and 2.4b (Gravity and Geoid in South America), respectively. The capacity building and knowledge transfer activities are coordinated by the WGII. The interaction with the IGS is done by the IGS Regional Associate Analysis Centre for SIRGAS (IGS RNAAC SIR). Efforts and results of these WGs are also reported to the PAIGH Cartography Commission. The interaction between SIRGAS and UN-GGIM: Americas is founded in the WG Geodetic Reference Frame for the Americas (GRFA-WG), which promotes and provides mechanisms for capacity development and knowledge transfer in the field of Geodesy among the Nations of the Americas. The main goal is to cooperate in the implementation of the UN Resolution about a "Global Geodetic Reference Frame for sustainable development" (A/RES/69/2663) adopted in 2015. To optimise resources and harmonise the SIRGAS and GRFA-WG activities, the president and vice-president of SIRGAS are the co-chairs of the GRFA-WG. Thus, SIRGAS is the meeting point for policy, science, technology, and capacity building in geodesy in the Americas.

3   Advances in the Physical Reference Frame

As mentioned, one of the main goals of SIRGAS is to establish a unified physical reference frame that ensures consistency between gravity observations, geoid model, and physical heights. Surface (terrestrial, airborne, shipborne) gravity values are the main input for the computation of levelling-based geopotential numbers (i.e., physical heights) and the high-frequency signals of the geoid. In turn, the disturbing potential determined for the geoid modelling is also needed for the calculation of geopotential numbers in the IHRF (see e.g., Sánchez et al. 2021). For this reason, SIRGAS seeks to ensure consistency between the gravity reference (Sect. 3.1), the geoid model (Sect. 3.2), and the IHRF coordinates (Sect. 3.3).

3.1   Reference Frame for Terrestrial Gravimetry

The gravimetric reference frame within SIRGAS is mostly based on local absolute gravity networks determined mainly by Micro-g LaCoste A10 gravity meter measurements

(Blitzkow et al. 2018). Today, most of the countries have absolute gravity networks (Fig. 1), which are usually densified by relative gravimeter measurements. Current goals are to identify areas with few observations and to distribute and set new stations more homogeneously in order to support the establishment of the IHRF (Sánchez et al. 2021) and the precise determination of the geoid.

SIRGAS is also involved in establishing the International Terrestrial Gravity Reference System (ITGRS) and Frame (ITGRF; Wziontek et al. 2021) on a regional level. One of the key aspects of the ITGRF is the demand for reference stations that provide a precise gravity reference supporting frame accessibility at any time. In this regard, the Argentinean-German Geodetic Observatory (AGGO) located close to La Plata, Argentina (Fig. 1), plays a fundamental role as it provides continuous gravity measurements using a superconducting gravimeter (SG). These measurements were complemented with absolute gravity measurements performed with a FG5 gravity meter between 2019 and 2022. The combination of both allowed for the computation of a gravity reference function for the station (Antokoletz et al. 2020). According to Wziontek et al. (2021), these characteristics, and the available infrastructure, allow AGGO to be a core station of the ITGRF.

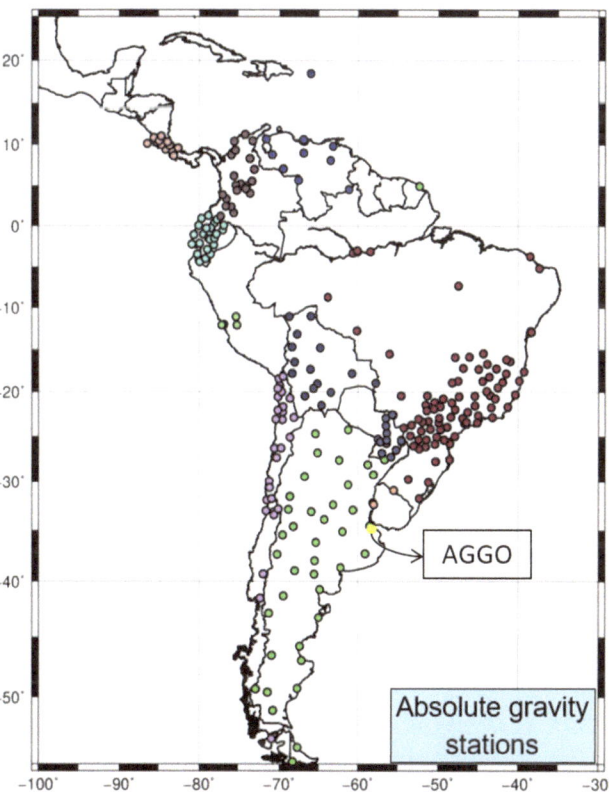

Fig. 1 Distribution of absolute gravity stations along Latin America (as of Sep, 2022). Different colours correspond to the network belonging to different countries. In yellow the station AGGO

In this context, one of SIRGAS' challenges is to evaluate the quality of the existing absolute gravity measurements in order to ensure its compatibility with the standards and recommendations given for the ITGRF (Wziontek et al. 2021). Regional comparisons at reference stations like AGGO will play a key role, since all gravity meters in the region must participate in these comparison campaigns. Other activities include (a) training and capacity building in gravimetry with the aim of homogenising field procedures and processing standards of absolute and relative gravity measurements; (b) constant support to the national agencies in charge of the gravimetric reference frames; and (c) compilation of detailed documentation and metadata of the existing absolute gravity data.

3.2 Recent Improvements in the Modelling of the Geoid

The most recent geoid and quasi-geoid models for South America, called GEOID2021 and QGEOID2021 (de Matos et al. 2021a, b) respectively, were calculated thanks to the collaboration of several South American organizations, especially national mapping agencies, private companies and universities. These models cover the area between 15°N and 60°S latitude and 100°W and 30°W longitude, with a 5′ grid resolution. The comparison between the estimated geoid heights and the GPS/levelling data at 4,464 points in Argentina (2,931), Chile (176), Colombia (464), Ecuador (703) and Venezuela (190) shows differences with RMS values ranging from 34 cm for Argentina to 92 cm for Ecuador. The comparison between height anomalies and GPS/levelling data at 1,108 points in Brazil shows differences with a RMS of about 41 cm. Looking at the RMS it is possible to verify the convergence of the geoid and quasi-geoid models in relation to the GPS/levelling points. While levelling points are linked to the local vertical data of the different countries, the geoid and quasi-geoid models are linked to the equipotential surface of the Earth's gravity field with geopotential value $W_0 = 62,636,853.4\ m^2\ s^{-2}$. This can explain the differences in the comparison. Besides that, the zero degree term added to the geoid model was equal to −17 cm, where it was considered that the normal potential U_0 was different from W_0. The grids for both models are available on the website of the International Service for the Geoid (ISG; Reguzzoni et al. 2021; de Matos et al. 2021a, b).

3.3 Standardisation of Physical Heights

In the last 25 years, SIRGAS has been actively working on the unification of vertical datums and the determination of a unified height system for the region. Since 2015,

when the IAG defined the International Height Reference System (IHRS, see Drewes et al. 2016; Ihde et al. 2017), SIRGAS focused efforts to establish a regional densification of the IHRF and supported member states through workshops, schools, and webinars. In the region, 19 stations distributed over 10 countries were selected to compose the IHRF network. These stations are materialised by continuously operating GNSS stations and are integrated into the SIRGAS reference frame. Besides that, some of them are co-located with space geodesy and gravimetric techniques (Fig. 2).

It is recommended that regional unified height systems are based on geopotential numbers as different physical heights (orthometric or normal heights) are in use and they may introduce artificial errors in the connection of levelling networks at the borders between neighbouring countries. In this sense, SIRGAS provides training and capacity building to the national agencies responsible for the geodetic reference frames. To date, three member states have completed this task and three others are close to finish (Fig. 2). Furthermore, SIRGAS has emphasized the importance of international levelling connections (Fig. 2), gravity measurements and

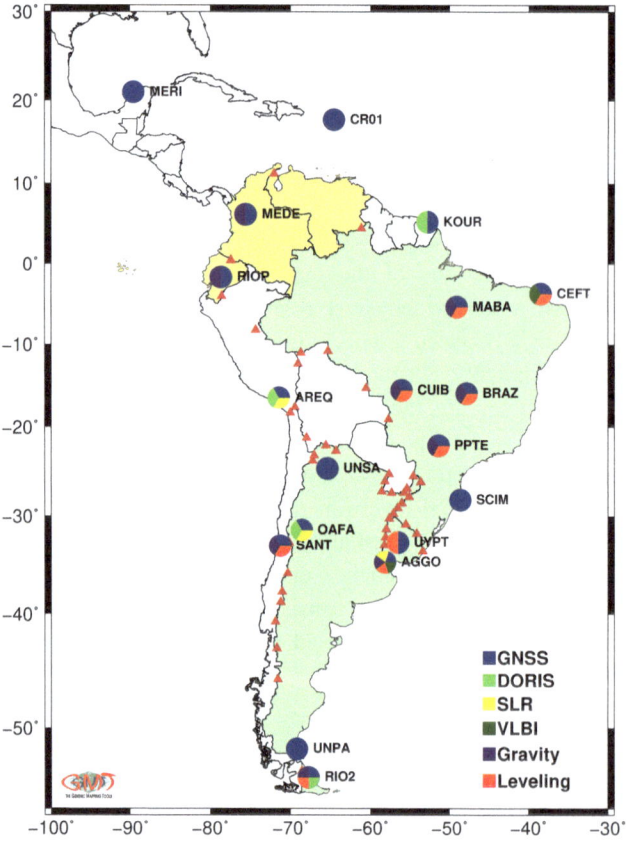

Fig. 2 Distribution of IHRF stations in Latin America (as of Sep, 2022), co-located with space techniques, gravity, and levelling. Triangles indicate the international levelling connections. Countries with vertical networks adjusted in terms of geopotential numbers are depicted in green and in yellow those in process

levelling connections at the IHRF stations. Two technical guides were developed: "Guidelines to select IHRF stations" and "Guidelines for gravimetric measurements around IHRF stations", both available at https://sirgas.ipgh.org/. Additional ongoing activities are (a) station selection for national densifications of the IHRF, and (b) the determination of geopotential numbers at the Latin American IHRF stations (more details in Tocho et al. 2020; Guimarães et al. 2022a, b; Silva et al. 2022). The present challenges in this regard are the evaluation of discrepancies between different computation methods and the quality assessment in the determination of geopotential numbers.

4   Status of the Geometric Reference Frame

The current realization of SIRGAS is a network of 500 continuously operating GNSS stations (Fig. 3). From these stations, 109 belong to the IGS global network; the rest belong to the national reference frames. All SIRGAS stations track GPS, 89% of them track GLONASS, 39% Galileo, and 30% Beidou.

The SIRGAS reference stations are classified in core stations (core network, SIRGAS-C) and national densification stations (national networks, SIRGAS-N). All stations follow the same operational criteria and are analysed on a weekly basis in agreement with the standards of the International Earth Rotation and Reference Systems Service (IERS, Petit and Luzum 2010) and the IGS (Johnston et al. 2017). Currently, 10 SIRGAS analysis centres (SIRGAS-AC) process the GNSS data. Each station is included in at least three individual solutions. The SIRGAS-ACs generate weekly loosely constrained solutions (LCS) for station positions and Zenith Path Delays (ZPD) hourly estimates. The station positions' LCS are combined by the SIRGAS combination centres to generate a unified solution of the reference frame (Sect. 4.1). The ZPD estimates are combined by the SIRGAS analysis centre for the neutral atmosphere (Sect. 4.2). The weekly combinations are the input for the determination of reference frame multi-year solutions (Sect. 4.3), which are the basis for the calculation of SIRGAS velocity models (Sect. 4.4). Table 1 summarizes present and former SIRGAS analysis centres. Figure 4 depicts the data flow within the SIRGAS reference frame analysis.

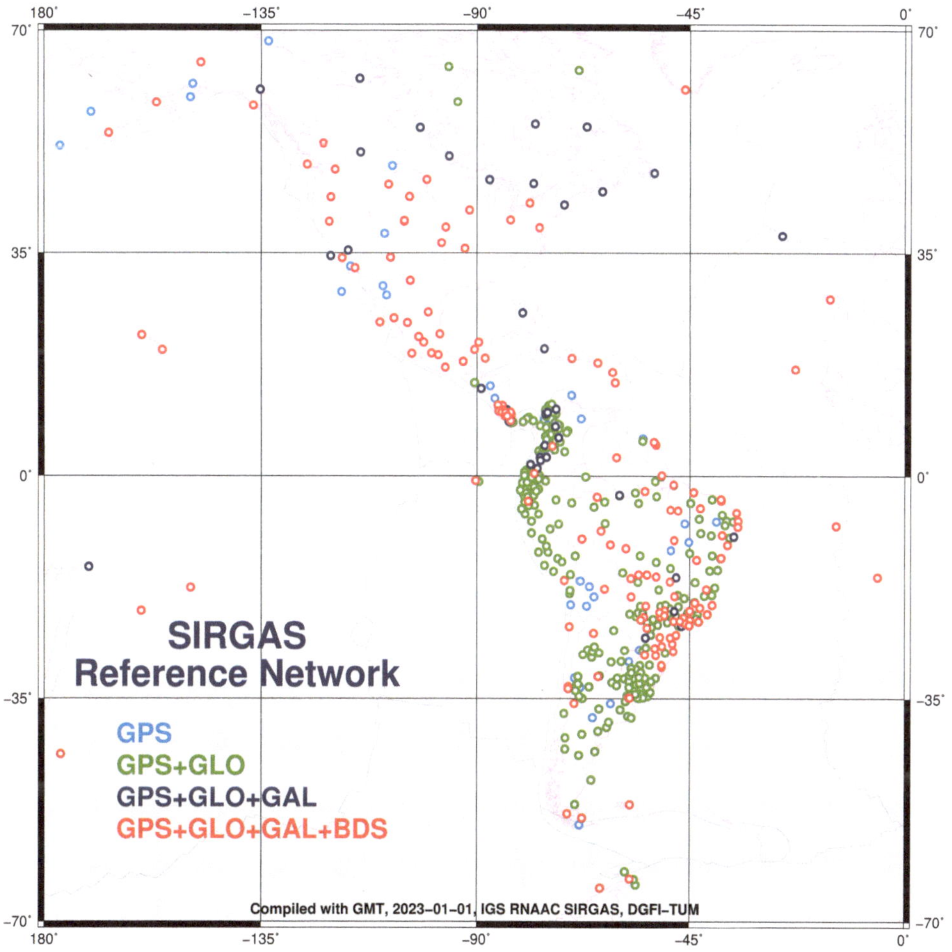

Fig. 3 SIRGAS reference network (as of Sep, 2022)

Table 1 Active and former SIRGAS analysis centres (as of Sep 2022). CIMA acts as the SIRGAS analysis centre for the neutral atmosphere since Nov. 2019. DGFI-TUM acts as the IGS regional network associate analysis centre for SIRGAS (IGS RNAAC SIR) since June 1996

Analysis centre	Country	Agency	Software	Coordinate solutions		Tropospheric estimates	
				Since	To	Since	To
DGF	Germany	Deutsches Geodätisches Forschungsinstitut, Technischen Universität München (DGFI-TUM)	BSW52[a]	1996-06-30	Present	2014-04-27	Present
ECU	Ecuador	Instituto Geográfico Militar (IGM-Ec)	BSW52	2010-01-01	Present	2014-12-21	Present
IBG	Brazil	Instituto Brasileiro de Geografia e Estatistica (IBGE)	BSW52	2008-08-31	Present	2014-04-27	Present
IGA	Colombia	Instituto Geográfico Agustín Codazzi (IGAC)	BSW52	2008-08-31	Present	2014-12-21	Present
CHL	Chile	Instituto Geográfico Militar (IGM-Cl)	BSW52	2013-01-01	Present	2014-04-27	Present
URY	Uruguay	Instituto Geográfico Militar (IGM-Uy)	BSW52	2010-01-01	Present	2014-04-27	Present
USC	Chile	Universidad de Santiago de Chile (USACH)	BSW52	2019-01-01	Present	2019-05-01	Present
GNA	Argentina	Instituto Geográfico Nacional (IGN-Ar)	GG[b]	2011-01-01	Present	2022-01-01	Present
INE	Mexico	Instituto Nacional de Estadística y Geografía (INEGI)	GG	2011-01-01	Present	–	
PER	Peru	Instituto Geográfico Nacional (IGN-Pe)	GG	2022-01-01	Present	–	
CIM	Argentina	Centro de Ingeniería Mendoza, Argentina (CIMA)	BSW52	2008-08-31	2021-12-31		
LUZ	Venezuela	Universidad de Zulia	BSW52	2010-01-01	2019-02-09	2014-12-14	2019-02-09
UNA	Costa Rica	Universidad Nacional de Costa Rica	BSW52	2014-01-01	2018-12-31	2014-01-01	2018-12-31

[a]BSW52: Bernese GNSS Software, version 5.2 (Dach et al. 2015)

[b]GG: GAMIT/GLOBK: GNSS at MIT/Global Kalman filter (Herring et al. 2015, 2018)

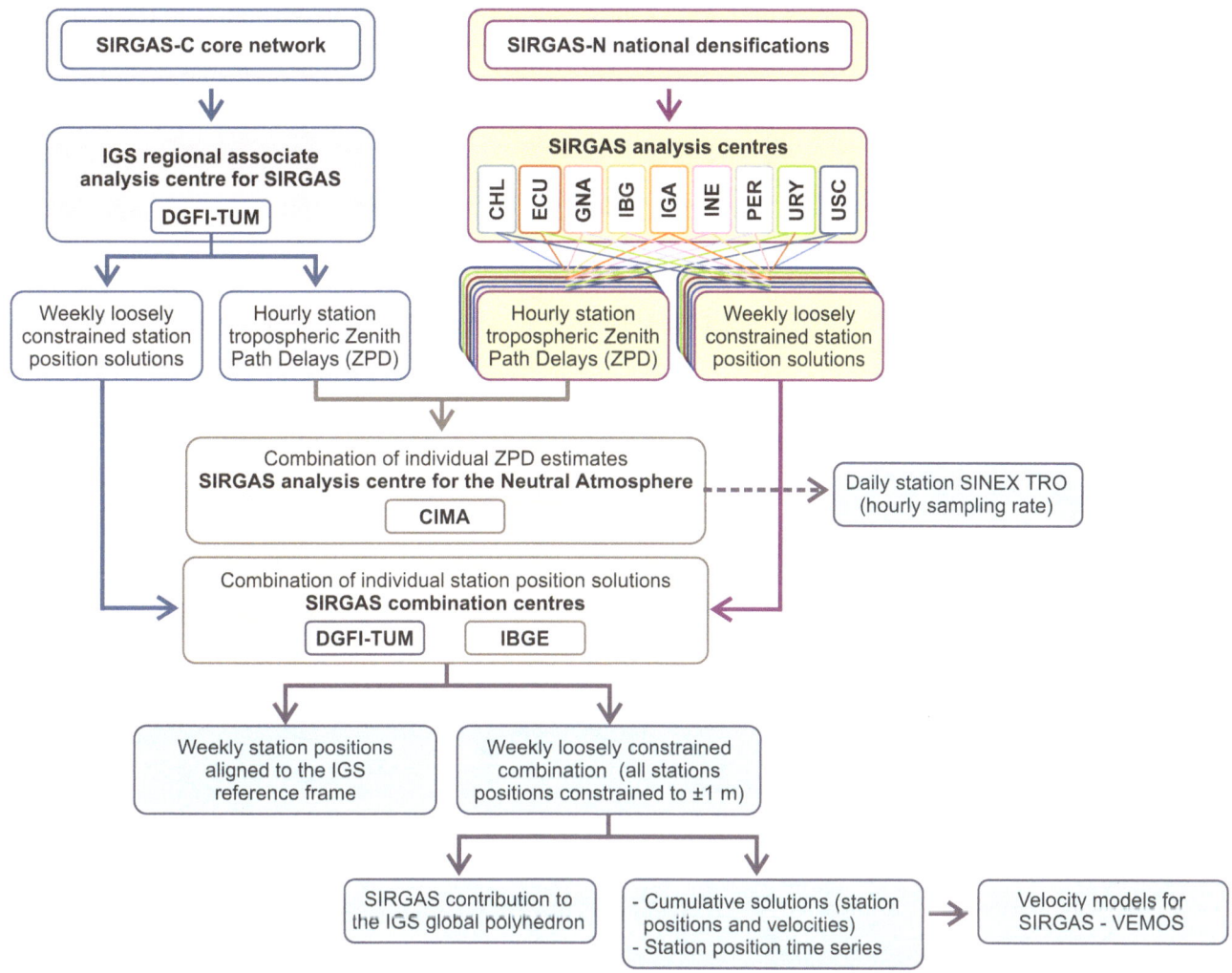

Fig. 4 Data flow in the analysis of the SIRGAS reference frame (adapted from Sánchez et al. 2022). Please see Table 1 for the SIRGAS-AC acronyms

4.1 Operational and Reprocessed SIRGAS Weekly Station Positions

In the weekly analysis of the SIRGAS reference frame, the IGS final satellite orbits, satellite clocks, and Earth orientation parameters (Johnston et al. 2017) are included as known parameters (see Tarrío et al. 2021). Thus, the SIRGAS weekly solutions are based on the models and standards valid at the time of computation and refer to the IGS reference frame in use during that specific time. Updated models, better processing standards or improved IGS reference frame solutions are directly reflected in the quality of the SIRGAS coordinates. As an example, Table 2 summarises the weekly station position repeatability and the consistency with the IGS weekly solutions of the SIRGAS positions referring to different IGS reference frames.

To ensure the long-term reliability of the SIRGAS reference frame, the complete GNSS data series are homogeneously reprocessed to refer all weekly normal equations to a unified set of standards and to the same reference frame. The first SIRGAS reprocessing, Repro1, comprised GNSS data from 2000-01-02 to 2008-08-30 and its main goals were to consider absolute corrections for the phase centre variations of the GNSS antennae and to refer positions and velocities to the IGS05 reference frame (see Sánchez and Seitz 2011).

The DGFI-TUM recently reprocessed all the GNSS data from de SIRGAS Reference Network and a set of globally distributed IGS stations, covering the time span between January 2000 and December 2021 (see Sánchez et al. 2022). This Repro2 refers to the IGS14/IGb14 reference frame (Rebischung and Schmid 2016; Griffiths 2019). In total, 537 SIRGAS and 128 IGS stations (with 88 in the IGS14/IGb14 reference frame) were reanalysed (Fig. 5). The normal equations obtained in Repro2 were the input for the computation of a new DGFI-TUM reference frame solution called

Table 2 Mean RMS values of the weekly SIRGAS station position repeatability and after comparing the SIRGAS station positions with the weekly coordinates of the IGS stations. The last row presents the values obtained from Repro2 (more details in Sánchez et al. 2022)

IGS reference frame	From	To	Weekly station position repeatability [mm]		Compatibility of weekly SIRGAS reference frame solutions with the IGS reference frame [mm]	
			N/E	Up	N/E	Up
IGS05	2000-01-02	2011-04-16	2.3	4.5	2.8	6.0
IGS08/IGb08	2011-04-17	2017-01-28	1.8	3.2	1.8	3.5
IGS14/IGb14	2020-05-17	2022-11-26	1.0	3.2	0.8	2.6
IGS14/IGb14	2000-01-02	2022-11-26	1.0	3.0	0.8	2.6

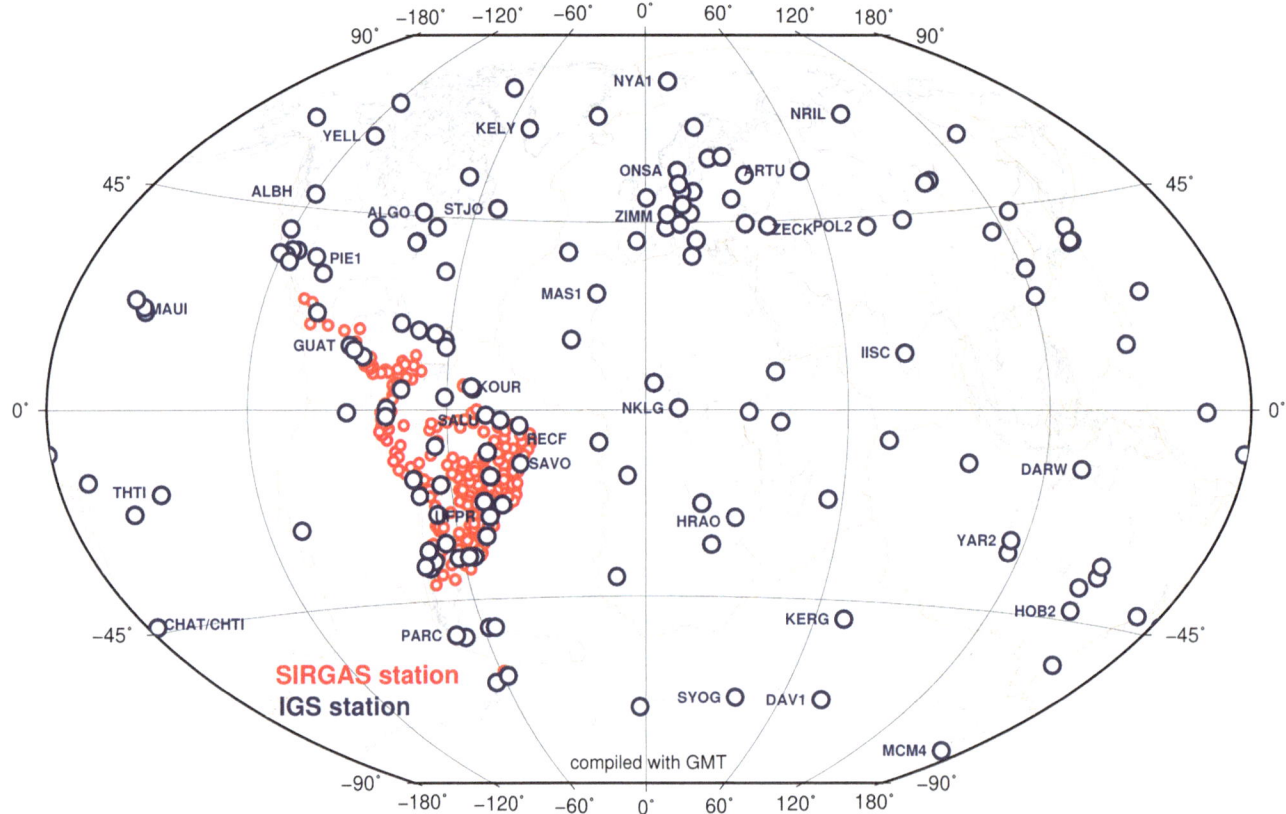

Fig. 5 GNSS network included in the latest SIRGAS data reprocessing. Labels identify the reference stations utilised for the geodetic datum realisation (adapted from Sánchez et al. 2022)

SIRGAS2022 (see Sect. 4.3). The Repro2 normal equations are available for combination with solutions from other SIRGAS-ACs to realize a SIRGAS-wide reference frame.

4.2 Combined Tropospheric Zenith Path Delays

The ZPDs estimated by the SIRGAS-ACs (see Table 1) are combined to generate the ZPD_{SIR} values in hourly sampling rates. This combination is performed on a weekly basis by CIMA, since Nov. 2019 (Mackern et al. 2020). The methodology is described in Mackern et al. (2022). Three or more individual solutions are needed to obtain

statistical controls over the combined values of ZPD_{SIR}. Figure 6 shows significant progress towards this goal, mainly since 2019.

The ZPD_{SIR} precision was calculated using the mean annual Standard Deviation (SD) for each station. Table 3 summarises the results of the last precision analysis carried out for 2021.

The final ZPD_{SIR} have been validated (Mackern et al. 2020) with respect to final IGS' ZTD products (at 15 IGS stations) and with respect to computed ZTD at radiosonde stations (10 sites). This study shows that the ZPD_{SIR} agree with the corresponding values obtained by the IGS (mean RMSE 6.8 mm; mean bias −1.5 mm) as well as those from radiosondes (mean RMSE 7.5 mm; mean bias −2 mm).

Fig. 6 Number of stations with 1, 2, 3 or more individual ZPD solutions before combination

Table 3 Precision ZPD analysis carried out for 2021 (Mackern et al. 2022)

	Mean RMS < 1 mm	1.1 mm < Mean RMS <3 mm	3.1 mm < Mean RMS <6 mm
Number of stations	309	113	150
Percent of stations	54	20	26

4.3 SIRGAS2022: The Latest DGFI-TUM Reference Frame Solution for SIRGAS

Due to the occurrence of seismic events in the SIRGAS region, the SIRGAS reference frame cumulative solutions require frequent updates (e.g., Seemüller et al. 2011; Sánchez and Seitz 2011; Sánchez and Drewes 2016, 2020). The latest DGFI-TUM reference frame cumulative solution, called SIRGAS2022, is based on the Repro2 normal equation series up to December 2021. The normal equations from January 2022 to April 2022 were obtained from the weekly combination of individual solutions from the SIRGAS-ACs and are all based on the weekly IGS14/IGb14 normal equations (Fig. 7). A description of the processing and analysis methodology can be found in Sánchez et al. (2022).

SIRGAS2022 (Fig. 8) contains 587 stations with 1,389 occupations. The station positions refer to the IGb14 and are given at the epoch 2,015.0. Their accuracy is estimated to be ±0.8 mm in N/E and ± 1.4 mm in U at the reference epoch. The accuracy of the velocities is assessed to ±0.6 mm/year in N/E and ± 1.0 mm/year in U.

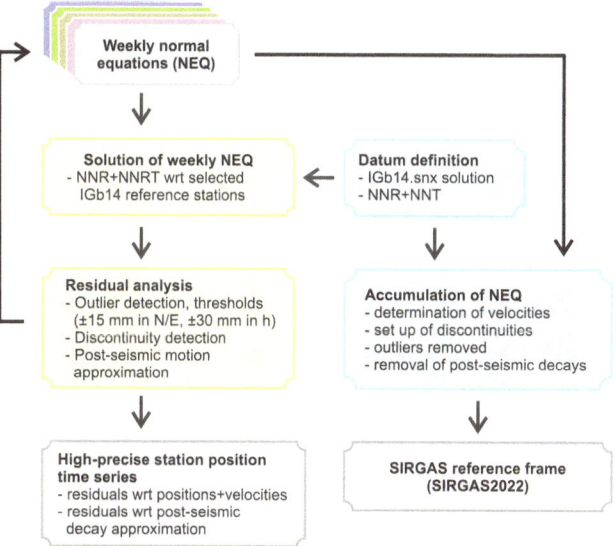

Fig. 7 Analysis steps in the determination of SIRGAS2022 (adapted from Sánchez et al. 2022)

Fig. 8 SIRGAS2022 horizontal velocities

4.4 VEMOS: Overall Velocity Models for the Entire SIRGAS Region

The constant velocities determined in the computation of the SIRGAS reference frame cumulative solutions are the input for the prediction of velocity grids over the entire SIRGAS region (Fig. 9). They are needed to interpolate station motions in regions where no SIRGAS stations are in operation and serve as the basis for the analysis of regional surface deformations. The VEMOS models represent mean yearly horizontal surface displacements for a period of data used for the model (Table 4). A new updated version of VEMOS, including the latest processing results, is in preparation.

5 Final Remarks

SIRGAS is a well-established comprehensive regional geodetic reference frame and widely used in practical and scientific applications. The routine analysis of the SIRGAS reference frame is in accordance with the new models, standards, and procedures defined by the IERS and the IGS. The accuracy of the weekly SIRGAS station positions is 1.0 mm in N/E and about 3.0 mm in the vertical component. The accuracy of the latest DGFI-TUM reference frame solution SIRGAS2022 is estimated to be ± 0.8 mm in N/E and ± 1.4 mm in U for the station positions at the reference epoch and ± 0.6 mm/year in N/E and ± 1.0 mm/year in

Fig. 9 VEMOS2017 (adapted from Drewes and Sánchez 2020)

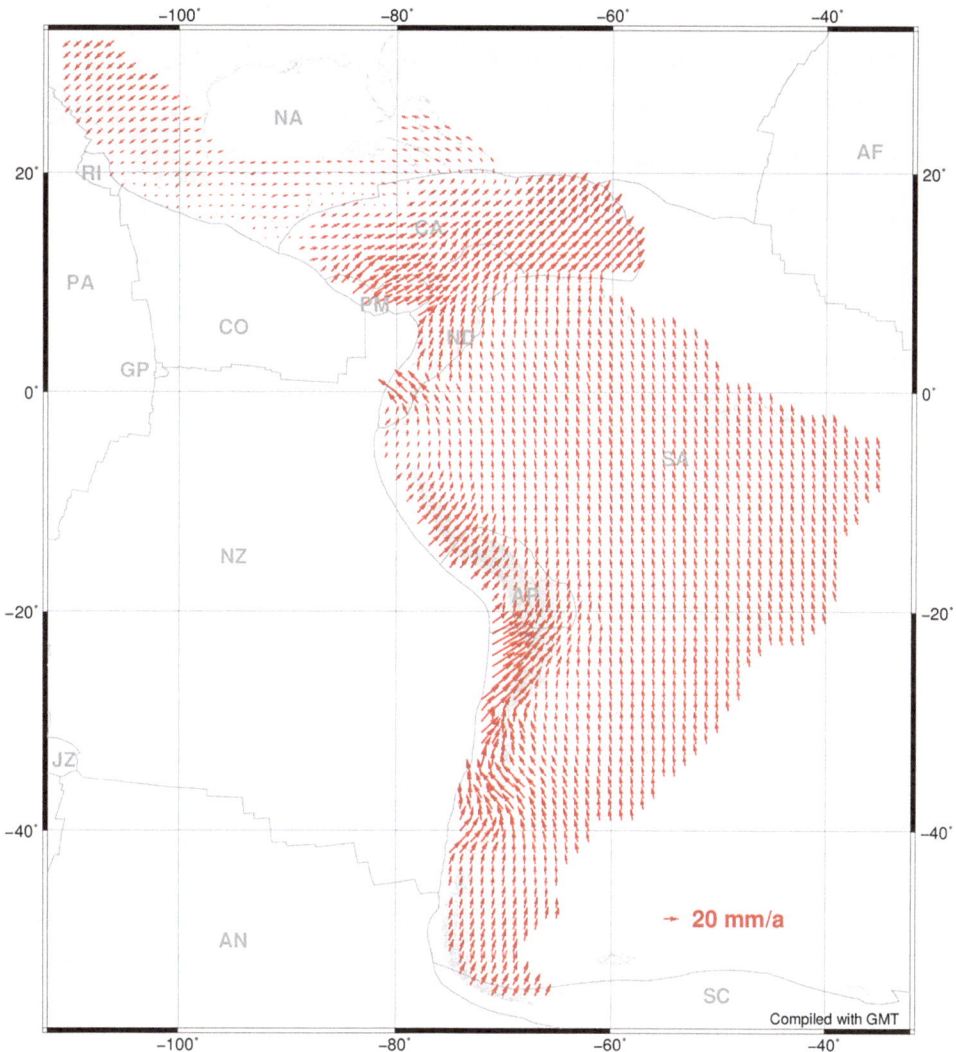

Table 4 SIRGAS velocity models (more details in https://sirgas.ipgh.org/en/products/vemos/)

VEMOS	Reference frame	Observation period included		Reference
		From	To	
VEMOS2003	ITRF2000	May 1995	April 2001	Drewes and Heidbach (2005)
VEMOS2009	ITRF2005	January 2000	June 2009	Drewes and Heidbach (2012)
VEMOS2015	IGb08	March 2010	April 2015	Sánchez and Drewes (2016)
VEMOS2017	IGS14	January 2014	January 2017	Drewes and Sánchez (2020)

U for the velocities. Main challenges in the determination of the reference frame are the modelling of seismic and post-seismic effects and strong seasonal signals observed in the Amazon basin. A strategic priority of SIRGAS is the advancement in the establishment of a physical reference frame to support gravimetry, physical heights and geoid determination with an accuracy similar to that of the geometric reference frame. This is yet a difficult challenge to overcome. Current SIRGAS efforts are aimed at collecting the necessary data and linking the different national agencies through training and knowledge transfer. The joint work between SIRGAS and UN-GGIM: Americas highlights the importance of geodetic reference frames as a strategic tool for sustainable development. Yet, governmental support is needed to obtain human, technical, and financial resources to continue the development of SIRGAS. This governmental support can only be achieved by each of the members of SIRGAS, and we strive to provide the mechanisms to support raising the necessary awareness.

Acknowledgements The SIRGAS activities are possible thanks to the active support of more than two hundred colleagues contributing to the WGs, to capacity building activities, operating GNSS stations, operating SIRGAS analysis centres, etc. This support and that provided by the IAG and PAIGH to the geodetic reference activities in the region are highly appreciated. We would like to thank Editor-in-Chief Jeff Freymueller and two anonymous reviewers for their insightful comments that helped improve this work.

Data Availability All SIRGAS products are freely available at https:// sirgas.ipgh.org. and ftp.sirgas.org.

References

Antokoletz ED, Wziontek H, Tocho CN et al (2020) Gravity reference at the Argentinean–German Geodetic Observatory (AGGO) by co-location of superconducting and absolute gravity measurements. J Geod 94:81. https://doi.org/10.1007/s00190-020-01402-7

Blitzkow D, Oliveira Cancoro de Matos AC, Moraes Bjorkstrom I et al (2018) Absolute gravity network in South America-Comparisons. EGU General Assembly Conference Abstracts (p. 7815). https://ui.adsabs.harvard.edu/abs/2018EGUGA..20.7815B

Brunini C, Sánchez L, Drewes H, Costa SMA, Mackern V, Martinez W, Seemüller W, Da Silva AL (2012) Improved analysis strategy and accessibility of the SIRGAS Reference Frame, International Association of Geodesy Symposia Series, vol 136. Springer, Heidelberg, pp 3–10. https://doi.org/10.1007/978-3-642-20338-1_1

Dach R, Lutz S, Walser P, Fridez P (2015) Bernese GNSS Software Version 5.2. Astronomical Institute, University of Bern. https://boris.unibe.ch/72297/

de Matos ACOC, Blitzkow D, Guimarães GN, Silva VC (2021a) The South American gravimetric quasi-geoid: QGEOID2021. V. 1.0. GFZ Data Services. https://doi.org/10.5880/isg.2021.005

de Matos ACOC, Blitzkow D, Guimarães GN, Silva VC (2021b) The South American gravimetric GEOID: GEOID2021. V. 1.0. GFZ Data Services. https://doi.org/10.5880/isg.2021.006

Drewes H, Sánchez L (2020) Velocity model for SIRGAS 2017: VEMOS2017. Deutsches Geodätisches Forschungsinstitut der Technischen Universität München, PANGAEA. https://doi.org/10.1594/PANGAEA.912350

Drewes H (2022) Historical development of SIRGAS. J Geodetic Sci 12(1):120–130. https://doi.org/10.1515/jogs-2022-0137

Drewes H, Heidbach O (2005) Deformation of the South American crust estimated from finite element and collocation methods, International Association of Geodesy Symposia Series, vol 128. Springer, Heidelberg, pp 544–549. https://doi.org/10.1007/3-540-27432-4_6

Drewes H, Heidbach O (2012) The 2009 horizontal velocity field for South America and the Caribbean, International Association of Geodesy Symposia Series, vol 136. Springer, Heidelberg, pp 657–664. https://doi.org/10.1007/978-3-642-20338-1_81

Drewes H, Kaniuth K, Voelksen C, Alves Costa SM, Souto Fortes LP (2005) Results of the SIRGAS campaign 2000 and coordinates variations with respect to the 1995 South American geocentric reference frame, International Association of Geodesy Symposia Series, vol 128. Springer, Heidelberg, pp 32–37. https://doi.org/10.1007/3-540-27432-4_6

Drewes H, Kuglitsch F, Ádám J, Rózsa S (2016) Geodesist's handbook 2016. J Geod 90:907. https://doi.org/10.1007/s00190-016-0948-z

Griffiths J (2019) Combined orbits and clocks from IGS second reprocessing. J Geod 93:177–195. https://doi.org/10.1007/s00190-018-1149-8

Guimarães GN, Blitzkow D, Silva VC, Matos ACOC, Inoue MEB, Oliveira SL (2022a) New gravimetric infrastructure in Southeast Brazil: from absolute gravity network to a geoid model. J Surv Eng 148:3. https://doi.org/10.1061/(ASCE)SU.1943-5428.0000393

Guimarães GN, Blitzkow D, Matos ACOC, Silva VC, Inoue MEB (2022b) The establishment of the IHRF in Brazil: current situation and future perspectives. Rev Bras Cartogr 74(3):651–670. https://doi.org/10.14393/rbcv74n3-64949

Herring TA, King RW, Floyd MA, McClusky SC (2015) GLOBK: global Kalman filter VLBI and GPS analysis program, Reference Manual, Release 10.6, http://geoweb.mit.edu/gg/docs/GLOBK_Ref.pdf

Herring TA, King RW, Floyd MA, McClusky SC (2018) GAMIT: GPS Analysis at MIT, Reference Manual, Release 10.7. http://geoweb.mit.edu/gg/docs/GAMIT_Ref.pdf

Ihde J, Sánchez L, Barzaghi R, Drewes H, Foerste C, Gruber T, Liebsch G, Marti U, Pail R, Sideris M (2017) Definition and proposed realization of the International Height Reference System (IHRS). Surv Geophys 38(3):549–570. https://doi.org/10.1007/s10712-017-9409-3

Johnston G, Riddell A, Hausler G (2017) The international GNSS service. Springer handbook of global navigation satellite systems, pp 967–982. https://doi.org/10.1007/978-3-319-42928-1

Mackern MV, Mateo ML, Camisay MF, Morichetti PV (2020) Tropospheric products from high-level GNSS processing in Latin America. In: Freymueller JT, Sánchez L (eds) Beyond 100: the next century in geodesy, International Association of Geodesy Symposia, vol 152. Springer, Cham. https://doi.org/10.1007/1345_2020_121

Mackern MV, Mateo ML, Camisay MF, Rosell PA (2022) Quality control of SIRGAS ZTD products. J Geodetic Sci 12(1):42–54. https://doi.org/10.1515/jogs-2022-0136

Petit G, Luzum B (2010) IERS conventions (IERS technical note; no. 36). IERS conv. Cent. 179

Rebischung P, Schmid R (2016) IGS14/igs14.atx: a new framework for the IGS products. Fall Meeting of the American Geophysical Union, San Francisco, USA, December 2016. https://mediatum.ub.tum.de/doc/1341338/file.pdf

Reguzzoni M, Carrion D, De Gaetani CI, Albertella A, Rossi L, Sona G, Batsukh K, Toro Herrera JF, Elger K, Barzaghi R, Sansó F (2021) Open access to regional geoid models: the International Service for the Geoid. Earth Syst Sci Data 13:1653–1666. https://doi.org/10.5194/essd-13-1653-2021

Sánchez L, Drewes H (2016) Crustal deformation and surface kinematics after the 2010 earthquakes in Latin America. J Geodyn. https://doi.org/10.1016/j.jog.2016.06.005

Sánchez L, Drewes H (2020) Geodetic monitoring of the variable surface deformation in Latin America, International Association of Geodesy Symposia Series, vol 152. Springer, Heidelberg. https://doi.org/10.1007/1345_2020_91

Sánchez L, Seitz M (2011) Station positions and velocities of the SIR11P01 multi-year solution, epoch 2005.0. Deutsches Geodätisches Forschungsinstitut der Technischen Universität München, PANGAEA. https://doi.org/10.1594/PANGAEA.835098 [In supplement to: Sánchez L; Seitz M (2011): Recent activities of the IGS Regional Network Associate Analysis Centre for SIRGAS (IGS RNAAC SIR) - Report for the SIRGAS 2011 General Meeting August 8–10, 2011. Heredia, Costa Rica. DGFI Report, 87, 48 pp. hdl:10013/epic.43995.d001]

Sánchez L, Ågren J, Huang J, Wang YM, Mäkinen J, Pail R, Barzaghi R, Vergos GS, Ahlgren K, Liu Q (2021) Strategy for the realisation of the International Height Reference System (IHRS). J Geod 95:33. https://doi.org/10.1007/s00190-021-01481-0

Sánchez L, Drewes H, Kehm A, Seitz M (2022) SIRGAS reference frame analysis at DGFI-TUM. J Geodetic Sci 12(1):92–119. https://doi.org/10.1515/jogs-2022-0138

Seemüller W, Sánchez L, Seitz M (2011) The new multi-year position and velocity solution SIR09P01 of the IGS Regional Network Associate Analysis Centre (IGS RNAAC SIR), International Association of Geodesy Symposia Series, vol 136. Springer, pp 675–680. https://doi.org/10.1007/978-3-642-20338-1_110

Silva VC, Blitzkow D, Almeida FGV, Matos ACOC, Guimarães GN (2022) Computation and analysis of geopotential number in São Paulo, Brazil. Earth Sci Res J 26(2):107–118. https://doi.org/10.15446/esrj.v26n2.100645

SIRGAS Project Committee (1997) SIRGAS final report; working groups I and II IBGE, Rio de Janeiro, 96 p

Tarrío JA, Costa S, da Silva A, Inzunza J (2021) Processing guidelines for the SIRGAS analysis centres, SIRGAS Working Group I. https://doi.org/10.35588/dig.g3.2021

Tocho CN, Antokoletz ED, Piñón DA (2020) Towards the realization of the International Height Reference Frame (IHRF) in Argentina. In: Freymueller JT, Sánchez L (eds) Beyond 100: the next century in geodesy, International Association of Geodesy Symposia, vol 152. Springer. https://doi.org/10.1007/1345_2020_93

Wziontek H, Bonvalot S, Falk R, Gabalda G, Mäkinen J, Pálinkáš V, Rülke A, Vitushkin L (2021) Status of the international gravity reference system and frame. J Geod 95:7. https://doi.org/10.1007/s00190-020-01438-9

A Review of Space Geodetic Technique Seasonal Displacements Based on ITRF2020 Results

Xavier Collilieux, Zuheir Altamimi, Paul Rebischung, Maylis de La Serve, Laurent Métivier, Kristel Chanard, and Jean-Paul Boy

Abstract

The new release of the International Terrestrial Reference Frame, ITRF2020, differs from ITRF2014 by the addition of parametric functions describing annual and semi-annual displacements for every station. ITRF2020 coordinates are now described with piece-wise linear functions, occasional exponential and logarithmic functions modelling post-seismic displacements and the newly provided seasonal parameters. The paper first shortly presents the ITRF2020 seasonal parameters provided both in the Center of Mass (CM) and in the Center of Fig. (CF) frames. The station-specific seasonal displacements determined by the four space geodetic techniques (DORIS, GNSS, SLR, VLBI) are then reconstructed from the ITRF2020 results in the CF frame. The estimated seasonal signals are shown to agree generally within their uncertainties at co-location sites if a realistic noise model is considered.

Keywords

ITRF2020 · Non-tidal loading displacements · Seasonal displacements · Space geodetic techniques · Terrestrial reference frame

1 Introduction

The International Terrestrial Reference Frame (ITRF) is widely used for societal and science applications. It is composed of the coordinates of a primary network of stations that sample the Earth's surface. These coordinates are monitored by space geodetic techniques, namely Doppler Orbitography and Radiopositioning Integrated by Satellite (DORIS), Global Navigation Satellite Systems (GNSS), Satellite Laser Ranging (SLR) and Very Long Baseline

Interferometry (VLBI). The official ITRF products are computed by combining altogether the coordinates estimated from these four techniques. The homogenization of their reference frames is carried out in this process by estimating Helmert parameters (Altamimi et al. 2023). The estimation of these parameters is made possible by adding the relative position vectors of the instrument reference points at co-location sites that host several techniques, the so-called local ties. For the combination to be optimal, it is essential to monitor how coordinates of co-located stations agree at those fundamental sites.

Seasonal displacements at ITRF co-location sites have been investigated in various studies in the past, based on ITRF2008 input data (Collilieux et al. 2007; Altamimi and Collilieux 2010) or homogeneously reprocessed series (Tesmer et al. 2009). While attempts were made to evaluate the error introduced by reference frame alignment, the so-called network effect, no specific methodology was carried out to mitigate it. Indeed, in order to compute coordinates from a network of stations in a well-defined reference frame,

X. Collilieux (✉) · Z. Altamimi · P. Rebischung · M. de La Serve ·
L. Métivier · K. Chanard
Université Paris Cité, Institut de physique du globe de Paris, CNRS, IGN, Paris, France

ENSG-Géomatique, IGN, Marne-la-Vallée, France
e-mail: xavier.collilieux@ensg.eu

J.-P. Boy
Institut Terre et Environnement Strasbourg (ITES, UMR7063:
Université de Strasbourg, CNRS, ENGEES), Strasbourg, France

© The Author(s) 2023
J. T. Freymueller, L. Sánchez (eds.), *Gravity, Positioning and Reference Frames*,
International Association of Geodesy Symposia 156, https://doi.org/10.1007/1345_2023_216

it is necessary to apply a Helmert transformation which includes transformation parameters (translation, rotation and scale). The coordinate time series derived with this method suffer from periodic errors which depend on the network distribution and on the magnitude of periodic displacements in the time series (Collilieux et al. 2012). It is thus of utmost importance to mitigate these errors for the purpose of comparing seasonal displacements observed by the different space geodetic techniques.

In the scope of the ITRF2014 processing (Altamimi et al. 2016), seasonal coordinate variations were estimated for each technique but not combined. A rigorous combination was proposed by Collilieux et al. (2017, 2018) in order to express seasonal parameters in the same reference frame and thus mitigate network effect errors. The availability of 6 years of additional data in ITRF2020 is an opportunity to revisit this comparison. Indeed, a larger set of co-location sites is now available with longer and more overlapping position time series.

For the first time, seasonal coordinate variations, rigorously combined, have been included in the ITRF2020 products (Altamimi et al. 2022, 2023). However, as will be explained in Sect. 2, they cannot be used to assess inter-technique agreement at co-location sites. In this paper, we propose in Sect. 3 a method to compute technique-specific seasonal displacements in the ITRF2020 Center of Fig. (CF) frame. Then, we compare them while accounting for the time-correlated nature of the noise processes in station coordinate time series. The results are introduced in Sect. 4 and discussed in Sect. 5.

2 Input Data

2.1 ITRF2020 Seasonal Parameters

The original data used in this study are those submitted for ITRF2020 (Moreaux et al. 2023; Rebischung 2022; Hellmers et al. 2022; Pavlis and Luceri 2022) by the International Association of Geodesy (IAG) technique services: the International DORIS Service (IDS), the International GNSS Service (IGS), the International VLBI Service for Geodesy and Astrometry (IVS) and the International Laser Ranging Service (ILRS). Station coordinate time series with their full variance-covariance information have been combined during the ITRF2020 computation to estimate station positions at the reference epoch (2015.0), velocities and seasonal coordinate variations in addition to Earth Orientation Parameters (Altamimi et al. 2023). Thus, constant annual and semi-annual displacements along each component (East, North and Up) have been estimated for each station over its whole data span. The amplitudes of the cosine and sine terms at

these two frequencies are hereafter referred to as "seasonal parameters".

In the ITRF2020 computation process, the estimated seasonal parameters have been equated within co-location sites at the 0.1 mm level except where seasonal displacements of the different techniques were found to be inconsistent. In this case, they were only loosely equated as described in (Altamimi et al. 2023). This explains why different seasonal parameters have been published for certain pairs of co-located stations in ITRF2020, but also why they are generally equal. In any case, this does not mean that the coordinate residuals of the ITRF2020 combination are not free from seasonal variations.

The ITRF2020 seasonal parameters have been estimated in the Center of Mass (CM) frame as estimated from SLR data. However, the averaged station displacements in the CM frame is non-zero due to geocenter motion. Thanks to the seasonal geocenter motion model estimated by Rebischung et al. (2022), the ITRF2020 seasonal parameters could also be brought to the CF frame (Blewitt 2003). As no net translational motion exists in the CF frame, this frame is the most relevant for the seasonal displacement comparisons presented in this paper. Indeed, leaving geocenter motion included in the seasonal displacements would artificially increase their level of agreement.

2.2 Station Selection

Only stations with sufficient data span will be discussed, since short position time series are known to yield unreliable seasonal displacement estimates (Blewitt and Lavallée 2002). Thus, stations with at least 150 points for DORIS (weekly), SLR (weekly) and VLBI (daily sessions) have been investigated. With this criterion, all selected VLBI series span longer than 3 years. Moreover, the SLR station coordinates estimated before 1993.0 – without Lageos II observations – were excluded since they exhibit significantly larger scatter. This led to the complete exclusion of only one SLR station: the older Arequipa station (7907). As GNSS solutions are provided by the IGS on a daily basis, a minimum of 1000 points has been considered for GNSS stations. The selection of stations used in this study is shown in Fig. 1. It includes 180 GNSS, 121 DORIS, 45 SLR and 45 VLBI stations distributed over 111 distinct co-location sites. 20 sites include three techniques or more. A few vast co-location sites were split into two sub-sites when inter-station distances were exceeding 2 km and at least one GNSS station was available for every sub-site. Besides, remote GNSS stations (>2 km) were excluded if a closer GNSS station was available.

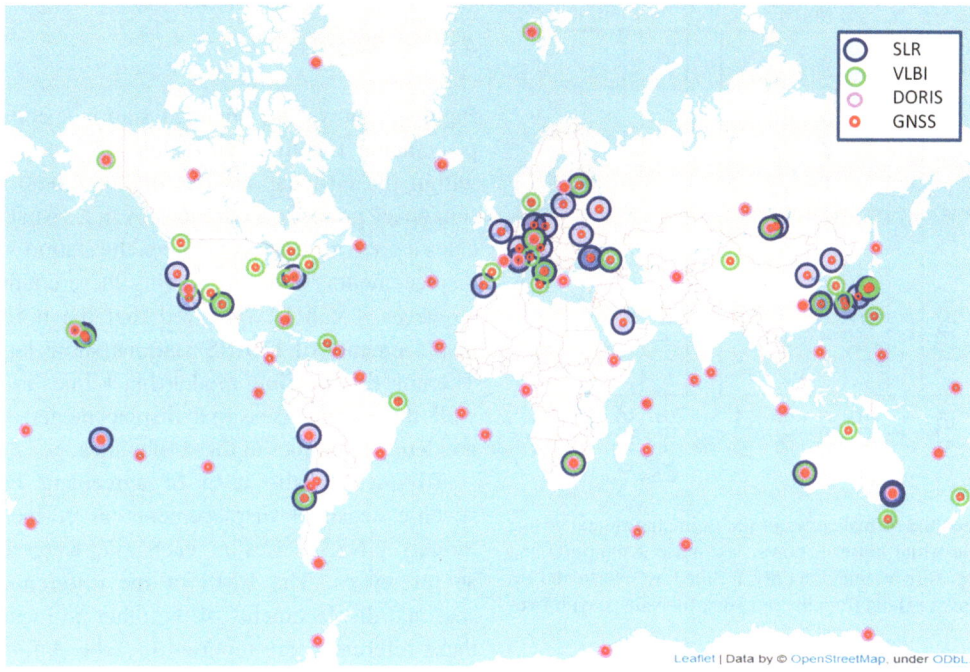

Fig. 1 Network of stations used in this study

For comparison with seasonal coordinate variations of geodetic stations, the non-tidal loading deformation model computed by Boy (2021) has been considered. It is based on ERA5 atmospheric pressure (Hersbach et al. 2020), TUGO-m induced barotropic ocean response to pressure and winds (update of Carrère and Lyard 2003), and on ERA5 soil-moisture and snow loading. Daily average displacements have been computed over the period 01/01/1994 to 01/01/2021.

3 Methodology

In order to obtain station-specific seasonal parameters expressed in a common frame from the ITRF2020 results, three steps were followed:

1. Estimation of annual and semi-annual variations in the ITRF2020 residual time series of each individual station. In this process, possible outliers may have been filtered out.
2. Addition of ITRF2020 seasonal parameters (in the CF frame) to the residual seasonal variations from step 1.
3. Re-evaluation of seasonal parameter formal errors.

The advantages of this three-step method are that it is easy to carry out, and ensures that the obtained station-specific seasonal variations are expressed in the same reference frame by benefiting from the ITRF2020 combination carried out by Altamimi et al. (2023). Indeed, the published ITRF2020 seasonal parameters are expressed in the same reference

frame and the residual seasonal signals estimated in step 1 are free from residual translation, rotation and scale components since the ITRF2020 combination model includes transformation parameters.

In order to carry out step 3, a station- and component-specific noise model, composed of variable white noise and power-law noise (Williams 2003), has been adjusted to each ITRF2020 residual coordinate time series. The variable white noise variance factor, as well as the power-law noise variance factor and spectral index were estimated by restricted maximum likelihood following Gobron et al. (2021) and de la Serve et al. (2023). The estimation model also included annual and semi-annual signals, of which the a posteriori formal errors were extracted and will be used in the following as estimates of the precision of the station-specific seasonal parameters from step 2. Correlations between cosine and sine terms are neglected in the following but are smaller than 0.1 (absolute value) in 99% of cases.

Figure 2 shows the ratio between the formal errors of the annual cosine terms under variable white noise + power-law noise assumption and under variable white noise only assumption. In average, the ratio is larger than 1.0 for the four techniques. GNSS estimated parameters are the most impacted with a median value close to 4.0 (3.9 for East and 4.2 for North and Up components) but also with significantly larger differences between stations. SLR, VLBI and DORIS formal error changes are more homogenous while there are still differences on station by station basis. The median values for the three techniques lie between 1.2 and 1.5 for the horizontal components. It is 1.9 for the SLR vertical

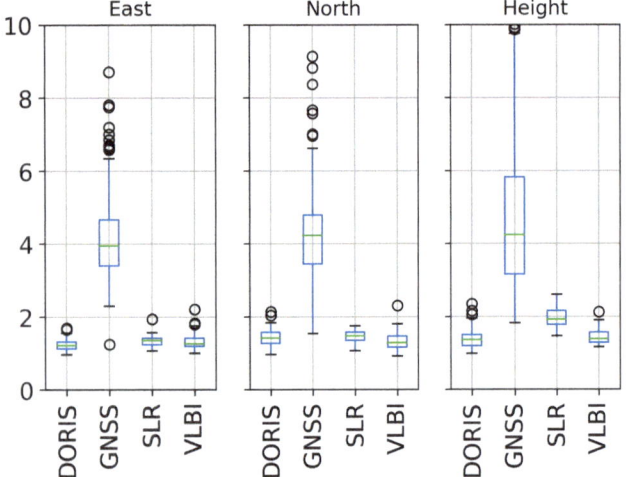

Fig. 2 Ratio between the formal errors of the estimated annual cosine terms under variable white noise + power-law noise assumption and under variable white noise assumption only, for the East, North and Up components. The boxes extend from the ratio quartile values, with lines at the medians

component against 1.4 for DORIS and VLBI. Looking at these numbers, it is clear that accounting for time-correlated noise in coordinate series analysis impacts inter-technique coordinate series comparison.

4   Results

Figure 3 shows the obtained station-specific seasonal displacements together with their 95% confidence intervals within the co-location sites of our selection that host the four space geodetic techniques. A first visual inspection indicates a good agreement between the station-specific seasonal displacements when considering these confidence intervals. However it can be observed that the horizontal seasonal displacements of DORIS stations show larger amplitudes. The non-tidal loading model (black lines in Fig. 3) matches well the vertical seasonal displacements observed by the geodetic techniques at these four sites.

To quantify the level of agreement between station-specific seasonal displacements at co-location sites, the longest GNSS series at each site were arbitrarily taken as references. The RMS of the differences between the seasonal displacements of the other co-located stations and these references are reported in Table 1 for each technique and component.

As can be observed in Table 1, the best agreement between seasonal displacements of the longest GNSS series and the other techniques is found for VLBI, especially for the horizontal components. The RMS values are smaller

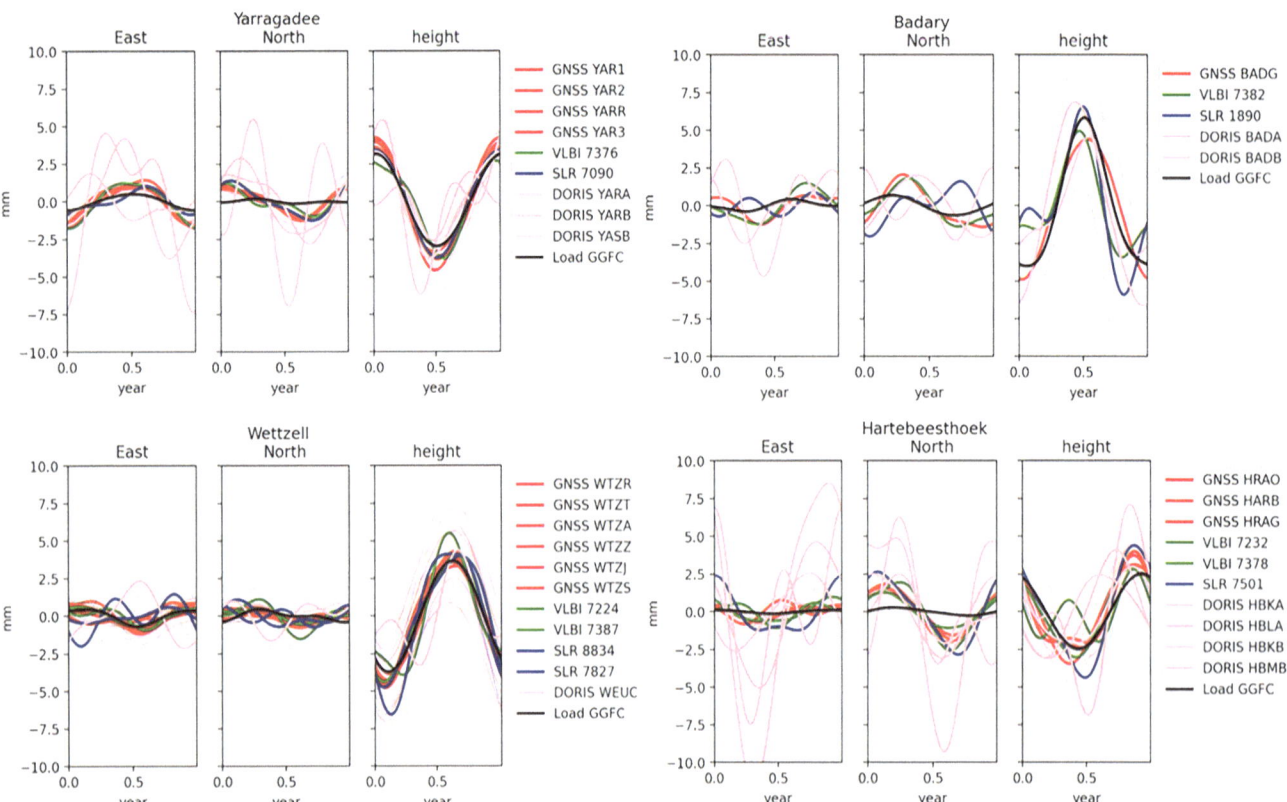

Fig. 3 Station-specific seasonal displacements (in the CF frame) within four co-location sites. The light curves represent 95% confidence intervals

Table 1 Median RMS and minimum/maximum RMS (between brackets) of the differences between (1) the seasonal displacements of the longest GNSS time series of each co-location site and (2) the seasonal displacements of other stations within the same co-location site or non-tidal loading model

	East (mm)	North (mm)	Up (mm)
DORIS	1.9 (0.4, 7.6)	1.7 (0.4, 5.9)	1.8 (0.3, 9.8)
SLR	1.0 (0.3, 9.0)	1.4 (0.3, 4.9)	1.6 (0.3, 7.3)
VLBI	0.5 (0.1, 1.2)	0.4 (0.1, 3.3)	1.4 (0.2, 5.7)
GNSS	0.3 (0.0, 2.8)	0.3 (0.0, 1.3)	0.6 (0.1, 3.3)
NT-loading	0.4 (0.1, 1.9)	0.5 (0.1, 1.6)	1.2 (0.1, 4.1)

Table 2 Percentage of "ratio statistics" larger than 3.0 at co-location sites. Between brackets: percentage of "ratio statistics" larger than 3.0 corresponding to RMS of seasonal differences larger than 2.0 mm

	East	North	Up
DORIS	11.6% (9.9%)	26.4% (15.7%)	8.3% (6.6%)
SLR	2.3% (2.3%)	0.0% (0.0%)	9.1% (6.8%)
VLBI	4.5% (0.0%)	11.1% (2.2%)	6.7% (6.7%)
GNSS	20.0% (1.5%)	15.4% (0.0%)	1.5% (0.0%)

than 2.0 mm for 72% of the SLR, 68% of the VLBI stations and 57% of the DORIS stations for the vertical. However, such RMS values do not consider the uncertainties of the estimated seasonal displacements.

To quantify the level of agreement between station-specific seasonal displacements in a more statistically meaningful way, the ratios of the maximum absolute values of the seasonal displacement differences to their formal errors have been computed. The formal errors are based on the noise models adjusted to the ITRF2020 residual time series in step 3. Figure 4 shows the distribution of these ratios, hereafter referred to as "ratio statistics", for each technique and component. Values larger than 3.0 point to

significant inconsistencies between seasonal displacements in the longest GNSS series and in other co-located series.

As reported in Table 2, more than 90% of the SLR, VLBI and DORIS vertical seasonal displacements agree with GNSS within the 3σ level. The SLR and VLBI stations with ratio statistics larger than 3.0 in vertical are respectively Changchun (7237), Fort Davis (7080), Arequipa (7403), Tidbinbilla (7843) and Shanghai (7227), Chichijima (7347), Warkworth (7377). The DORIS stations with ratio statistics larger than 3.0 are Kitab (KIUB), Krasnoyarsk (KRAB), Cibinong (CIDB), Libreville (LIBB), Palmeira (SALB), Arta observatory (DJIA, DJIB), Goldstone (GONC), Fairbanks (FAIA) and Miami (MIAB).

In the horizontal components, the DORIS seasonal displacements show higher proportions of significant discrep-

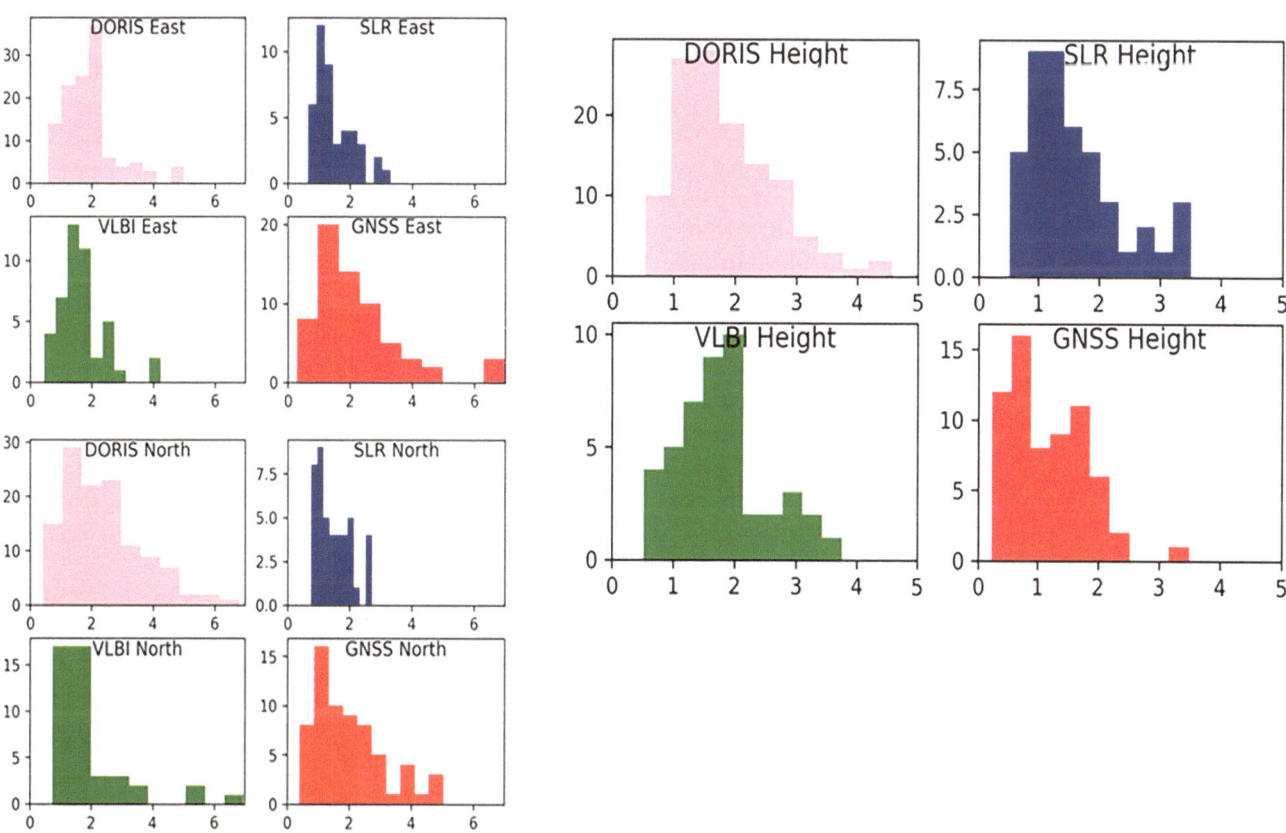

Fig. 4 Distributions, for each technique and component, of the "ratio statistics" introduced in the text

ancies with GNSS, especially along the North component. 15–20% of the selected co-located GNSS station pairs also have ratio statistics larger than 3.0, although the RMS of the corresponding seasonal displacement differences are smaller than 2.0 mm, except in Fairbanks (station pair FAIV/FAIR). The horizontal seasonal displacements of the Irkutsk SLR station (1891) and Warkworth VLBI station (7377) finally also differ by more than 2.0 mm RMS from the co-located GNSS stations.

5   Discussion

We reported an overall good agreement between technique-specific seasonal displacements, and pointed out co-location sites where the seasonal displacements sensed by the different techniques are statistically inconsistent. However, as previously mentioned, constant seasonal displacements were estimated, whereas the Earth's deformations are not strictly periodic, such as the deformations caused by non-tidal loading effects. Moreover, space geodetic data are not regularly sampled. For example, the SLR technique is weather-dependent, while VLBI observation sessions are not continuous. Besides, the observation periods of co-located stations do not necessarily overlap since new instruments can be installed after others are decommissioned. Differences between the constant seasonal displacements adjusted to the series of co-located stations with different time spans are thus expected.

We evaluated the magnitude of this "sampling effect" by performing simulations based on the non-tidal loading model introduced above. The RMS of the differences between seasonal displacements estimated from continuous loading time series over the full time interval 01/01/1994 to 01/01/2021 and seasonal displacements estimated from loading time series re-sampled at the same epochs as the space geodesy observations have been computed and reported in Table 3. The largest differences have been found for SLR observations. This could be explained by the weather dependent availability of SLR data, which causes a coordinate bias related to atmospheric loading, the so-called blue sky effect (Otsubo et al. 2004). Differences between the vertical sea-

sonal displacements are always smaller than 1.0 mm RMS except for the Mendeleevo SLR (7814) station (1.5 mm RMS). Only 13% of the SLR stations and 11% the VLBI stations show seasonal displacement differences larger than 0.5 mm RMS (6% for DORIS and 2% for GNSS). Differences between the horizontal seasonal displacements are generally smaller than 0.1 mm RMS. Overall, the evaluated magnitude of the "sampling effect" is much smaller than the differences between station-specific seasonal displacements reported in Sect. 4, see values in Table 1. This indicates that these differences are likely mainly due to errors in the space geodetic station position time series.

Only a small number of sites was found with statistically significant differences between seasonal displacements (i.e., ratio statistics larger than 3.0). Our re-evaluation of the seasonal parameter formal errors based on time-correlated noise models adjusted to the ITRF2020 residual time series likely contributes to this result. While we expect these models to provide more realistic formal errors than the white noise model used in the past by Collilieux et al. (2017, 2018), further work is needed to improve the modelling of space geodetic technique noise.

The overall good consistency of technique-specific seasonal displacements supports the choice made to equate them in the ITRF2020 combination – except in case of notable discrepancies – and ensures good confidence in the published ITRF2020 seasonal parameters. In particular, the good agreement between SLR and GNSS seasonal displacements made it possible to transfer reliably the SLR origin to the ITRF2020 seasonal parameters of GNSS stations and thus express them in the CM frame.

Acknowledgements This study contributes to the IdEx Université de Paris ANR-18-IDEX-0001. This work is partly funded by the Centre National d'Etudes Spatiales (CNES), under the TOSCA grant. All data used in this paper are available at Altamimi et al. (2022). We are grateful to the editor in chief Jeff Freymueller and to the two anonymous reviewers who commented on this manuscript.

References

Altamimi Z, Collilieux X (2010) Quality assessment of the IDS contribution to ITRF2008. Adv Space Res 45(12):1500–1509. https://doi.org/10.1016/j.asr.2010.03.010

Altamimi Z, Rebischung P, Métivier L, Collilieux X (2016) ITRF2014: a new release of the International Terrestrial Reference Frame modeling nonlinear station motions. J Geophys Res 121(B8):6109–6131. https://doi.org/10.1002/2016JB013098

Altamimi Z, Rebischung P, Collilieux X, Métivier L, Chanard K (2022) ITRF2020 [Data set]. IERS ITRS Center Hosted by IGN and IPGP. https://doi.org/10.18715/IPGP.2023.LDVIOBNL

Altamimi Z, Rebischung P, Collilieux X, Métivier L, Chanard K (2023) ITRF2020: an augmented reference frame refining the modeling of nonlinear station motions. J Geod 97:47. https://doi.org/10.1007/s00190-023-01738-w

Table 3 Median RMS, 95% quantile and maximum RMS between seasonal time series estimated from continuous load time series (Boy 2021) [a] over the time interval 01/01/1994 to 01/01/2021 and estimated from sampled time series [a] by space geodesy

	East	North	Up
DORIS	0.03, 0.08, 0.13	0.03, 0.10, 0.19	0.17, 0.51, 0.93
SLR	0.03, 0.09, 0.16	0.03, 0.10, 0.13	0.17, 0.76, 1.51
VLBI	0.03, 0.07, 0.11	0.04, 0.07, 0.09	0.20, 0.57, 0.67
GNSS	0.01, 0.05, 0.13	0.01, 0.05, 0.11	0.09, 0.31, 0.65

[a] Computed from daily load values

Blewitt G (2003) Self-consistency in reference frames, geocenter definition, and surface loading of the solid earth. J Geophys Res 108(B2).
https://doi.org/10.1029/2002JB002082

Blewitt G, Lavallée D (2002) Effect of annual signals on geodetic velocity. J Geophys Res 107:2145. https://doi.org/10.1029/2001JB000570

Boy J-P (2021) Contribution of GGFC to ITRF2020. Technical report
EOST/IPGS. http://loading.u-strasbg.fr/ITRF2020/ggfc.pdf

Carrère L, Lyard F (2003) Modeling the barotropic response of the
global ocean to atmospheric wind and pressure forcing - comparisons
with observations. Geophys Res Lett 30:1275. https://doi.org/10.
1029/2002GL016473

Collilieux X, Altamimi Z, Coulot D, Ray J, Sillard P (2007) Comparison of very long baseline interferometry, GPS, and satellite
laser ranging height residuals from ITRF2005 using spectral and
correlation methods. J Geophys Res 112(B12403). https://doi.org/
10.1029/2007JB004933

Collilieux X, van Dam T, Ray J, Coulot D, Métivier L, Altamimi
Z (2012) Strategies to mitigate aliasing of loading signals while
estimating GPS frame parameters. J Geod 86:1–14. https://doi.org/
10.1007/s00190-011-0487-6

Collilieux X, Altamimi Z, Rebischung P, Métivier L, Chanard K (2017)
Analysis of the seasonal parameters estimated in the ITRF2014
processing. Presented at IAG-IASPEI Joint Scientific Assembly,
Kobe

Collilieux X, Chanard K, Altamimi Z, Rebischung P, Métivier L,
Ray J, Coulot D (2018) Comparison of the seasonal displacement
parameters estimated in the ITRF2014 processing, what can we
learn? Presented at 42nd COSPAR Scientific Assembly, Pasadena

de La Serve M, Rebischung P, Collilieux X, Altamimi Z, Métivier
L. (2023) Are there detectable common aperiodic displacements
at ITRF co-location sites? J Geod 97:79. https://doi.org/10.1007/
s00190-023-01769-3

Gobron K, Rebischung P, Van Camp M, Demoulin A, de Viron O (2021)
Influence of aperiodic non-tidal atmospheric and oceanic loading

deformations on the stochastic properties of global GNSS vertical
land motion time series. J Geophys Res 126. https://doi.org/10.1029/
2021JB022370

Hellmers H, Modiri S, Bachmann S, Thaller D, Bloßfeld M, Seitz M,
Gipson J (2022) Combined IVS contribution to the ITRF2020. In:
International Association of Geodesy Symposia. Springer, Heidel-
berg. https://doi.org/10.1007/1345_2022_170

Hersbach H, Bell B, Berrisford P et al (2020) The ERA5 global
reanalysis. Q J R Meteorol Soc 146:1999–2049. https://doi.org/10.
1002/qj.3803

Moreaux G, Lemoine FG, Capdeville H, Otten M, Štěpánek P, Saunier
J, Ferrage P (2023) The international DORIS service contribution
to ITRF2020. Adv Space Res 72(1):65–91. https://doi.org/10.1016/
j.asr.2022.07.012

Otsubo T, Kubo-oka T, Gotoh T, Ichikawa R (2004) Atmospheric blue
sky effects on SLR station coordinates. In: Proceedings of 14th ILRS
workshop, San Fernando, pp 69–74

Pavlis EC, Luceri V (2022) The ILRS contribution to ITRF2020. Tech.
Rep. https://itrf.ign.fr/en/solutions/ITRF2020

Rebischung P (2022) The IGS contribution to ITRF2020. Tech. Rep.
https://itrf.ign.fr/en/solutions/ITRF2020

Rebischung P, Altamimi Z, Collilieux X, Métivier L, Chanard K (2022).
ITRF2020 seasonal geocenter motion model, presented at REFAG
2022, this meeting

Tesmer V, Steigenberger P, Rothacher M, Boehm J, Meisel B (2009)
Annual deformation signals from homogeneously reprocessed VLBI
and GPS height time series. J Geod 83:973–988. https://doi.org/10.
1007/s00190-009-0316-3

Williams SDP (2003) The effect of coloured noise on the uncertainties
of rates estimated from geodetic time series. J Geod 76:483–494.
https://doi.org/10.1007/s00190-002-0283-4

Validation of Reference Frame Consistency of GNSS Service Products

Lennard Huisman and Huib de Ligt

Abstract

In Global Navigation Sattelite System (GNSS) point positioning the coordinate reference frame of the positioning results is determined by the reference frame of the used GNSS service product. These products include broadcast ephemeris, precise orbits, clocks, biases, and reference station observations. Consistency in the reference frame is crucial for analyzing coordinate differences and velocities in earth science and geomatics applications. National agencies calculate coordinates for GNSS reference stations to ensure reference frame consistency within a country, however this approach is not suitable for providers covering multiple countries.

This contribution will introduce two new approaches for reference frame validation of GNSS service products and their relation with the International Association of Geodesy's Reference Frame Sub-Commission for Europe (EUREF) densification guidelines, including results of a first prototype assessing the consistency of a cross-border GNSS RTK service with the EUREF Permanent Network (EPN) reference frame ETRF2000 and consistency of a GNSS PPP service with the International GNSS Service reference frame IGb14.

Keywords

ETRF2000 · GNSS · PPP · Reference frame · RTK

1   Introduction

To obtain GNSS positions from GNSS code and phase observations, information is needed on the state of the satellites, such as the satellite positions and the clock offset to a reference time. For high precision applications, the user also needs information on error sources that affect the observations, such as signal biases and atmospheric delays.

GNSS service products provide users with the necessary information for GNSS point positioning. The GNSS service products can be divided into two groups: the State Space Representation (SSR) products and Observations Space Representation (OSR) products as described by Wübbena et al (2001, 2005) and the RTCM standard SC 10403.2 (2013). The SSR products provide information on the state of individual GNSS error sources, while the OSR products provide observations or corrections to observations that can be used by the user to eliminate common error sources of the user observations and the OSR product.

The coordinate reference frame of GNSS point positioning results is determined by the reference frame of the GNSS reference stations used to generate GNSS service products. In earth science applications and geomatics, consistency between the reference frame of point positioning results is of importance in the analysis of coordinate differences and velocities. In many countries, for example Belgium (Bruyn-

L. Huisman (✉)
Rijksdriehoeksmeting, Kadaster, Apeldoorn, The Netherlands

Department of Geoscience and Remote Sensing, Faculty of Civil Engineering and Geosciences, Delft University of Technology, Delft, The Netherlands
e-mail: lennard.huisman@kadaster.nl

H. de Ligt
Rijksdriehoeksmeting, Kadaster, Apeldoorn, The Netherlands
e-mail: huib.deligt@kadaster.nl

© The Author(s) 2023
J. T. Freymueller, L. Sánchez (eds.), *Gravity, Positioning and Reference Frames*,
International Association of Geodesy Symposia 156, https://doi.org/10.1007/1345_2023_232

inx et al 2018), Canada (Bond et al 2018), Great Britain (Edwards et al 2010) and the Netherlands (van Willigen and Salzmann 2002), the consistency of GNSS services is achieved through consistent computation of GNSS reference station coordinates by national authorities. Implicitly, these computations usually also result in a quality check of the reference station observations. In case of low quality, the station observations will be rejected in the coordinate computation.

The approach of computing reference station coordinates for each country's individual realization of the reference frame is not suitable for GNSS service providers that provide GNSS service products for multiple countries. These providers can only select one coordinate for each station when operating a cross-border network. This holds for some GNSS real-time kinematic (RTK) service providers, but especially for other GNSS point positioning techniques, such as single point positioning (SPP), precise point positioning (PPP) and PPP-RTK where a global network of reference stations is used to compute the GNSS service products. Also, the approach has some drawbacks for both providers and users, especially as the consistency of the reference frame that is provided by the GNSS service product to the end user is not validated, but only the input data. In a series of interviews, not included in this contribution, we found that users in the Netherlands actually expect that the delivered GNSS service products are validated by the national authority and not only the input coordinates of the GNSS reference stations.

This contribution introduces two new approaches for reference frame validation of GNSS service products, capable of validating the consistency of global but also regional and local GNSS service products and reference frames.

Section 2 provides a short overview of existing approaches to validate the reference frame of GNSS services. Section 3 introduces the new methods for reference frame validation of GNSS services. In Sect. 4 results are presented of a prototype implementation of the approaches for a cross-border GNSS RTK service and a GNSS PPP service. Section 5 gives an outlook on next steps.

2   Existing Validation Approaches

This section describes the approach of coordinate computation for GNSS reference stations in Sect. 2.1 and two approaches for quality control of GNSS service products that are capable of validating the reference frame the user obtains with these products. Section 2.2 describes the Physical quality control and Sect. 2.3 describes the circular quality control approach.

2.1   Coordinate Computation

In the coordinate computation approach, the coordinates of GNSS reference stations are computed using a consistent approach, for example based on the guidelines for densifications of the EUREF permanent network (EPN) in Europe (Legrand et al 2021). GNSS service providers use GNSS reference stations for which coordinates are computed; these stations can be operated by a national agency or by operators themselves. For example, in Great Britain, providers rely mainly on the GNSS reference stations provided by the Ordnance Survey, while in the Netherlands, providers operate their own stations.

In case of coordinate computation, the chain from GNSS reference stations' observations to user positions can be summarized as follows. Coordinates are computed, and possibly monitored, such as in Belgium (Bruyninx et al 2018) and Canada (Bond et al 2018), by a national agency. A GNSS service provider uses the observations and computed coordinates of the GNSS reference stations to create GNSS service products. The end user processes its own observations in combination with the service products to do point positioning for the user location. The obtained position will be in the reference frame of the GNSS service station, provided that this is handled correctly by the GNSS service provider when creating the service products. The latter is not validated in this approach.

2.2   Physical Quality Control

To validate the quality of GNSS service products, a straightforward approach is to perform point positioning at known physical points. For example, Kadaster (2003), Edwards et al (2010), Wang et al. (2010), Cina et al (2015), Sedell (2015) and NavCert (2016) use surveys at known points to assess the quality of GNSS service products. When the reference frame of the GNSS service products and the known points is the same, or the relation between the reference frames is known, such a physical quality control can also validate the reference frame of the GNSS service products. The reference frame can be validated by a comparison of the known coordinates and the point positioning results. These procedures have proven to provide good insight into the quality of service products, but cannot be easily automated and are very labour-intensive, due to a significant amount of fieldwork.

It is, of course, also possible to use permanent stations for such a quality control, as done by Janssen (2013). However, to avoid the risk that an issue with a permanent station remains undetected, this approach requires a network of stations independent of the stations used by the GNSS service provider. This is especially important for OSR based GNSS service products, as the individualized observations of the GNSS service products will be highly correlated with the reference stations' original observations.

In addition, it is also important to be aware of the source of the coordinates of the known points. For example, in the Netherlands, the coordinates of the known points used by the procedure of Kadaster (2003) and NavCert (2016) are obtained using observations and a GNSS service product from the national authority. In this case, the procedure is a comparison between two independent surveys using two different GNSS products. As a result, this procedure is not an independent validation of the GNSS service product and is also affected by the precision of the local setup of equipment at the known point of the two independent surveys.

2.3 Circular Quality Control

The European Position Determination System (EUPOS) is a collaboration between public agencies that provide GNSS services. Within EUPOS, GNSS reference products are validated by monitoring the coordinates of the GNSS reference stations using OSR GNSS service products (Droščak and Smolik 2015). The monitoring takes place in the form of 2-min sessions. The sessions vary using different distances and directions; the individualized OSR products are requested by the monitoring system for distances of 2, 11 and 20 km from the GNSS reference station in different directions with an interval of 15 degrees. Using the open-source software RTKLIB (Takasu and Yasuda 2010), the coordinates of the GNSS reference station are calculated using the observation of the OSR products. The reference frame is validated by a comparison of the known and computed coordinates.

An advantage of this approach is that permanent GNSS reference stations are used for the quality control; hence, no additional measurements need to be carried out in the field. Another advantage is that any organization can monitor the GNSS service products as long as they have access to the GNSS service products and the GNSS reference station data. This access can be made available easily by GNSS service providers, as it requires the same communication protocols as used by customers of the services. A disadvantage of the method is that in EUPOS, most of the reference stations used for checking the product are also used for creating the GNSS product. Because systematic errors in the coordinates of GNSS stations propagate in the GNSS product, such errors will not be detected by this method. In this contribution, this approach is labelled the circular quality control, as the output data is validated with the input data,

making it a circular flow of data. In the current EUPOS implementation, the distance for the monitoring is limited to 20 km from the nearest station, which means that nationwide monitoring of the GNSS services is currently not possible when the distance between reference stations is larger than 40 km.

3 New Validation Methods

Section 2 described current approaches that are used to ensure a consistent reference frame of GNSS services, as well as methods that are used to validate the reference frame of GNSS service products. These current approaches for validation have limitations, as they are either not fully independent from data that was used to generate the GNSS service product, can be labour-intensive, or are actually a relative validation compared to another GNSS service product. To overcome these limitations, we introduce two approaches for validation of the reference frame provided by the GNSS reference products: the Grid check approach, described in Sect. 3.1 and an approach that is complementary to the grid check, the Systematic quality control, described in Sect. 3.2.

3.1 Grid Check

The grid check approach was developed to validate the reference frame of OSR GNSS service products. A nationwide OSR based GNSS service can usually provide data for any location. In real-time kinematic positioning, implementations of such services are known as the Virtual Reference Station (VRS), Pseudo Reference Station (PRS) or individualized Master Auxiliary (iMAX) approaches (Wübbena et al 2005; Takac and Zelzer 2008). In the grid check, observations for a grid of virtual stations are collected, using the OSR based GNSS service, covering the area for which the service products will be validated. The coordinate computation approach described in Sect. 2.1 is then used to compute coordinates for the virtual stations. The validation consists of a comparison between the computed coordinates and the coordinates for which the virtual stations were created. This difference is expected to be zero if the reference frame of the service product is consistent with the reference frame in which the virtual stations are requested. The advantage of this approach is that the grid check can be performed independently from the reference stations used to generate the service products; no field measurements are required and the existing coordinate computation approach can be re-used, which adds to the consistency of the validation. A disadvantage of this approach is that it can only be applied to OSR based GNSS service products.

3.2 Systematic Quality Control

To overcome the limitation of only validating OSR based GNSS service products with the grid check, the complementary approach of the systematic quality control was developed. This approach introduces the concept of a grid-checked GNSS service. The grid-checked service is an OSR based GNSS service that is validated using the grid check approach and serves as a benchmark for other services. The grid-checked service generates virtual stations that can be used to validate OSR and SSR based GNSS service products, such as broadcast ephemeris, post-processing kinematic services and PPP services. In this approach, the coordinates for the virtual stations are computed using the OSR or SSR based service products. The validation consists of a comparison between the computed coordinates and the coordinates for which the virtual stations were created. When the service product is provided in a different reference frame than the grid-checked service, a coordinate transformation is required. The advantage of this approach is that it can be used for both OSR and SSR based GNSS service products at any distance from reference stations within the coverage of the grid-checked service. A disadvantage of the systematic quality is the dependency on a grid-checked GNSS service.

Actually, this concept is very similar to the concept of the physical control survey described in Sect. 2.2. The difference is that the relative control is not done by fieldwork, but the actual GNSS service products are directly used. Besides eliminating the labour-extensive fieldwork it also eliminates the uncertainty introduced by local setups and only compares GNSS service providers, which is the actual goal of the validation for both the physical and the systematic quality control. The validation of OSR based service products is also very similar to the validation done by the circular quality control described in Sect. 2.3.

4 Results

This section gives results on the implementation of the grid check and systematic quality control. The implementation is based on a prototype.

4.1 Grid Check for a Cross-Border GNSS Service Provider

This section shows the result of the grid check of the GNSS service product of a GNSS service provider. The service is based on reference stations in Belgium, France and the Netherlands. The reference frame for the GNSS service product is ETRF2000 at epoch 2010.0. Data for the grid

of virtual stations were collected from real-time streams and converted to RINEX files. As the data collection was limited to 500 parallel streams to limit the load on the GNSS service, the resulting grid resolution was 35 km. To avoid extrapolation outside the coverage of the active GNSS service, no grid points were used with a distance of more than 50 km to the nearest reference station. A European digital elevation model was used to obtain heights for the virtual stations in the grid at ground level. The coordinate computation was done following EUREF guidelines for EPN densifications (Legrand et al 2021) using the Bernese 5.2 software (Dach et al 2015). Reference station selection was done using the EPN densification online tool. Figure 1 shows the horizontal differences between the computed coordinates and the coordinates for which virtual stations were requested from the service. Table 1 gives metrics for the differences.

In general the validation shows coordinate differences at the millimetre level. In the northwest of Belgium, an outlier is visible of several centimetre in the southwest direction. This outlier actually identified an error in the used coordinates of the nearest reference station of the same magnitude. This example shows how the grid check is able to identify such errors, that directly affect the users point positioning in the same magnitude. In the south of France, systematic errors are visible. The reason for these effects is under investigation. One hypothesis is that station coordinates are incorrect in this region; however, the effect covers an area with multiple reference stations. Another hypothesis is that the atmospheric modelling in the mountain areas of the Pyrenees and Alps regions is less precise than in areas with smaller height differences.

4.2 Systematic Quality Check for OSR and SSR Based GNSS Service Products

This section shows the result of the systematic quality control of real-time and post-processing GNSS service products within the Netherlands. Following the concept introduced in Sect. 3.2, the Netherlands Positioning Service (NETPOS), the nationwide network RTK service operated by the Dutch Kadaster (NSGI 2022), was used as the grid-checked service. To confirm the reference frame of NETPOS, the grid check was performed on a grid of 25 km×25 km. Section 4.2.1 shows results for a real-time OSR service and Sect. 4.2.2 shows results for two post-processing SSR services.

4.2.1 Real-Time OSR Service

To validate the reference frame of a real-time OSR service, a real-time virtual user was created using NETPOS. RTKLIB was used with the observations of the virtual user and the individualized OSR products of the service to compute

Fig. 1 Horizontal coordinate differences for the grid check of a cross border OSR based GNSS service product

Table 1 Metrics for the validation of ETRF2000 coordinates for a cross-border GNSS service provider using the Grid check approach

	North [mm]	East [mm]	Up [mm]
RMS	8	7	17
95-percentile	16	13	37
99-percentile	25	18	51

coordinates for the virtual user. The coordinate computation was done in sessions of 10 min. The coordinate computation and validation are very similar to the computation done by Droščak and Smolik (2015) in the circular quality control approach described in Sect. 2.3 as our prototype also uses RTKLIB. To minimize the load on the GNSS service provider, a maximum of four simultaneous sessions was run. Each session would randomly select a location for the virtual user from the grid of 25 km×25 km mentioned in Sect. 4.2. The systematic quality control was run for one month resulting in approximately 160 sessions per grid point. The number of sessions differs per point as the points were randomly selected in each session. The validation was done by comparing the computed coordinates of the virtual user obtained with the GNSS service product, with the known coordinates for which the virtual user was generated. Figure 2a shows the results for the validation, metrics for the results are included in Table 2.

The results for the systematic quality control of the real-time OSR service show that the average coordinate difference for the sessions on a single point is below 7.5 mm in the horizontal component and below 15 mm in height, within the borders of the Netherlands. The standard deviation is less than 2 mm for the horizontal components and 7.1 mm for the height. Given the specification of 20 mm in planimetry and 30 mm in height for this service, users can rely on this service to provide them coordinates in the national reference system within the specifications of the service.

4.2.2 Post-Processing SSR Services

To show that the systematic quality control can also be applied to post-processing services, the Canadian NRCAN CSRS-PPP service (NRCAN 2022) and the Trimble RTX post-processing service (TrimbleRTX 2022) were validated using the NETPOS grid-checked network. These online services allow users to upload a RINEX file and will deliver an extensive report on the processing results including coordinates. In this case, both services provided coordinates in the ITRF2014 reference frame at the epoch of observation. For this validation, a single session of 24 h of observations for virtual users on the

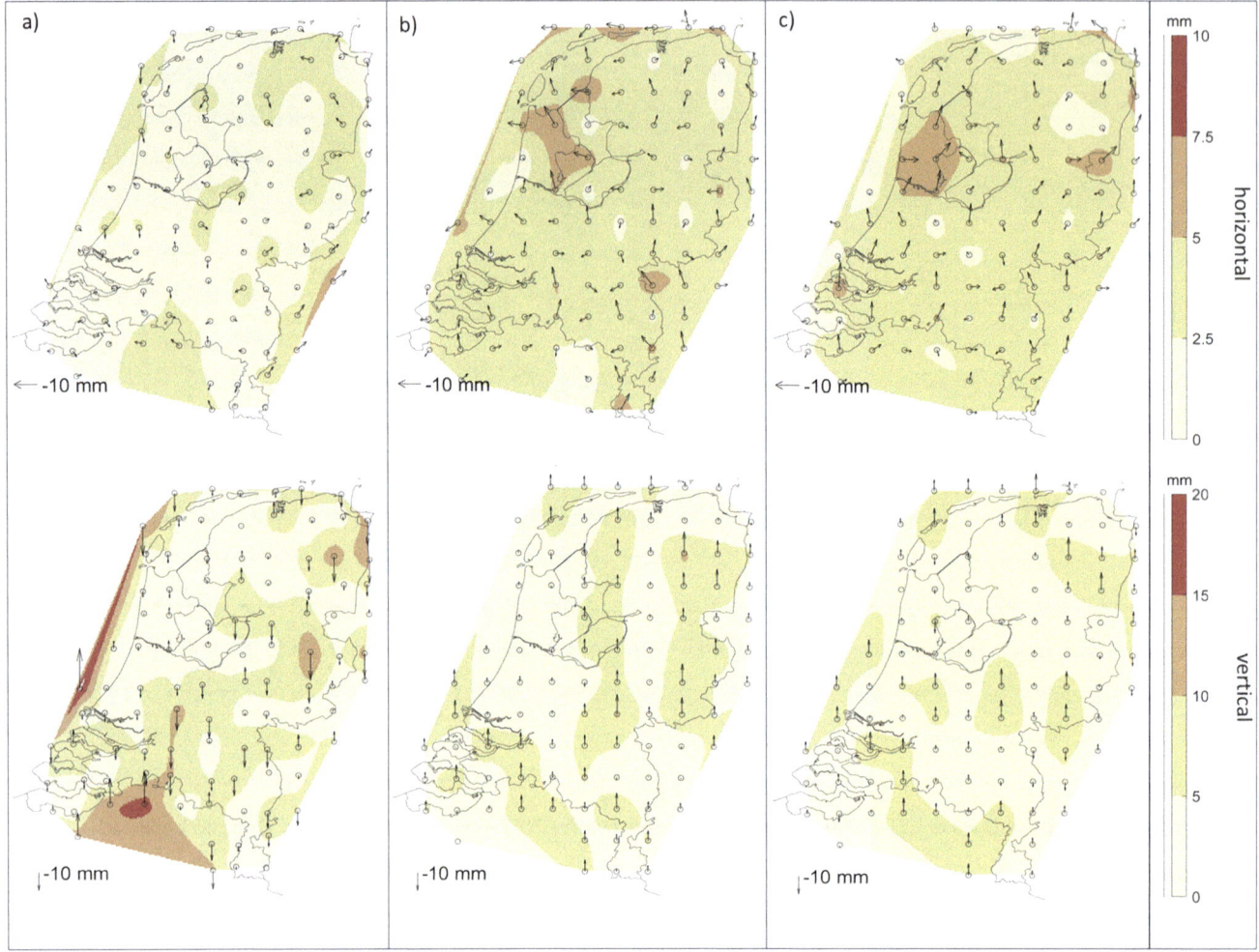

Fig. 2 Coordinate differences for the systematic quality control of three different GNSS service products. Top row shows the horizontal differences, bottom row shows the vertical differences

Table 2 RMS error for the validation of coordinates for OSR and SSR GNSS service products using the Systematic quality control approach

	North [mm]	East [mm]	Up [mm]
Real-time OSR service	1.6	1.9	7.1
Post-processing Trimble RTX	2.8	2.6	4.6
Post-processing NRCAN PPP	2.6	3.0	5.0

same grid of 25 km x 25 km was used for the grid check of NETPOS. The coordinates were validated by comparing the ITRF2014 coordinates obtained for the post-processing services and the known coordinates of the virtual stations. Figure 2b and 2c show the results for the validation, metrics for the results are included in Table 2.

The results for the systematic quality control of the post-processing services show that the services have a good agreement with each-other and the ITRF2014 reference frame at the epoch of observations. The average coordinate difference for the sessions on a single point is below 7.5 mm in the horizontal component and below 10 mm in height. The standard deviation of the differences is less than 3 mm for the horizontal components and about 5 mm for the height. The results show that for these services, users can rely on this service to provide them coordinates in the specified reference system at the centimetre level.

5   Outlook

To ensure the consistency of the reference frames of GNSS service products, current approaches have limitations. These limitations are that the approaches either do not validate the end-user product (coordinate computation), are not fully independent from data that was used to generate the GNSS service product (circular quality control), can be labour-intensive or are actually a relative validation compared to another GNSS service product (physical quality control). To overcome these limitations we introduced two approaches, the grid check and the complementary systematic quality control, for validation of the reference frame provided by the GNSS reference products. Implementation of the approaches in a prototype show that the approaches can be applied to both OSR and SSR based GNSS service products to validate the reference frame.

In this contribution, the systemic quality control approach was evaluated using an independent grid-checked service as a benchmark. In this approach, the GNSS service of the national agency was validated with the grid check and served as the grid-checked service for the systematic quality control. Other services, in this case commercial and public (open) services, were then validated using the virtual OSR based observation generated with this grid-checked network. It can be argued that the dependency on a grid-checked service is a disadvantage of the approach, as it requires an independent infrastructure. We think this disadvantage is relatively small when the national agency already operates a national GNSS service, but are also investigating other approaches. For example, when it is assumed that the majority of the GNSS services are providing the correct reference frame, the approach could be to perform a relative comparison between GNSS services to be validated.

Currently, there are no standards or guidelines for the validation of GNSS service products. Each national agency has its own implementation and interpretation of existing approaches. The described approach can serve as a standardized methodology and it is shown that it can be applied cross-border and for different types of GNSS service products. We are seeking collaboration with other national agencies and service providers to further develop the method. Scientific challenges are the optimum density of the grid and the development of a robust positioning algorithms that can handle different types of GNSS service products.

References

Bond J, Donahue B, Craymer M, Banham G (2018) Nrcan's compliance program for high accuracy, gnss services: ensuring compatibility with the canadian spatial reference system. Geomatica **72**(4), 101–111. https://doi.org/10.1139/geomat-2019-0001

Bruyninx C, Bergeot N, Chatzinikos M, Chevalier J-M, Fabian A, Legrand J, Pottiaux E, Voet P, Doncker FD (2018) EUREF 2018 National Report of Belgium. http://www.euref.eu/symposia/2018Amsterdam/05-03-p-Belgium.pdf

Cina A, Dabove P, Manzino AM, Piras M (2015) Network real time kinematic (NRTK) positioning – description, architectures and performances. Satellite Positioning - Methods, Models and Applications, pp 23–45. https://doi.org/10.5772/59083

Dach R, Lutz S, Walser P, Fridez P (eds) (2015) Bernese GNSS Software Version 5.2. Astronomical Institute, University of Bern, Switzerland

Droščak B, Smolik K (2015) Monitoring tool for EUPOS countries network RTK quality. http://www.euref.eu/symposia/2015Leipzig/05-06-Droscak.pdf

Edwards SJ, Clarke PJ, Penna NT, Goebell S (2010) An examination of network rtk gps services in great britain. Sur Rev **42**(316), 107–121. https://doi.org/10.1179/003962610X12572516251529

Janssen V (2013) Investigation of virtual rinex data quality. In: International Global Navigation Satellite Systems Society Symposium 2013, pp 1–11

Kadaster (2003) RDNAP kwaliteitsmeting van real time GPS-diensten. Kadaster

Legrand J, Bruyninx C, Altamimi Z, Caporali A, Kenyeres A, Lidberg M (2021) Guidelines for EUREF densifications. Royal Observatory of Belgium (ROB). https://doi.org/10.24414/ROB-EUREF-Guidelines-DENS

NavCert (2016) NavCert Certification Marks. Online: http://www.navcert.de/1/certification/certification-marks/

NSGI (2022) Netherlands Positioning Service. [Online: Visited 13 February 2023]. https://www.netpos.nl

NRCAN (2022) NRCAN CSRS-PPP post-processing service. [Online: Visited 13 February 2023]. https://webapp.csrs-scrs.nrcan-rncan.gc.ca/geod/tools-outils/ppp.php

RTCM (2013) RTCM STANDARD 10403.2 for Differential GNSS. Radio Technical Commision for Maritime Services, Arlington, Virginia, USA

Sedell D (2015) Network-RTK: A comparative study of service providers currently active in Sweden. School of Architecture and the Built Environment, Royal Institute of Technology (KTH), Stockholm, Sweden

Takac F, Zelzer O (2008) The relationship between network RTK solutions MAC, VRS, PRS, FKP and i-max. In: Proceedings of ION GNSS, pp 348–355

Takasu T, Yasuda A (2010) Kalman-filter-based integer ambiguity resolution strategy for long-baseline RTK with ionosphere and troposphere estimation. In: Proceedings of the 23rd International Technical Meeting of the Satellite Division of the Institute of Navigation (ION GNSS 2010), pp 161–171

TrimbleRTX (2022) Trimble RTX post-processing. [Online: Visited 13 February 2023]. https://www.trimblertx.com/

van Willigen CW, Salzmann MA (2002) Certificering van gps-referentiestations. Geodesia **44**(1), 4–8

Wang C, Feng Y, Higgins M, Cowie B (2010) Assessment of commercial network rtk user positioning performance over long inter-station distances. J Global Position Syst **9**(1), 78–89

Wübbena G, Bagge A, Schmitz M (2001) Rtk networks based on geo++ gnsmart-concepts, implementation, results. In: Proceedings of ION GPS 2001, pp 11–14

Wübbena G, Schmitz M, Bagge A (2005) Ppp-rtk: precise point positioning using state-space representation in rtk networks. In: Proceedings of ION GNSS 2005, vol 5, pp 13–16

Intra-Technique Combination of VLBI Intensives and Rapid Data to Improve the Temporal Regularity and Continuity of the UT1-UTC Series

Lisa Klemm, Daniela Thaller, Claudia Flohrer, Anastasiia Walenta, Dieter Ullrich, and Hendrik Hellmers

Abstract

The difference UT1-UTC is the most variable quantity among the Earth Orientation Parameters (EOP) with significant unpredictable variation. It can be measured only with the quasi-space-fixed technique VLBI. The IVS organizes two different VLBI observation campaigns: The bi-weekly 24-hour Rapid campaigns and the daily 1-hour Intensive sessions. As a result, two independent UT1-UTC time series are estimated and published as official IVS EOP-S and EOP-I products. These have different strengths and weaknesses in terms of continuity and accuracy, but both are characterized by irregular temporal resolution. We present the current activities of BKG towards a combined processing of VLBI Intensive and Rapid data in one common adjustment. In this way, we unify the strengths of both sessions and generate a UT1-UTC time series characterized by a daily, continuous and temporally regular resolution, e.g., at 12:00 UTC. We achieved a significant improvement in accuracy of 35% lower WRMS values compared to the regular session-wise Intensive-only solution. By using a continuous EOP parameterization, the accuracy is almost at a constant level and less dependent on the irregularity of the VLBI observation period. The processing is based on homogenized, datum-free NEQs which allow a rigorous combination on the normal equation level instead of the observation level. Based on the improved combination method, we intend to set up a new operational VLBI EOP product at BKG. Its characteristics make it suitable as an input for EOP prediction algorithms.

Keywords

Combination · dUT1 · EOP · VLBI

1 Introduction

The Earth Orientation Parameters (EOP) describe the rotational part of the transformation between the Celestial Reference Frame (CRF) and the Terrestrial Reference Frame (TRF). They are represented by five components: the pole coordinates x_p and y_p, the celestial pole offsets δX and δY, and the difference dUT1 between Universal Time UT1 and Coordinated Universal Time UTC (Thaller 2008; Bloßfeld 2015). The knowledge of accurate EOP plays an important role for various applications. This includes, for example, precise positioning and satellite navigation, precise orbit determination and Earth system monitoring, e.g. climate change studies (Gambis and Luzum 2011). The time component dUT1 is the most variable component among the EOP. It is dominated by significant and unpredictable variations which can only be measured with the quasi-space-fixed space-geodetic technique Very Long Baselines Inteferometry (VLBI) (Dermanis and Mueller 1978; Artz et al 2011; Thaller 2008). Since the VLBI observation and correlation process is not fully automated, continuous 24-hour observations are not yet possible (Nothnagel et al 2017).

L. Klemm (✉) · D. Thaller · C. Flohrer · A. Walenta · D. Ullrich · H. Hellmers
Federal Agency for Cartography and Geodesy (BKG), Frankfurt am Main, Germany
e-mail: Lisa.Klemm@bkg.bund.de

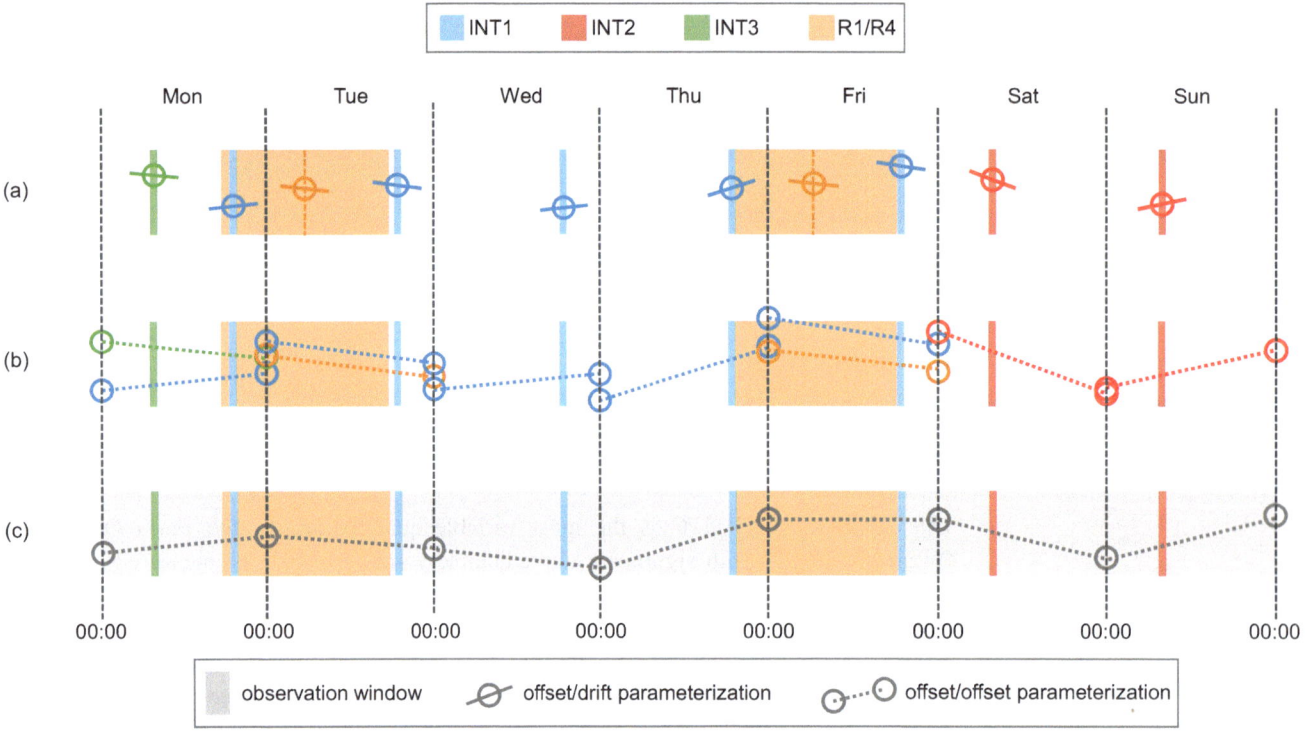

Fig. 1 Weekly distribution of the regular VLBI legacy sessions used for combination and their EOP parameterization: (**a**) initial offset/drift parameterization of input NEQs, (**b**) transformed piece-wise linear offset parameterization with two offsets per day, (**c**) seven-day piece-wise linear offset parameterization after combination

Therefore, the International VLBI Service for Geodesy and Astrometry (IVS) organizes two different geodetic VLBI session types with a limited observation time and a subset of radio telescopes. The 24-hour IVS-Rapid (RAP) observation campaigns are conducted every Monday (R1) and Thursday (R4) with an observation period of 24 hours. With a global network of up to ten antennas, these sessions are suitable for determining all five EOP components, but they have a rather long latency of up to two weeks until the final products in form of SINEX (Solution INdependent EXchange Format) files are available (Nothnagel et al 2017). The IVS additionally organizes daily single- (or triple-) baseline sessions with a large east-west extension and one hour duration. The so-called Intensives (INT) sessions are suitable only for the daily monitoring of the highly variable dUT1 (Robertson et al 1985; Leek 2015). The latency of the dUT1 estimates based on the VLBI INT sessions is about two days or less (Nothnagel et al 2017). Figure 1a illustrates the weekly session distribution of both VLBI observation campaigns.

At this stage, the two VLBI observation campaigns are analyzed separately and two independent EOP time series are estimated and published as official IVS EOP products[1]:

- The Session EOP product (EOP-S) is a series of EOP results, one for each 24-hour geodetic session. The esti-

mates are characterized by a high degree of accuracy. However, the temporal resolution is not daily and irregular.

- The Intensive EOP product (EOP-I) is characterized by a daily but not regular temporal resolution, since it is referred to the mid-session epochs which vary from one Intensive to the next one. The high accuracy is limited to the observation interval of one hour per day.

The objective of our study is the development of a method in which data from both VLBI campaigns are combined in a common adjustment. In this way, we aim to generate a dUT1 time series characterized by a daily, continuous, and temporally regular resolution, with estimates e.g. at 12:00 UTC. By combining the VLBI data of the last seven consecutive days and estimating EOP as continuous piece-wise linear polynomials, we use the past days' information to stabilize the estimated parameters and minimize random deviations. We expect a higher level of accuracy of the dUT1 time series, which becomes less dependent on the duration and timing of the observations. The use of a regular parameter epoch facilitates the comparison of the dUT1 time series of other techniques with the VLBI series. Furthermore, the EOP series can be used as input for EOP prediction algorithms, which usually require a time series with a daily and regular resolution.

[1]https://ivscc.gsfc.nasa.gov/products-data/products.html.

For this purpose, we combine the data of the VLBI legacy S/X campaigns available within approx. two weeks (i.e. INT 1/2/3 and R1/R4) on the normal equation (NEQ) level. The combination at the NEQ level represents the second most rigorous combination method. It can be considered as a good approximation of the combination at the observation level, being the most rigorous combination approach, if the modeling and parameterization of common parameter have been handled in the same way for each input NEQ system (Thaller 2008; Seitz 2009; Schmid 2009; Rothacher et al 2011; Bloßfeld 2015). The solution is more consistent than the typical parameter-level combination approach typically used to generate the official IERS EOP products (Luzum et al 2001; Bizouard et al 2019). The advantages of combining at NEQ level rather than at the parameter level are, for example, that the correlations between parameters are taken into account and it is ensured that the same underlying reference frame is used.

2 Data Input and Combination Methodology

For the combination, we used NEQs in the period between 2014 and 2022 of the BKG IVS Analysis Center (AC), which were provided as SINEX files at the BKG IVS Data Center (DC) (Engelhardt et al 2021). The VLBI RAP NEQs contained all components of the EOP and their temporal derivatives (except for the drift parameter of the celestial pole), station coordinates, and radio source coordinates. The EOP were parameterized at the mid-session epochs. The parameterization of the VLBI INT NEQs was the same, except for the pre-eliminated celestial pole offsets and radio source coordinates. The weekly distribution of the respective sessions and their initial EOP parameterization is shown in Fig. 1a. In preparation for the combination, the NEQs first underwent a parameter transformation step, as depicted in Fig. 1b. The parameterization of the EOP was transformed from the offset/drift representation at mid-session epochs to a piece-wise linear representation consisting of two offsets at 00:00 and 24:00 UTC. A linear epoch transformation was used for the polar motion components. For the time component dUT1 and its negative temporal derivative the Length-of-Day (LOD), it should be noted that the short-time tidal variations were first subtracted according to the IERS Conventions before the reduced parameters $UT1_R$ and LOD_R can finally be linearly transformed (Petit and Luzum 2010; Bloßfeld 2015). Since the 24-hour RAP campaigns usually start on Mondays and Thursdays at 17:00 and 18:30 UTC, respectively, the observation window of these sessions contains two consecutive days. The default parameterization of the official IVS products does not allow to separate the observations of both days in post-processing. Therefore, the

Table 1 Summary of the a priori values and the respective constraints used for the combined solutions

		a priori	7d-INT	7d-INT+RAP
EOP	dUT1	IERS 14 C04	None	None
	x_p, y_p	IERS 14 C04	Fixed	Loose
	$\delta X, \delta Y$	IERS 14 C04	Fixed	None
Station coordinates	X, Y, Z	ITRF14	Fixed	Minimum
Source coordinates	α, δ	ICRF3	Fixed[a]	Fixed

[a] Parameter group is pre-eliminated

second day with the majority of observations was chosen for converting the EOP from mid-session epochs to two offsets at 00:00 and 24:00 UTC (See Fig. 1b). In addition, the a priori values of the parameters were transformed to consistent reference series. For this purpose, we used the official products of the IERS, i.e. the *IERS 14 C04* for the EOP (Bizouard et al 2019) and the *ITRF2014* time series for the station coordinates (Altamimi et al 2016) (See Table 1). If no ITRF coordinates were available, we use the coordinates listed in the SINEX files as parameter a priori values.

In the next step, the homogenized constraint-free NEQs of seven consecutive days were stacked to one NEQ system. As shown in Fig. 1c, the EOP were finally parameterized as continuous 7-day piece-wise linear polynomials with offsets every 24 hours at the day boundaries.

In this study, we compared two different combination approaches:

- **7d-INT:** The combination of the VLBI INT data of seven consecutive days with a latency of about two days.
- **7d-INT+RAP:** The combination of VLBI INT and VLBI R1/R4 data of seven consecutive days with a latency of about two weeks.

The analysis of both session types was performed with the same software using identical parameterization and background models. Therefore, no rescaling of NEQs was required before combination. We generated an EOP time series by using a sliding window approach. The window was shifted by one day over the daily NEQs which were combined into a 7-day NEQ system. This procedure was iterated over the entire series of daily NEQs.

For solving the datum-free session-wise and multi-day INT-only NEQs (1d-INT, 7d-INT) all EOP parameters except dUT1 had to be fixed to their a priori values (See Table 1). However, the advantage of the multi-day solution 7d-INT in comparison to the sessions-wise solution 1d-INT is that no constraining on the LOD parameter is required, since relative drift information is obtained by stacking consecutive sessions. Due to the sparse network all station coordinates were fixed to their a priori values (ITRF14). The radio source coordinates were pre-eliminated before generating the corresponding SINEX file and fixed to the a priori values (ICRF3) (Charlot et al 2020). The 24-hour VLBI sessions are appropriate for the determination of all five EOP, so no additional conditions were required. For the combined intra-technique solution 7d-INT+RAP, the pole coordinate information of the 24-hour sessions within the continuous polynomial is not sufficient to estimate high-quality pole coordinates over the entire seven-day period. At this point, it was necessary to apply supporting loose constraints with a threshold of 0.1 mas. This constraints stabilizes the pole coordinate estimates based on INT data only. For the session-wise solution 1d-RAP as well as for the intra-technique combined multi-day solution 7d-INT+RAP no-net-rotation (NNR) and no-net-translation (NNT) conditions were applied on the a priori coordinates of a subset of well-defined and stable VLBI stations. The radio source coordinates are fixed to their a priori values (ICRF3). After applying the constraints, summarized in Table 1, all these datum-free NEQ systems could be solved for the parameters to be estimated.

For the combination processing we use the *Combination and Solution* package of the *DGFI Orbit and Geodetic parameter estimation Software* (DOGS-CS), developed and maintained at DGFI-TUM (Deutsches Geodätisches Forschungsinstitut, Technische Universität München) (Gerstl et al 2004).

3　Resulting UT1-UTC Series

Dealing with multi-day time series allows to compare estimated values from the center of a multi-day session window with EOP values from the boundaries. Therefore, for each 7-day solution, we generated seven subseries by extracting the estimates at 00:00 and 24:00 UTC of the same day d from each multi-day solution. The analysis day d ranges from 0 to -6 and represents the analyzed day within the polynomial, where day $d = 0$ is the rightmost and day $d = -6$ the leftmost day on the time axis. For validation purposes, the dUT1 estimates at 00:00 and 24:00 UTC were interpolated to noon epochs, i.e., 12:00 UTC. We analyzed the Weighted Root Mean Square (WRMS) of the residuals of the estimated dUT1 values w.r.t. the *IERS Bulletin A* series, interpolated at the same validation epochs (Luzum and Gambis 2014). The weighting factors for the WRMS were the reciprocal values of the individual dUT1 variance.

The WRMS values of the dUT1 differences at 12:00 UTC of both combination approaches 7d-INT and 7d-INT+RAP are summarized in Table 2. For validation purposes, we had additionally listed the WRMS level of the session-wise VLBI-INT solution 1d-INT as well. In the following, we will not go into more detail about the results of the session-wise and multi-day INT-only solutions (1d-INT, 7d-INT), as they are covered in an other publication. These can be found in Lengert et al (2022).

Table 2 Comparison of the different dUT1 solution types w.r.t. *IERS Bulletin A*. The WRMS of the differences computed at 12:00 UTC epochs in µs

Analysis day d	−6	−5	−4	−3	−2	−1	0
1d-INT	–	–	–	–	–	–	23.2
7d-INT	20.8	19.2	18.0	17.3	17.8	19.3	19.6
7d-INT+RAP	15.7	15.7	15.2	15.1	15.0	15.5	15.8

In summary, improvement in all WRMS values can be achieved when INT data and 24-hour VLBI data are combined with continuously parameterized EOP over 7-days. (See Table 2). The largest reduction in WRMS values of more than 8 µs compared to the session-wise INT-only (1d-INT) solution is obtained for the three middle days of the polynomial $d = -4$ to $d = -2$. As expected, the addition of the 24-hour VLBI data stabilized the estimates over the entire polynomial and significantly flattened the symmetric, parabolic behavior of the WRMS of the 7d-INT solution with minima on the middle day. The WRMS values of the 7d-INT+RAP solution ranged at nearly equal level (from 15.0 µs to 15.8 µs), with the lowest values for the middle days. This corresponds to an improvement in WRMS values compared to seven-day INT-only (7d-INT) solution of 3.8 µs and 5.1 µs for the boundary days $d = 0$ and $d = -6$, respectively. By using a continuous EOP parameterization, the accuracy was almost constant and less dependent on the irregularity of the VLBI observation periods.

4　Conclusion and Outlook

This paper presents the combined processing of VLBI INT and RAP data in a joint adjustment. The aim is to combine the strengths of both session types to estimate a dUT1 time series characterized by a daily, continuous and temporally regular resolution. We achieved a significant improvement in accuracy as evidenced by 35% lower WRMS values of the dUT1 residuals compared to the regular session-wise INT solution. By using a continuous EOP parameterization, the accuracy was almost at a constant level and less dependent on the irregularity of the VLBI observation period. The combination processing was based on homogenized, datum-free NEQs provided via SINEX files from the BKG IVS-AC, which allowed a combination on the NEQ level. We used the DOGS-CS software, developed and maintained at DGFI-TUM. Based on the improved combination method, we intend to set up a new operational VLBI EOP product at BKG, whose characteristics facilitates the comparability of different dUT1 time series with the VLBI series. In addition, the new VLBI EOP series is suitable as input data for EOP prediction algorithms.

There are still some challenges for the future work. The current datum definition of a combined 7-day solution assumes that the antennas of the INT session network are also included in the station network of the 24-hour sessions. This is usually the case, but there are some exceptions. If one or even two INT stations are not included in the 24-hour session, the short observation time of usually one hour may not be sufficient to estimate stable station coordinates. This has a direct effect on the accuracy and stability of the estimated EOP. In the next step we will investigate systematically the affected sessions and improve our datum definition.

In order to estimate a daily, continuous and regular dUT1 series, the daily and rapid availability of input data, especially of VLBI INT sessions, is a mandatory requirement. The series of the daily SINEX files of the legacy (S/X) VLBI INT campaigns has some gaps in the past. The reasons are manifold and can be found throughout the entire VLBI processing chain, i.e., from observation to analysis. However, in the last two years, an increasing number of VGOS INT campaigns has been conducted in addition to the legacy (S/X) INT sessions. As a result, the INT series is nowadays almost without gaps and there are even more than one INT sessions available per day. In the near future, we plan to extend the VLBI intra-technique combination by adding the new VGOS INT data.

References

Altamimi Z, Rebischung P, Métivier L, Collilieux X (2016) ITRF2014: A new release of the International Terrestrial Reference Frame modeling nonlinear station motions. J Geophys Res Solid Earth 12(1):6109–6131. https://doi.org/10.1002/2016JB013098

Artz T, Bernhard L, Nothnagel A, Steigenberger P, Tesmer S (2011) Methodology for the combination of sub-daily Earth rotation from GPS and VLBI observations. J Geodesy 86:221–239. https://doi.org/10.1007/S00190-011-0512-9

Bizouard C, Lambert S, Gattano C, Becker O, Richard JY (2019) The IERS EOP 14C04 solution for Earth orientation parameters consistent with ITRF 2014. J Geodesy 93:621–633. https://doi.org/10.1007/s00190-018-1186-3

Bloßfeld M (2015) The key role of Satellite Laser Ranging towards the integrated estimation of geometry, rotation and gravitational field of the Earth. PhD thesis, Dissertation der Ingenieurfakultät Bau Geo Umwelt der Technischen Universität München

Charlot P, Jacobs CS, Gordon D, Lambert S, Witt AD, Böhm J, Fey AL, Heinkelmann R, Skurikhina E, Titov O, Arias EF, Bolotin S, Bourda G, Ma C, Malkin Z, Nothnagel A, Mayer D, Macmillan DS, Nilsson T, Gaume R (2020) The third realization of the International Celestial Reference Frame by very long baseline interferometry. Astron Astrophys 644:159. https://doi.org/10.1051/0004-6361/202038368

Dermanis A, Mueller II (1978) Earth rotation and network geometry optimization for very long baseline interferometers. Bull Geodesique 52:131–158. https://doi.org/10.1007/BF02521695

Engelhardt G, Girdiuk A, Goltz M, Ullrich D (2021) BKG VLBI analysis center. International VLBI service for geodesy and astrometry 2019+2020 Biennial report, edited by D. Behrend, K. L. Armstrong, and K. D. Baver, NASA/TP-2020-219041 (2021)

Gambis D, Luzum B (2011) Earth rotation monitoring, UT1 determination and prediction. Metrologia 48(4):165. https://doi.org/10.1088/0026-1394/48/4/S06

Gerstl M, Kelm R, Müller H, Ehrnsperger W (2004) DOGS-CS Kombination und Lösung großer Gleichungssysteme. Deutsches Geodätisches Forschungsinstitut (DGFI)

Leek J (2015) The application of impact factors to scheduling VLBI Intensive sessions with twin telescopes. PhD thesis, Rheinische Friedrich-Wilhelms-Universität Bonn. https://hdl.handle.net/20.500.11811/6226

Lengert L, Thaller D, Flohrer C, Hellmers H, Girdiuk A (2022) On the improvement of combined EOP series by adding 24-h VLBI sessions to VLBI intensives and GNSS data, pp 1–8. Springer, Berlin, Heidelberg. https://doi.org/10.1007/13452022175

Luzum B, Gambis D (2014) Explanatory supplement to IERS bulleting A and bulletin B/ C04. ftp://hpiers.obspm.fr/iers/bul/bulbnew/bulletinb.pdf

Luzum BJ, Ray JR, Carter MS, Josties FJ (2001) Recent improvements to IERS Bulletin A combination and prediction. GPS Solut 4(3):34–40

Nothnagel A, Artz T, Behrend D, Malkin Z (2017) International VLBI Service for Geodesy and Astrometry: Delivering high-quality products and embarking on observations of the next generation. J Geodesy 91:711–721. https://doi.org/10.1007/s00190-016-0950-5

Petit G, Luzum B (eds) (2010) IERS technical note, no. 36. International Earth Rotation and Reference Systems Service, Central Bureau. Verlag des Bundesamts für Kartographie und Geodäsie, Frankfurt am Main. http://www.iers.org/TN36

Robertson DS, Carter WE, Campbell J, Schuh II (1985) Daily Earth rotation determinations from IRIS very long baseline interferometry. Nature 316(6027):424–427

Rothacher M, Angermann D, Artz T, Bosch W, Drewes H, Gerstl M, Kelm R, König D, König R, Meisel B, Müller H, Nothnagel A, Panafidina N, Richter B, Rudenko S, Schwegmann W, Seitz M, Steigenberger P, Tesmer S, Tesmer V, Thaller D (2011) GGOS-D: Homogeneous reprocessing and rigorous combination of space geodetic observations. J Geodesy 85:679–705. https://doi.org/10.1007/s00190-011-0475-x

Schmid R (2009) Zur Kombination von VLBI und GNSS. PhD thesis, Dissertation der Fakultät für Bauingenieur- und Vermessungswesen der Technischen Universität München

Seitz M (2009) Kombination geodätischer Raumbeobachtungsverfahren zur Realisierung eines terrestrischen Referenzsystems. PhD thesis, Dissertation der Fakultät für Forst-, Geo- und Hydrowissenschaften der Technischen Universität Dresden

Thaller D (2008) Inter-technique combination based on homogeneous normal equation systems including station coordinates, Earth orientation and troposphere parameters. PhD thesis, Scientific Technical Report STR 08/15, Deutsches GeoForschungsZentrum. https://doi.org/10.2312/GFZ.b103-08153

Automatic Determination of the SLR Reference Point at Côte d'Azur Multi-Technique Geodetic Observatory

Julien Barnéoud, Clément Courde, Jacques Beilin, Madec Germerie-Guizouarn, Damien Pesce, Maurin Vidal, Xavier Collilieux, and Nicolas Maurice

Abstract

The Satellite Laser Ranging (SLR) station known as GRSM-7845 in the International Laser Ranging Service (ILRS) is hosted by the Observatoire de la Côte d'Azur (OCA) located in Caussols, France. Its reference point is the intersection of the telescope axes, which is supposed to be static. Measuring devices and a data processing chain were set up to automatically determine this point, more quickly and accurately than traditional local survey. In order to use an indirect approach (circular fitting), circular and motorized prisms were fixed on the station to be always visible during the telescope rotation. A software package was developed to control the telescope, the dome and the total station motions for fully automatic measurements. In addition to providing an easy determination of the cross-axis for local ties, this system will allow to study the potential motion of the telescope's axes intersection throughout the year.

Keywords

Automation · Instrumentation · Local tie · Metrology · SLR

1 Introduction

The *Observatoire de la Côte d'Azur* (OCA) hosts permanent geodetic stations in Grasse area ("Grasse co-location site", Caussols, France). The relative positions of the reference points of these instruments, hereafter named local tie vectors, are essential for the International Terrestrial Reference Frame (ITRF) construction and should be known at one millimeter accuracy (Altamimi et al. 2017; Poyard et al. 2017). More specifically, the satellite laser ranging (SLR) station GRSM-7845 which belongs to the International Laser Ranging Service (ILRS) network performs daily distance measurements. It is one of the few telescopes in the world capable of laser ranging on the Moon (Lunar Laser Ranging) (Chabé et al. 2020). Its reference point is the intersection of the telescope axes.

Currently, local tie vectors are determined once a year during a multi-technique local survey (Pesce 2013; Poyard 2009). However, this is a time-consuming operation during which the telescope cannot perform satellite measurements. Moreover, it requires specific metrology accessories and trained surveyors. Thus, this paper describes measuring devices and data-processing chain set up to automatically determine the reference point of the SLR station.

J. Barnéoud (✉) · X. Collilieux
Université de Paris, Institut de Physique du globe de Paris, CNRS, IGN, Paris, France

ENSG-Géomatique, IGN, Marne-la-Vallée, France
e-mail: julien.barneoud@ign.fr

C. Courde · M. Vidal · N. Maurice
Université Côte d'Azur, CNRS, Observatoire de la Côte d'Azur, IRD, Géoazur, Caussols, France

J. Beilin
ENSG-Géomatique, IGN, Marne-la-Vallée, France

M. Germerie-Guizouarn · D. Pesce
Institut National de l'Information Géographique et Forestière (IGN), Saint-Mandé, France

© The Author(s) 2023
J. T. Freymueller, L. Sánchez (eds.), *Gravity, Positioning and Reference Frames*,
International Association of Geodesy Symposia 156, https://doi.org/10.1007/1345_2023_223

2 Measuring Devices

2.1 Methodology

The SLR reference point is determined by an indirect approach (Dawson et al. 2007). For this, reflector targets are fixed on the SLR telescope. During the telescope rotation, these reflectors draw circle arcs. They are shot at different telescope angles by a motorized total station, located on the roof of the building, about forty meters away from the SLR station (Fig. 1). Moreover, several circular prisms are set up all around the SLR station, on concrete pillars and buildings. The normal vector to the circle plane that passes through the center of the circle defines the axis of rotation. By performing these measurements for both elevation and azimuth axes, the SLR reference point can be determined as the axe's intersection (Fig. 2). Physically, the axes do not necessarily intersect: this sub-millimeter distance is called "axis offset". The reference point is therefore defined as the orthogonal projection of the horizontal axis (elevation axis) onto the vertical axis (azimuth axis).

Measuring devices were developed to achieve automatic measurements, namely pendular prisms and a motorized corner cube. The four steps of field measurements are the following:

1. Retrieve meteorological data (to apply corrections later in post-processing).
2. Shoot all reference prisms shown in Fig. 1.
3. Shoot the pendular prism in several telescope elevation positions. For this, the telescope is positioned in front of the total station, see Sect. 2.2.
4. Shoot the motorized corner cube in several telescope azimuthal positions. The telescope is positioned at 90 degrees of elevation during these sessions, see Sect. 2.3.

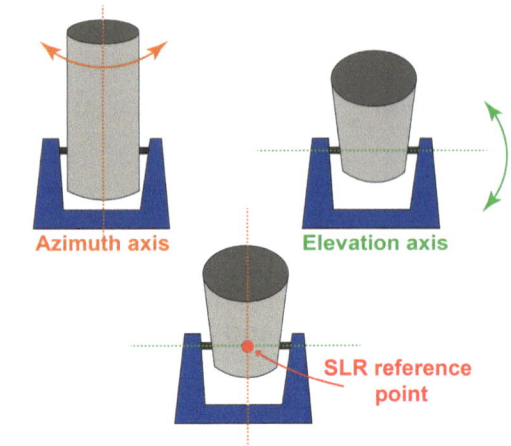

Fig. 2 SLR reference point defined as the intersection of elevation and azimuth axes

Measurements with the total station are performed on face left and face right, with intermediate closures to ensure validity of measurements.

2.2 Elevation Axis Measuring Devices

Several prisms are placed on telescope to determine elevation circles (Fig. 3). They have counterweights to always face the total station. At the beginning of the measurements, the telescope is positioned at an elevation of 5 degrees facing the total station. The total station shoots the pendular prisms in the two circles every 5 degrees of telescope elevation, up to the limit of 90 degrees (example of three steps in Fig. 4).

For a faster center determination, just one pendular prism can be targeted (i.e. one circle of elevation). In this case, the elevation axis is defined as the normal vector to the circle plane that passes through the center of this single circle. To

Fig. 1 Overview of the Grasse co-location site configuration during a local tie survey

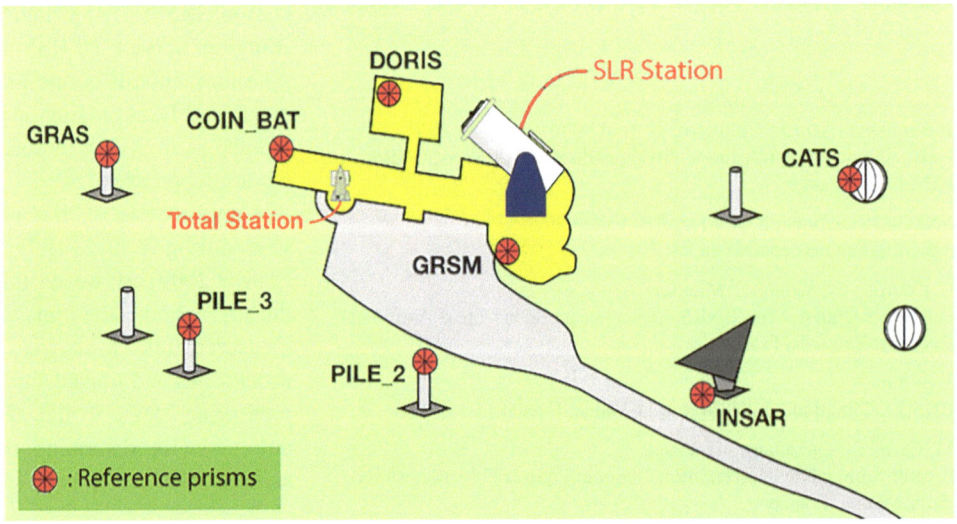

Fig. 3 Setup of a pendular prism on the telescope (left) and global view during rotation along the elevation axis (right)

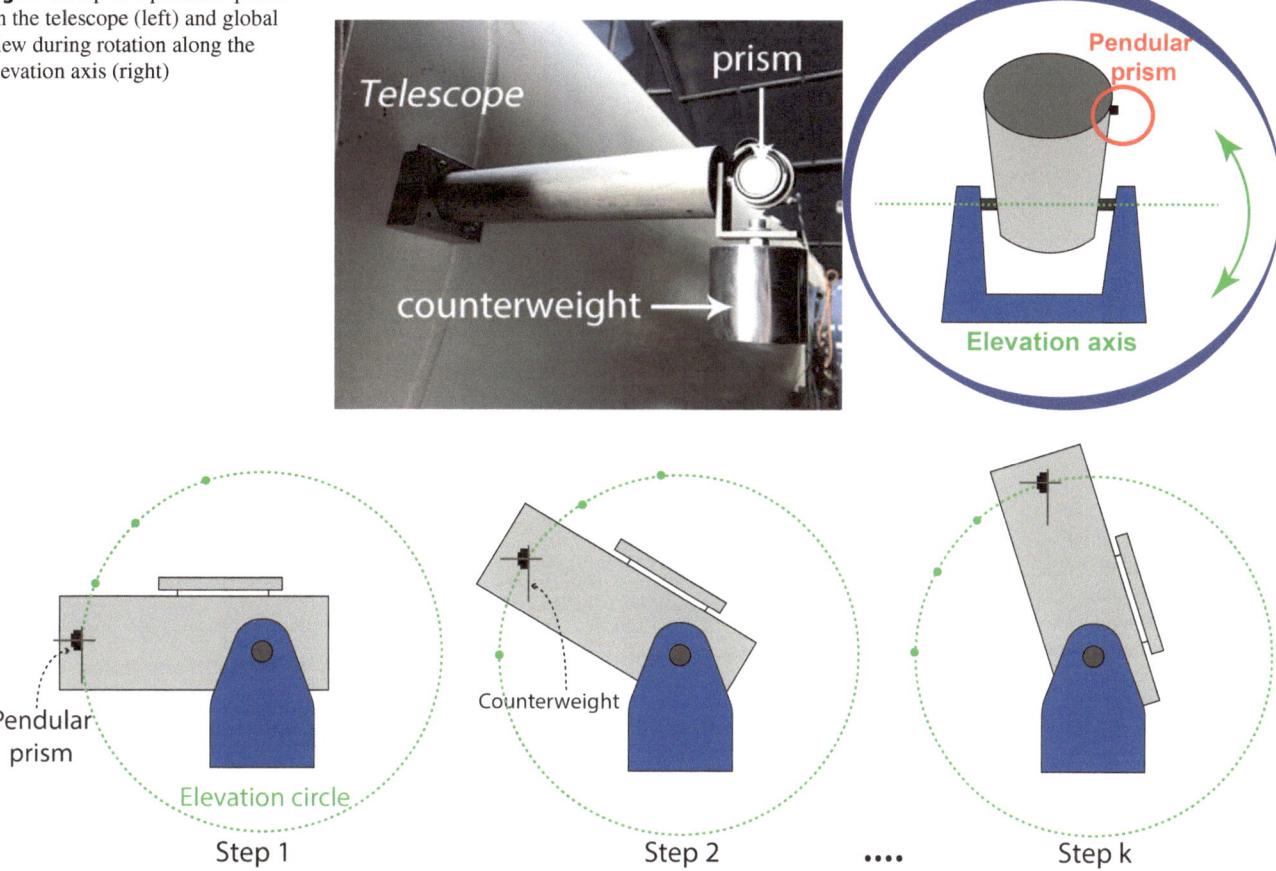

Fig. 4 Steps of elevation circle measurements. Thanks to counterweights, the pendular prism always faces the total station

improve the accuracy, up to three pendular prisms can be used (i.e. three circles determined). In this other case, the elevation axis passes through the center point of each circle.

2.3 Azimuth Axis Measuring Devices

When the telescope moves around its azimuth axis, a motorized corner cube is used as target. Indeed, this prism must rotate with telescope to be always visible from the total station: it is mounted on a stepper motor and driven by an Arduino microcontroller (Fig. 5). This device is fixed on a steel disc, located on the telescope fork. Figure 6 shows the device in four azimuthal positions of the telescope. Motor steps are automatically computed according to the number of points chosen by the user: it is adaptable, depending on the time available for measurement session or the required precision. A minimum of ten points is used to describe the circle.

Initially the prism is facing the telescope center thanks to an initialization sensor. This is the reference angle from which the motor steps will be calculated by trigonometry, to be directed toward the total station. As for the measurements of the elevation axis, points cannot be equally distributed on the azimuth circle since the telescope masks the prims when it is located behind (hidden area in Fig. 6).

3 Automation and First Measurement Tests

Now that we understand the principles of measuring axes and circles, it is important to coordinate the movements of the telescope, of its dome (to avoid masking prisms), of the total station and of the stepper motor. The *MeOCenter* software was developed to monitor all of them (Fig. 7).

From measurements to processing, it provides a complete determination of the SLR reference coordinates. In the software interface, the user can choose to determine a single axis or to have a complete determination of the center (azimuth and elevation axes). The total station and the stepper motor are respectively driven by a Raspberry Pi and Arduino microcontrollers. Arduino communication is provided by serial port via the Firmata protocol whereas the MeOCenter software communicates with the telescope, the dome and the total station via Sockets through the OCA computer network.

Fig. 5 Arduino circuit and setup (left) and global view during rotation along the elevation axis (right)

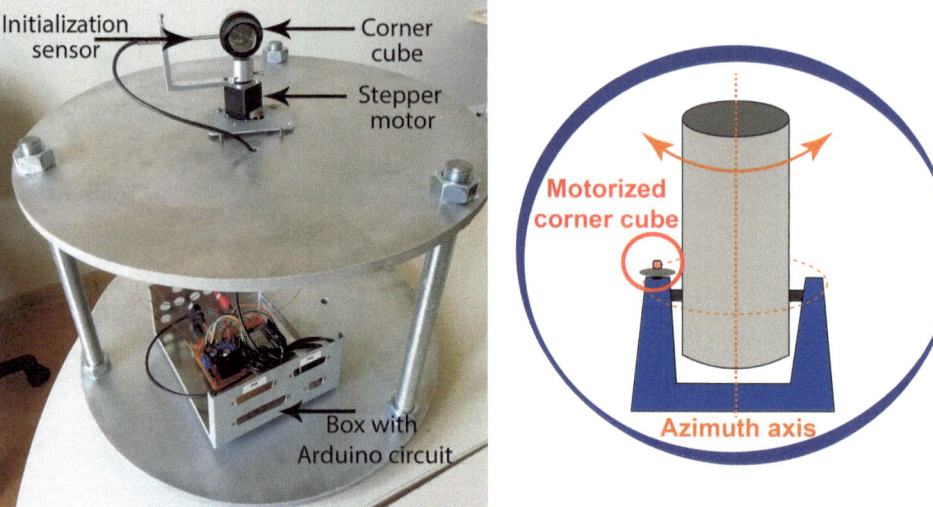

Fig. 6 Steps of azimuth circle measurements. With stepper motor, the corner cube is always visible from the total station after telescope rotation. The telescope is pointed upwards, in the zenith direction (top view, four steps shown)

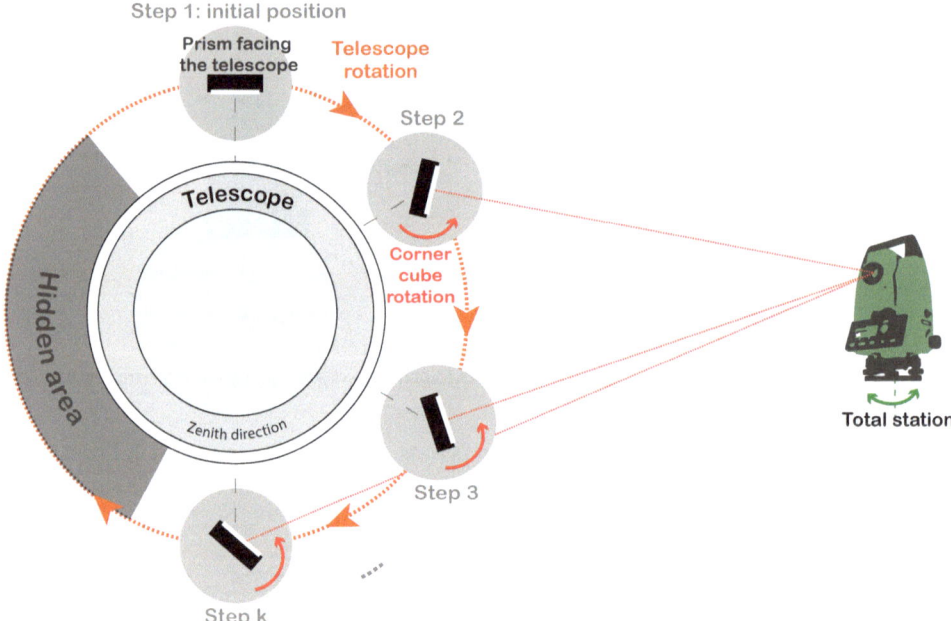

At the end of the measurement process, the angle and distance data provided by the total station are formatted in a text file. They are sent to a Linux server where the Comp3d5 software is installed, in order to calculate the point coordinates in a local projected coordinate system. Comp3D, developed by IGN-France, is a micro-geodesy compensation software that implements a global 3D least-squares adjustment of several topometric observation types (Pesce 2013). The coordinates computed are sent back to the *MeOCenter* client by Sockets. Then the parameters of elevation and azimuth circle axes are determined by least squares circular regressions of these points. Finally, the SLR reference point is estimated by the orthogonal projection of the elevation axis onto the azimuth.

Since the implementation of the automatic method on GRSM-7845 station, several test sessions have been per-

formed. As shown in Fig. 8, the four determinations agree at the sub-millimeter level. Now, measurements should be continued to monitor the reference point position with the change of seasons or after a maintenance operation of the telescope.

4   Conclusion

This study aimed to automatically determine the SLR reference point at the Grasse co-location site. The project combines mechanics, electronics and IT developments. Thanks to the developed devices and software package, a continuous monitoring of the telescope reference point is made possible. Nevertheless, a total station has to remain permanently on

Fig. 7 The different components of the MeOCenter software package: communication with telescope, cupola, total station and measuring devices (by sockets, serial)

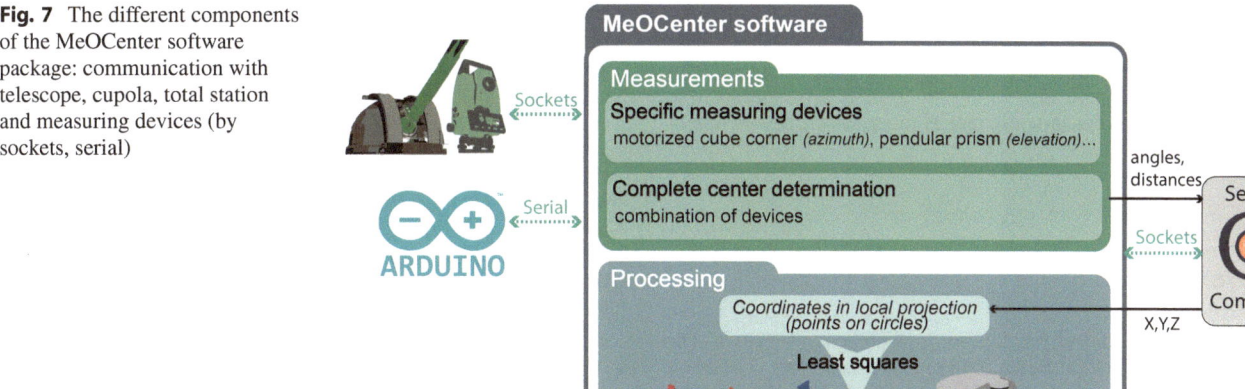

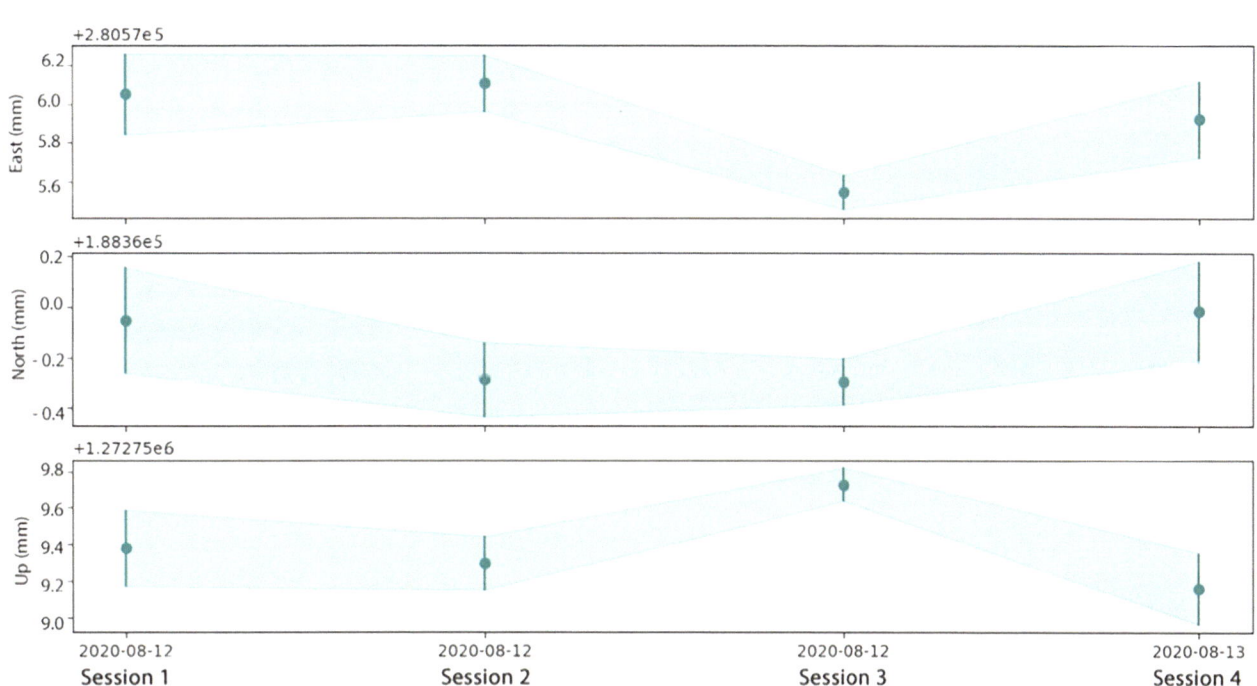

Fig. 8 Computed SLR reference point coordinates provided in a local coordinate system during different measurement sessions

the site to continue the measurements, which is not yet the case. However, this setup has been used during the last yearly local tie survey carried out in April 2021. More regular measurements are necessary to verify the SLR reference point position throughout the year, especially at seasonal time scale. As a perspective, several reflectors may be added to the GNSS and DORIS antennas of the Grasse site (see Figure 22 in Poyard et al. (2017)). Thus, it could be possible to perform an entire automated re-measurement of the local tie network. More generally, this system on the SLR station could be adapted to a VLBI telescope and set up at another co-location site.

Acknowledgments This study contributes to the IdEx Université de Paris ANR-18-IDEX-0001. A special thanks to the OCA team for their advice during the installation of measuring instruments and testings on the telescope. Thanks also to the group of ENSG students who worked on the design of in-house targets to enable the automation of the measurements.

References

Altamimi Z, Rebischung P, Métivier L, Collilieux X (2017) The International Terrestrial Reference Frame: lessons from ITRF2014. Rendiconti Lincei - Scienze Fisiche e Nat urali. https://doi.org/10.1007/s12210-017-0660-9

Chabé J, Courde C, Torre JM, Bouquillon S, Bourgoin A, Aimar M., et al. (2020). Recent progress in lunar laser ranging at Grasse laser ranging station Earth Space Sci 7(3):e2019EA000785

Dawson J, Sarti P, Johnston GM, Vittuari L (2007) Indirect approach to invariant point determination for SLR and VLBI systems: an assessment. J Geodesy 81:433–441

Pesce D (2013) ITRF Co-location Survey Observatoire de la Côte d'Azur Plateau de Calern (Grasse), France, https://itrf.ign.fr/docs/local-ties/reports/CR279_V1_PESCE_ITRFcolocationSurveyCalern.pdf

Poyard JC (2009) GRASSE ITRF co-location survey, IGN Service de Géodésie et Nivellement, https://itrf.ign.fr/docs/local-ties/reports/RT88_V1_POYARD_GrasseITRFColocationSurvey_ex.pdf

Poyard JC, Collilieux X, Muller JM, Garayt B, Saunier J (2017) IERS technical note 39 - IGN best practice for surveying instrument reference points at ITRF co-location sites, Frankfurt am Main: Verlag des Bundesamts für Kartographie und Geodäsie (ISBN 978-3-6482-129-5)

The K-Band (24 GHz) Celestial Reference Frame Determined from Very Long Baseline Interferometry Sessions Conducted Over the Past 20 Years

Hana Krásná, David Gordon, Aletha de Witt, and Christopher S. Jacobs

Abstract

The third realization of the International Celestial Reference Frame (ICRF3) was adopted in August 2018 and includes positions of extragalactic objects at three frequencies: 8.4 GHz, 24 GHz, and 32 GHz. In this paper, we present celestial reference frames estimated from Very Long Baseline Interferometry measurements at K-band (24 GHz) including data until June 2022. The data set starts in May 2002 and currently consists of more than 120 24h observing sessions performed over the past 20 years. Since the publication of ICRF3, the additional observations of the sources during the last four years allow maintenance of the celestial reference frame and more than 200 additional radio sources ensure an expansion of the frame. A study of the presented solutions is carried out helping us to understand systematic differences between the astrometric catalogs and moving us towards a better next ICRF solution. We compare K-band solutions (VIE-K-2022b and USNO-K-2022July05) computed by two analysts with two independent software packages (VieVS and Calc/Solve) and describe the differences in the solution strategy. We assess the systematic differences using vector spherical harmonics and describe the reasons for the most prominent ones.

Keywords

Celestial reference frame · K-band · Very long baseline interferometry

1   Introduction

The current International Celestial Reference Frame (ICRF3; Charlot et al. 2020) is the third realization of the International Celestial Reference System adopted by the International Astronomical Union (IAU) in August 2018. The ICRF3 is the first multi-wavelength radio frame since it contains positions of active galactic nuclei (AGN) observed with Very Long Baseline Interferometry (VLBI) at 2.3 and 8.4 GHz (S/X-band), 24 GHz (K-band), and 8.4 and 32 GHz (X/Ka-band). The three components differ as shown by several statistical indicators (e.g., data span, number of sources, coordinate uncertainty, error ellipse) and each of them faces different challenges. In 2018 IAU Resolution B2, "On The Third Realization of the International Celestial Reference Frame," (ICRF3 working group 2018) recommended that appropriate measures should be taken to both maintain and improve ICRF3. In response, this paper concentrates on the two main challenges in improving the accuracy of the celestial reference frame observed at K-band (K-CRF) which are (1) observations at a single frequency requiring an external ionospheric calibration and (2) the lack of a uniform global terrestrial network causing a non-optimal observation

H. Krásná (✉)
Department of Geodesy and Geoinformation, Technische Universität Wien, Vienna, Austria
e-mail: hana.krasna@tuwien.ac.at

D. Gordon
United States Naval Observatory, Washington, DC, USA

A. de Witt
South African Radio Astronomy Observatory, Cape Town, South Africa

C. S. Jacobs
Jet Propulsion Laboratory, California Institute of Technology, Pasadena, CA, USA

J. T. Freymueller, L. Sánchez (eds.), *Gravity, Positioning and Reference Frames*,
International Association of Geodesy Symposia 156, https://doi.org/10.1007/1345_2023_209

geometry. Our main goal is to assess systematic differences in the K-CRF solutions which are computed at two VLBI analysis centers: at TU Wien with VLBI software package VieVS (Böhm et al. 2018) and at the United States Naval Observatory (USNO) with Calc/Solve. We also compare these two frames to the ICRF3 using vector spherical harmonics (VSH) which provides information about systematic differences between pairs of astrometric catalogs and we investigate the possible reasons for the estimated differences.

2 Data and Solution Setup

2.1 Data Description

The celestial reference frames introduced in this paper are computed from $1.96 \cdot 10^6$ group delays observed at K-band in the VLBI sessions listed in Table 1. This data set was acquired mainly with the Very Long Baseline Array (VLBA) starting in May 2002 and it is available in the National Radio Astronomy Observatory Archive[1]. The first sessions belong to programs carried out by Lanyi et al. (2010) and Petrov et al. (2011). All sessions up to May 2018 are part of the current ICRF at K-band, ICRF3-K. The VLBA (Napier 1995), because its sites are limited to U.S. territory, does not allow observations of sources with declinations below $-46°$. Therefore, southern K-band sessions (KS) were organized starting in May 2014. The vast majority of southern observations are from single baseline sessions between the HartRAO 26m (South Africa) and the Hobart 26m (Tasmania, Australia) with the exception of one session involving the Tianma 65m (near Shanghai, China) and four sessions augmented with the Tidbinbilla 70m telescope (near Canberra, Australia). Of all the sources, 913 were observed in VLBA sessions, 328 were observed in southern hemisphere sessions, and 206 were observed in both types between $-46°$ and $+39°$ declination.

Table 1 Overview of sessions included in our solutions listed with recording rate

Time span	Session code	Data rate [Mbps]
Northern (VLBA) sessions		
05/2002–12/2008	BR079a-c, BL115a-c, BL122a-d, BL151a-b	128
06/2006–10/2006	BP125a-c	256
12/2015–10/2019	BJ083a-d, UD001a-x, UD009a-o	2048
11/2019–06/2022	UD009p-z, UD009aa-ah, UD015a-l	4096
Southern sessions		
05/2014–07/2016	KS1401, KS1601	1024
11/2016–02/2021	KS1603, KS1702-KS2102	2048

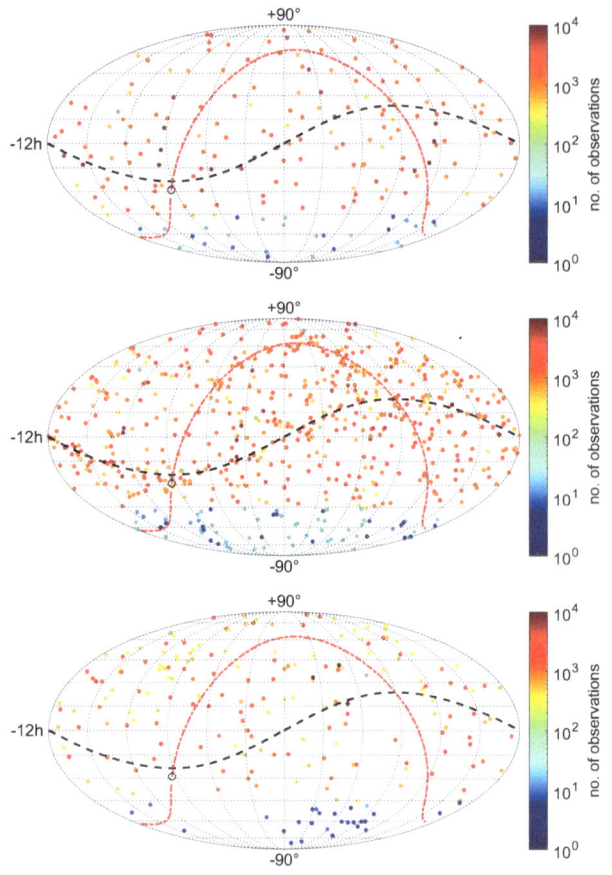

Fig. 1 Number of observations after the ICRF3-K data cutoff in May 2018 until June 2022. The sources are divided into three groups: ICRF3-K defining sources (top), ICRF3-K non-defining sources (middle), and sources not included in ICRF3-K (bottom)

In Fig. 1 we show the number of observations conducted after the ICRF3 K-band data cutoff on 5 May 2018 until June 2022 divided into three groups: (a) observations to ICRF3-K defining sources, (b) observations to ICRF3-K non-defining sources and (c) observations to sources which are not included in ICRF3-K. The consequence of using mainly the VLBA network for the K-band observations is the lack of observations of the deep south sources which is currently amplified by the technical problems of Hobart26 since March 2021. The low number of new observations (under 100) of the deep south sources since the ICRF3 release is seen in all three plots of Fig. 1.

2.2 Setup of Solutions

The treatment of the K-band VLBI observations in the VieVS solution (VIE-K-2022b) is similar to the S/X solution VIE2022b computed at the VIE Analysis Center[2] of the

---

[1]https://data.nrao.edu

[2]https://www.vlbi.at.

Table 2 Selected models and parametrization in VIE-K-2022b and USNO-K-2022July05. Values in parentheses represent the applied constraints. The abbreviation pwlo stands for piecewise linear offset

A priori modeling	VIE-K-2022b	USNO-K-2022July05
Ionosphere maps	CODE time series (Schaer 1999)	2 h average JPL maps
Ionospheric mf coefficients	MSLM, $k = 1$, $\Delta H = 56.7$ km, $\alpha = 0.9782$	2-D thin shell, MSLM
Hydrostatic delay	In situ pressure (Saastamoinen 1972)	In situ pressure (Saastamoinen 1972)
Hydrostatic + wet mf	VMF3 (Landskron and Böhm 2018)	VMF1 (Böhm et al. 2006)
Hydrostatic gradients	DAO (MacMillan and Ma 1997)	DAO (MacMillan and Ma 1997)
Precession/nutation model	IAU 2006/2000A	IAU 2006/2000A
Celestial pole offsets (CPO)	IERS Bulletin A, https://maia.usno.navy.mil/ser7/finals2000A.all	None
Parametrization		
Zenith wet delay	30 min pwlo (1.5 cm/30 min)	30 min pwlo (1.5 cm/h)
Tropo. grad.: VLBA	3 h pwlo (0.5 mm/3 h)	6 h pwlo (0.5 mm, 2 mm/day)
Tropo. grad.: KS	Fixed to a priori	Fixed to a priori
CPO: VLBA	24 h pwlo (0.1 μas/24 h)	Offset at midpoint of the session
CPO: KS	Fixed to a priori	Fixed to a priori
Weighting	Elevation-dependent (Gipson et al. 2008)	Baseline-dependent

International VLBI Service for Geodesy & Astrometry. A detailed description of the setup and applied theoretical models during the analysis are given in Krásná et al. (2022). In Table 2 we highlight models used in VIE-K-2022b and the USNO Calc/Solve solution USNO-K-2022July05[3] relevant to the presented investigations. While S/X frames calibrate the ionosphere directly from their dual-band data, K-band ionospheric effects require external calibration data. Specifically, K-band systems at the VLBA and the southern stations currently lack the complementary lower band needed for a dual-band ionospheric calibration, therefore the frequency-dependent delay coming from the dispersive part of the atmosphere has to be described by external models. In both K-band solutions presented here, ionospheric maps derived from Global Navigation Satellite System (GNSS) are applied. In VIE-K-2022b, global ionospheric maps provided by the Center for Orbit Determination in Europe (CODE; Schaer 1999)[4] are used with a time spacing of two hours from 05/2002 until 05/2014, and of one hour since that date. In USNO-K-2022July05, global ionospheric maps computed at the Jet Propulsion Laboratory (JPL) with two hours resolution are applied.

The alignment of the Terrestrial Reference Frame (TRF) is done by applying the No-Net-Translation (NNT) and No-Net-Rotation (NNR) conditions to the station position and velocity parameters in the global normal matrix. In VIE-K-2022b, the conditions are applied to all VLBA telescopes but one (MK-VLBA) with respect to the ITRF2020. In USNO-K-2022July05, the NNT/NNR condition is used w.r.t. a TRF solution based on ITRF2014 applied to all participating antennas except MK-VLBA (position discontinuity due to an Earthquake on June 15, 2006) and TIDBIN64 (limited number of observations).

The common practice for the rotational alignment of a new celestial reference frame to the current official one is to apply a three-dimensional constraint to the defining sources. In both solutions, ICRF3-SX is used as a priori celestial reference frame and the galactic acceleration correction is modeled with the adopted ICRF3 value of 5.8 μas/yr for the amplitude of the solar system barycenter acceleration vector for the epoch 2015.0. Datum definition of the CRFs is accomplished by the unweighted NNR (Jacobs et al. 2010) w.r.t. 287 (VIE-K-2022b) and 258 (USNO-K-2022July05) defining ICRF3-SX sources.

3   Results

We analyze the estimated VIE-K-2022b and USNO-K-2022July05 frames in terms of the vector spherical harmonics decomposition (VSH; Mignard and Klioner 2012; Titov and Lambert 2013; Mayer and Böhm 2020) w.r.t. ICRF3-SX which allows studying possible systematic differences between the catalogs. Prior to the comparison, outliers – defined as AGN with an angular separation greater than 5 mas from their ICRF3-SX position – were removed. In both solutions, there are four outlier sources: 0134+329 (3C48), 0316+162 (CTA21), 0429+415 (3C119), and 2018+295. Note that large position changes for 3C48 and CTA21 were found at X-band in observations made after the ICRF3 release and are reported by Frey and Titov (2021) and Titov et al. (2022). The number of remaining common sources is 993 in VIE-K-2022b and 995 in USNO-K-2022July05. The two sources (0227-542 and 0517-726) missing in VIE-K-2022b have 3 and 4 observations in USNO-K-2022July05. In VIE-K-2022b these observations were removed based on an outlier check of individual observations during the single session analysis.

---

[3]Latest version at https://crf.usno.navy.mil/data_products/RORFD/Quarterly/current//USNO_Kband_source_positions.iers.

[4]http://ftp.aiub.unibe.ch/CODE/.

Table 3 VSH parameters up to degree and order two for VIE-K-2022b and USNO-K-2022July05 w.r.t. ICRF3-SX (after eliminating four outliers from the solutions)

[µas]	VIE-K-2022b	USNO-K-2022July05
R_1	-1 ± 10	-4 ± 10
R_2	-8 ± 10	-16 ± 10
R_3	$+0 \pm 6$	-11 ± 6
D_1	-17 ± 9	-5 ± 9
D_2	-15 ± 9	$+9 \pm 9$
D_3	-4 ± 10	$+60 \pm 9$
$a_{2,0}^e$	-3 ± 12	-46 ± 11
$a_{2,0}^m$	-36 ± 7	$+1 \pm 7$
$a_{2,1}^{e,Re}$	-19 ± 10	-13 ± 10
$a_{2,1}^{e,Im}$	-21 ± 11	-26 ± 11
$a_{2,1}^{m,Re}$	-13 ± 11	$+13 \pm 10$
$a_{2,1}^{m,Im}$	-12 ± 11	-6 ± 11
$a_{2,2}^{e,Re}$	$+1 \pm 4$	$+3 \pm 4$
$a_{2,2}^{e,Im}$	$+8 \pm 4$	$+3 \pm 4$
$a_{2,2}^{m,Re}$	$+12 \pm 5$	$+23 \pm 5$
$a_{2,2}^{m,Im}$	$+6 \pm 5$	$+4 \pm 5$

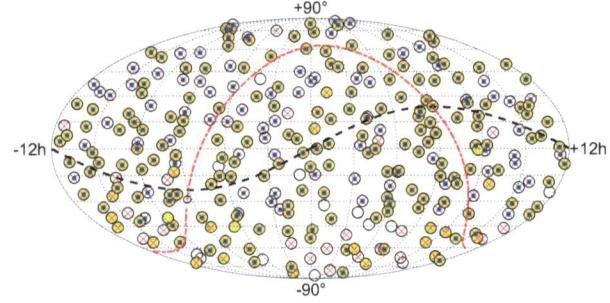

Fig. 2 Defining sources. The circles denote the 303 ICRF3-SX defining sources. The subgroup of 193 yellow circles depicts the ICRF3-K defining sources. Defining sources in VIE-K-2022b and USNO-K-2022July05 are red crosses and blue dots, respectively

The VSH are obtained with a least squares adjustment where the weight matrix contains inflated formal errors of the source coordinates. Similar to ICRF3-K, the formal errors of the source coordinates in both catalogs are inflated by a factor of 1.5, and a noise floor of 30 and 50 µas in quadrature is added to right ascension and declination, respectively. Table 3 summarizes the first order and second degree and order VSH, i.e., rotation (R_1, R_2, R_3), dipole (D_1, D_2, D_3), and ten coefficients (a) for the quadrupole harmonics of magnetic (m) and electric (e) type. All three rotation angles between the VIE-K-2022b and ICRF3-SX axes are within their formal errors and the angles do not exceed 8 µas. The largest angle (16 ± 10 µas) between USNO-K-2022July05 and ICRF-SX is around the y-axis (R_2). The selection of defining sources for the NNR constraint influences the mutual rotations of two catalogs (cf. Sect. 3.1 for more details). The three dipole parameters represent the distortion as a flow from a source to a sink located at two opposite poles. The D_3 term (-4 ± 10 µas in VIE-K-2022b and 60 ± 9 µas in USNO-K-2022July05) is susceptible to imperfect modeling of equatorial bulges in the ionospheric and tropospheric calibrations (cf. Sect. 3.2). The zonal quadrupole terms $a_{2,0}^e$ and $a_{2,0}^m$ reflect north-south asymmetries. Their values w.r.t. ICRF3-SX reach -3 ± 12 µas and -36 ± 7 µas in VIE-K-2022b, and -46 ± 11 µas and 1 ± 7 µas in USNO-K-2022July05, respectively (cf. Sect. 3.3).

3.1 Defining Sources

During the development of the ICRF3 a new set of sources observed at S/X-band was selected for defining the rotational alignment. This set of *defining* sources was based on three selection criteria in order to align the S/X-frame with its predecessor, the ICRF2 (Fey et al. 2015). These criteria were: (1) the overall sky distribution of the defining sources, (2) the position stability of the individual sources, and (3) the compactness of their structures (Charlot et al. 2020). For the alignment of the K-band reference frame ICRF3-K, a subset of 193 sources out of the set of 303 ICRF3-S/X defining sources – based mainly on the number of available K-band observations – was used. In Fig. 2 we show the distribution of the ICRF3-SX defining sources and highlight the ICRF3-K defining sources with yellow color. In the solutions VIE-K-2022b (red crosses) and USNO-K-2022July05 (blue dots) we take advantage of the additional observations gained after the ICRF3 release and choose the defining sources independently of the ICRF3-K ones. The current analysis of available sessions shows that there are no K-band observations of four ICRF3-S/X defining sources: 0044-846, 0855-716, 1448-648, 1935-692. This means, that 299 out of the 303 ICRF3-SX defining sources are observed in K-band (considering June 2022 to be the cutoff date for K-band observations). In VIE-K-2022b and USNO-K-2022July05 we apply different strategies for the selection of defining sources.

At TU Wien, we first computed a K-CRF solution from VLBA sessions only. We found 12 AGN (0038-326, 0227-369, 0316-444, 0437-454, 0743-006, 1143-245, 1606-398, 1929-457, 1937-101, 2036-034, 2111+400, 2325-150) among the 303 ICRF3-SX defining sources whose angular separation in this VLBA-only K-CRF solution is greater than 0.5 mas from their ICRF3-SX position and those are dropped from the NNR condition in VIE-K-2022b. All ICRF3-SX defining sources observed in the KS sessions only are kept in the NNR in VIE-K-2022b.

In USNO-K-2022July05 the following sources were excluded from the defining set: 0700-465, 0742-562, 0809-493 and 0918-534 since they show offsets of 0.5–1.5 mas from their ICRF3-SX positions in recent USNO

Table 4 Parameters of the ionospheric mapping function and the resulting VSH parameters D_3 and $a_{2,0}^e$

	k [−]	ΔH [km]	α [−]	D_3 [μas]	$a_{2,0}^e$ [μas]
MSLM	1	56.7	0.9782	-4 ± 10	-3 ± 12
SLM	1	0	1	-17 ± 10	$+2 \pm 12$
iono3	1	150.0	0.9782	15 ± 10	-10 ± 12
iono4	0.85	56.7	0.9782	42 ± 10	-15 ± 12

S/X solutions. An additional 41 sources, mostly in the deep south, were also excluded from the NNR condition because they had either very few or no observations.

The rotation angles in Table 3 show that the incorporation of the deep south sources in the alignment condition makes the adjustment more robust and keeps the estimated K-CRF solution slightly closer to the a priori one.

3.2   Ionospheric Mapping Function

The global ionosphere maps provide the Vertical Total Electron Content (VTEC). The conversion from VTEC to the Slant Total Electron Content (STEC) at an elevation angle (ϵ) of the VLBI observations at the telescope is done by the ionospheric mapping function (mf, M). In VIE-K-2022b we apply the thin shell ionospheric mf introduced by Schaer (1999) and recently discussed in detail by Petrov (2023):

$$M(\epsilon) = k \cdot \frac{1}{\sqrt{1 - \left(\frac{R_E}{R_E + H_i + \Delta H}\right)^2 \cdot \cos^2 \alpha \epsilon}}, \quad (1)$$

where k is a scaling factor, $R_E = 6371$ km stands for the Earth's base radius, $H_i = 450$ km is the height of the spherical single layer, ΔH represents an increment in the ionosphere height, and α is a correction factor to the elevation angle. In the default VIE-K-2022b solution we apply: $k = 1$, $\Delta H = 56.7$ km, $\alpha = 0.9782$ which is denoted as Modified Single-Layer Model (MSLM)[5] mapping function and claimed to be the best fit with respect to the JPL extended slab model mapping function. This parameter setting is recommended e.g. by Feltens et al. (2018), Wielgosz et al. (2018), and references therein. The standard Single Layer Model (SLM) mapping function is achieved with the parameters: $k = 1$, $\Delta H = 0$ km, and $\alpha = 1$. Following the discussion in Petrov (2023), we calculated two more solutions with different ionospheric mf parametrizations based on MSLM with different values of ΔH and k (i.e., iono3 and iono4) as summarized in Table 4.

In order to quantify the effect of the modified ionospheric mapping function on the K-CRF solution, we calculated

---

[5]http://ftp.aiub.unibe.ch/users/schaer/igsiono/doc/mslm.pdf.

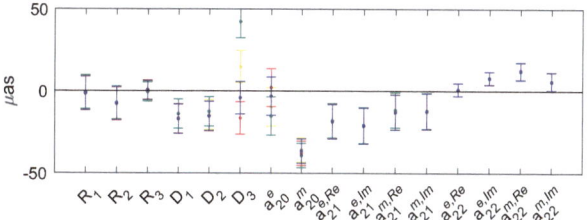

Fig. 3 Vector spherical harmonics of K-CRF solutions computed with ionospheric mf MSLM (blue, VIE-K-2022b), SLM (red), iono3 (yellow), and iono4 (green) w.r.t. ICRF3-SX

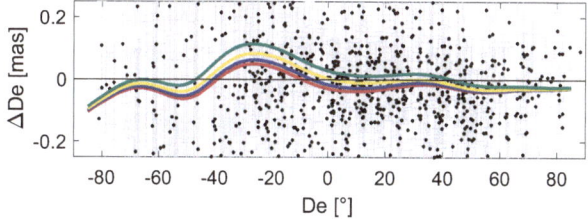

Fig. 4 Smoothed differences in declination from K-CRF solutions computed with ionospheric mf MSLM (blue, VIE-K-2022b), SLM (red), iono3 (yellow), and iono4 (green) w.r.t. ICRF3-SX. The black dots are differences in the declination of individual sources in VIE-K-2022b w.r.t. ICRF3-SX and their formal errors (in grey)

VSH for each solution w.r.t. ICRF-SX (Fig. 3). Changes in the three mf parameters (k, ΔH, α) influence the terms D_3 and $a_{2,0}^e$, which are sensitive to the equatorial bulge and north-south asymmetries, as mentioned earlier. The best fit to the ICRF3-SX is achieved with the MSLM mapping function applied in VIE-K-2022b where these parameters are negligibly small (-4 ± 10 μas and -3 ± 12 μas, respectively). On the other hand, in iono4 (where a scale factor $k = 0.85$ is applied to MSLM), the difference w.r.t. ICRF3-SX in D_3 and $a_{2,0}^e$ increases to 42 ± 10 μas and -15 ± 12 μas, respectively. In Fig. 4 we plot the differences in declination between the four discussed solutions w.r.t. ICRF3-SX over declination for individual sources. The smoothed curves are computed as moving averages with a Gaussian kernel and plotted with color coding identical to Fig. 3. The positive systematic difference in the declination estimates w.r.t. ICRF3-SX, appearing approximately between $-40°$ and $-10°$ declination, reaches its maximum of 63 μas for $-26°$ declination in VIE-K-2022b with applied MSLM mapping function (blue curve).

3.3   Systematic in Elevation Angles

Along with the ionospheric effects, the K-CRF suffers from an asymmetric observing network geometry with 99% of the data being from the all-northern VLBA. In Fig. 5 the percentage of observations from southern KS sessions for individual sources in VIE-K-2022b is shown. The logarithmic color

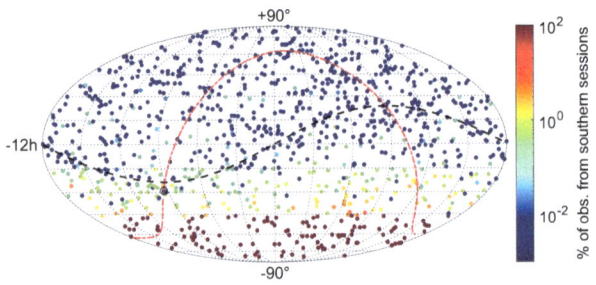

Fig. 5 Percentage of observations from KS sessions among the total number of K-CRF observations for individual sources in VIE-K-2022b

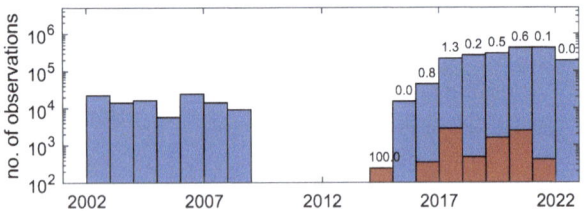

Fig. 6 The yearly distribution of the K-CRF observations used in our solutions. The numbers above the columns show the percentage of observations from southern sessions (KS, brown columns) w.r.t. the total number of observations (blue columns) during the individual year

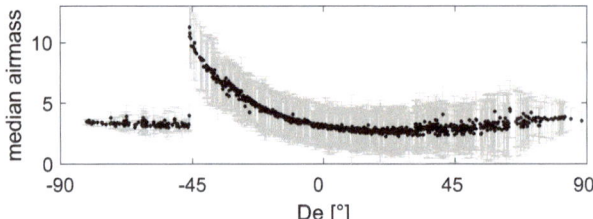

Fig. 7 Median airmass for individual sources computed over all their observations in VIE-K-2022b

scale highlights the fact that the number of observations to the sources with declination higher than $-45°$ builds only a tiny fraction of the total number of observations, although the mutual sky visibility between southern antennas HartRAO and Hobart allows observing sources up to approximately $30°$ declination. The mean percentage of observations from KS sessions for sources with declination between $-15°$ and $-45°$ (area with mainly yellow and green colors in Fig. 5) is 0.96%. The total number of K-CRF observations (blue color) and the number of observations from the KS (brown color) during individual years is plotted in Fig. 6. The numbers above the columns give the percentage of observations from KS w.r.t. the total number of observations within the individual year.

In order to explore the resultant elevation-dependent effects, we characterize the distribution of elevation angles at which the sources were observed. These distributions vary due to both the geometry of the VLBA network and the fact that we observe each source over a range of hour angles. First, we define a parameter called airmass in order to quantify the approximate total pathlength through the troposphere for each source – with the maximum at lower elevation angles. It is computed for each observation from the whole data set with the simplifying assumption of a flat slab atmosphere (ignoring the curvature of the atmosphere over a spherical Earth):

$$\text{airmass} = \frac{1}{\sin(\epsilon_1)} + \frac{1}{\sin(\epsilon_2)}, \qquad (2)$$

where ϵ is the elevation angle of the source at telescopes 1 and 2 of the baseline. Next, we compute the median value over the individual observations for each source and plot it with respect to the declination (Fig. 7) with the errors (in grey) obtained as standard deviations computed over the individual airmass values for the particular source. The systematic increase of the airmass parameter from $0°$ to $-45°$ declination can lead to an overestimation of the optimal data weights for VLBA observations in this declination range when the larger noise of observations conducted at low elevation angles is not considered. To partly account for the overweighting of the low elevation scans (which observe low declination sources in the mentioned area), elevation-dependent weighting (Eq. 3; Gipson et al. 2008) in VIE-K-2022b is applied. In the diagonal covariance matrix the measurement noise σ_m^2 is increased by the squared elevation-dependent noise terms for telescopes 1 and 2:

$$\sigma_{obs}^2 = \sigma_m^2 + \left(\frac{6\,\text{ps}}{\sin(\varepsilon_1)}\right)^2 + \left(\frac{6\,\text{ps}}{\sin(\varepsilon_2)}\right)^2. \qquad (3)$$

Hence, sources between $0°$ to $-45°$ declination obtain a lower weight in the least squares adjustment and the resulting distortion of the celestial reference frame is damped. For example, an observation conducted with two VLBA antennas at the elevation angles of $15°$ has an airmass value of 8 (Eq. 2) which corresponds in our data set to a source with a declination of about $-40°$ (Fig. 7). The additional noise added to the σ_m^2 of this observation in quadrature is 33 ps (Eq. 3) which decreases its weight in the solution.

4 Conclusion

Recent K-CRF solutions computed at TU Wien (VIE-K-2022b)[6] and USNO (USNO-K-2022July05) from single-frequency band VLBI observations (24 GHz) until June 2022 were assessed. The vector spherical harmonics were computed w.r.t. ICRF3-SX after eliminating four AGN as outliers. In VIE-K-2022b, all rotation values are lower than

---

[6]https://vlbi.at/data/analysis/ggrf/crf_vie2022b_k.txt.

8 μas and have significance at the level of their formal errors or less. With a single exception, all dipole and quadrupole terms are within 20 μas with a marginal significance of two times the formal error as maximum. The only quadrupole term above this limit is $a_{2,0}^m$ (-36 ± 7 μas). We discussed two major challenges which limit the accuracy of the current K-band VLBI solutions: external ionospheric corrections and the non-uniform observing network geometry – especially the lack of observations in the deep south. We show that the choice of ionospheric mapping function parameters influences the dipole, D_3, and quadrupole terms $a_{2,0}^e$. Because 99% of the data is observed with the all-northern VLBA sources between $0°$ and $-45°$ declination have a monotonic decrease in median elevation angle of observation making our solution vulnerable to atmospheric mis-modeling. We reduced sensitivity of the VIE-K-2022b solution to the effect of this observing geometry bias by computing elevation-dependent weighting to downweight low elevation observations. Future work will focus on improving the geometry of the K-band observing network, improving the modeling of atmospheric effects, and improving solution weighting schemes.

Declarations

Ethics Approval and Consent to Participate Not applicable.

Consent for Publication Not applicable.

Competing Interests There are no relevant financial or non-financial competing interests to report.

Funding We acknowledge our respective sponsors: SARAO/HartRAO is a facility of the National Research Foundation (NRF) of South Africa. Portions of this work were done at the Jet Propulsion Laboratory, California Institute of Technology under contract with NASA (contract no. 80NM0018D0004). Portions of this work were sponsored by the Radio Optical Reference Frame Division of the U.S. Naval Observatory. This work supports USNO's ongoing research into the celestial reference frame and geodesy.

Authors' Contributions HK wrote the manuscript, analyzed the VLBI data and created the VIE-K solutions. DG prepared the vgosDB databases and computed the USNO-K solution. AdW is the PI of the VLBI K-band group and the leader by planning of the VLBI K-band observations. CJ proposed the concept of the paper and contributed to the analysis of data. All authors contributed to regular discussions and interpretations of results. They read and commented the final paper.

Acknowledgements The authors appreciate comments provided by three anonymous reviewers. HK thanks Leonid Petrov (NASA GSFC) for fruitful discussions about single band astrometry. The authors gratefully acknowledge the use of the VLBA under the USNO's time allocation.

References

Böhm J, Werl B, Schuh H (2006) Troposphere mapping functions for GPS and very long baseline interferometry from European centre for medium-range weather forecasts operational analysis data. J Geophys Res Solid Earth 111(B2). https://doi.org/10.1029/2005JB003629

Böhm J, Böhm S, Boisits J, et al (2018) Vienna VLBI and satellite software (VieVS) for geodesy and astrometry. Publ Astron Soc Pac 130(986):044,503. https://doi.org/10.1088/1538-3873/aaa22b

Charlot P, Jacobs CS, Gordon D, et al (2020) The third realization of the international celestial reference frame by very long baseline interferometry. Astron Astrophys 644:A159. https://doi.org/10.1051/0004-6361/202038368

Feltens J, Bellei G, Springer T, et al (2018) Tropospheric and ionospheric media calibrations based on global navigation satellite system observation data. J Space Weather Space Clim 8:A30. https://doi.org/10.1051/swsc/2018016

Fey AL, Gordon D, Jacobs CS, et al (2015) The second realization of the international celestial reference frame by very long baseline interferometry. Astron J 150(2):58. https://doi.org/10.1088/0004-6256/150/2/58

Frey S, Titov O (2021) Change in the radio structure and position of the quasar CTA 21. Res Notes AAS 5(3):60. https://doi.org/10.3847/2515-5172/abf123

Gipson J, MacMillan D, Petrov L (2008) Improved estimation in VLBI through better modeling and analysis. In: Finkelstein A, Behrend D (eds) IVS GM Proceedings, pp 157–162. https://ui.adsabs.harvard.edu/abs/2008mefu.conf..157G/abstract

ICRF3 working group (2018) IAU resolution B2 on the third realization of the international celestial reference frame. In: Proceedings of the XXX IAU General Assembly. IAU, Vienna, Austria. https://www.iau.org/static/resolutions/IAU2018_ResolB2_English.pdf

Jacobs C, Heflin M, Lanyi G, et al (2010) Rotational alignment altered by source position correlations. In: Behrend D, Baver KD (eds) IVS GM Proceedings. NASA, pp 305–309. https://ivscc.gsfc.nasa.gov/publications/gm2010/jacobs2.pdf

Krásná H, Baldreich L, Böhm J, et al (2022) VLBI celestial and terrestrial reference frames VIE2022b. https://doi.org/10.48550/arXiv.2211.07338

Landskron D, Böhm J (2018) VMF3/GPT3: refined discrete and empirical troposphere mapping functions. J Geodesy 92(4):349–360. https://doi.org/10.1007/s00190-017-1066-2

Lanyi GE, Boboltz DA, Charlot P, et al (2010) The celestial reference frame at 24 and 43 GHz. I. Astrometry. Astron J 139(5):1695. https://doi.org/10.1088/0004-6256/139/5/1695

MacMillan DS, Ma C (1997) Atmospheric gradients and the VLBI terrestrial and celestial reference frames. Geophys Res Lett 24(4):453–456. https://doi.org/10.1029/97GL00143

Mayer D, Böhm J (2020) Comparing Vienna crf solutions to gaia-crf2. In: Freymueller JT, Sánchez L (eds) Beyond 100: The next century in geodesy. Springer International Publishing, Cham, pp 21–28. https://doi.org/10.1007/1345_2020_99

Mignard F, Klioner S (2012) Analysis of astrometric catalogues with vector spherical harmonics. Astron Astrophys 547:A59. https://doi.org/10.1051/0004-6361/201219927

Napier P (1995) VLBA Design. In: Zensus J, Diamond P, Napier P (eds) Very Long Baseline Interferometry and the VLBA, in ASP Conference Series, vol 82. Astronomical Society of the Pacific, pp 59–72. https://articles.adsabs.harvard.edu/pdf/1995ASPC...82...59N

Petrov L (2023) Single-band VLBI absolute astrometry. Astron J 165(4):183. https://doi.org/10.3847/1538-3881/acc174

Petrov L, Kovalev YY, Fomalont EB, et al (2011) The very long baseline array galactic plane survey - VGaPS. Astron J 142(2):35. https://doi.org/10.1088/0004-6256/142/2/35

Saastamoinen J (1972) Introduction to practical computation of astronomical refraction. Bull Geodesique 106:383–397. https://doi.org/10.1007/BF02522047

Schaer S (1999) Mapping and predicting the Earth's ionosphere using the global positioning system. PhD thesis, University of Berne

Titov O, Frey S, Melnikov A, et al (2022) Unprecedented change in the position of four radio sources. Mon Not R Astron Soc 512(1):874–883. https://doi.org/10.1093/mnras/stac038

Titov O, Lambert S (2013) Improved VLBI measurement of the solar system acceleration. Astron Astrophys 559:A95. https://doi.org/10.1051/0004-6361/201321806

Wielgosz P, Milanowska B, Krypiak-Gregorczyk A, et al (2018) Validation of gnss-derived global ionosphere maps for different solar activity levels: case studies for years 2014 and 2018. GPS Solut 25(103). https://doi.org/10.1007/s10291-021-01142-x

VGOS VLBI Intensives Between MACGO12M and WETTZ13S for the Rapid Determination of UT1-UTC

Matthias Schartner, Leonid Petrov, Christian Plötz, Frank G. Lemoine, Eusebio Terrazas, and Benedikt Soja

Abstract

In this work, we present a status update and preliminary results of the designated research and development VLBI Intensive program VGOS-INT-S, observed between MACGO12M and WETTZ13S for the rapid determination of the Earth's phase of rotation, expressed via UT1-UTC. Since 2021, 27 Intensive sessions have been observed successfully utilizing a special observation strategy alternating between high- and low-elevation scans for improved determination of delays caused by the neutral atmosphere. Between the end of January and mid of March 2022, VGOS-INT-S was among the most accurate Intensive programs. During this time, eight sessions were observed with an average formal error $\sigma_{\text{UT1-UTC}}$ of 3.1 µs and a bias w.r.t. IERS C04 of 1.1 µs. Later, the session performance decreased due to multiple technical difficulties.

Keywords

Intensives · IVS · VGOS · VLBI

1 Introduction

Among the space geodetic techniques, Very Long Baseline Interferometry (VLBI) is able to provide the most accurate and unbiased estimates of the angle of the Earth's orientation with respect to the rotation axis, expressed via UT1-UTC. Since 1984, regular observing campaigns have been launched for monitoring Earth orientation parameters (EOP)

including UT1-UTC. They are now coordinated by the International VLBI Service for Geodesy and Astrometry (IVS) (Nothnagel et al 2017). Most sessions of the IVS observing campaigns for EOP determination either run for 24 h to determine the full set of EOP, or for 1 h to determine solely UT1-UTC. The 24-h programs run 2–3 times a week with latency between observations and delivery of EOP estimates of about 15–20 days. The 1-h programs on average run 2–3 times a day and the latency between observations and delivery of UT1-UTC estimates is 1–3 days. For that reason, these campaigns are called *Intensives*. Nowadays, a number of VLBI Intensive programs dedicated to the estimation of UT1-UTC run in parallel.

VLBI observations commenced in 1967. Since then, the VLBI technique went through a number of upgrades. The most recent upgrade is called the VLBI Global Observing System (VGOS) (Niell et al 2018). The changes in that upgrade, relevant to the present study, are faster slewing speeds of 12° over azimuth and 6° over elevation, combined with an increased data rate of currently 8 Gbps distributed among four bands. The fast slewing rates reduce the time when the antenna is slewing and thus not recording signals.

Authors Leonid Petrov, Christian Plötz, Frank G. Lemoine, Eusebio Terrazas, and Benedikt Soja contributed equally to this work.

M. Schartner (✉) · B. Soja
Institute of Geodesy and Photogrammetry, ETH Zurich, Zurich, Switzerland
e-mail: mschartner@ethz.ch; soja@ethz.ch

L. Petrov · F. G. Lemoine · E. Terrazas
NASA Goddard Space Flight Center, Greenbelt, MD, USA
e-mail: leonid.petrov-1@nasa.gov; frank.g.lemoine@nasa.gov; chevo@utexas.edu

C. Plötz
Bundesamt für Kartographie und Geodäsie, Bad Kötzting, Germany
e-mail: Christian.Ploetz@bkg.bund.de

© The Author(s) 2023
J. T. Freymueller, L. Sánchez (eds.), *Gravity, Positioning and Reference Frames*,
International Association of Geodesy Symposia 156, https://doi.org/10.1007/1345_2023_222

The higher data rate allows for shorter observation times to reach the desired signal-to-noise ratio (SNR). Combined, this leads to a significantly increased number of scans per hour, up to 100, allowing for faster sampling of the atmosphere, which is considered one of the major error sources in VLBI.

There are around 200 extragalactic radio sources with sufficient brightness and compactness that are currently observed by VGOS radio telescopes and up to 100 sources can be observed in 1 h. Thus, the number of combinations of sources that can be selected for observations is extremely large. We are interested in developing new techniques for the generation of optimal observing plans, the so-called schedules, which provide UT1-UTC with minimum errors. The theoretical basis of the development of an optimal schedule was described in Schartner (2019).

In order to verify the optimized scheduling algorithm, we launched a research and development Intensive VLBI observing program, named VGOS-INT-S, on the 8418 km long baseline between MACGO12M and WETTZ13S in 2021. Station MACGO12M, also known as Mg, is located in Western Texas, USA, and station WETTZ13S, known as Ws, is located in Northeast Bavaria, Germany. Here, we outline the design of the observing program and discuss preliminary results.

2   Methods

The major error source in geodetic VLBI is mismodeling of the path delay in the neutral atmosphere. The a priori atmospheric path delay can be computed either using a regression model of surface atmospheric pressure and air temperature or by direct integration of equations of wave propagation through an inhomogeneous refractivity field derived from the output of numerical weather models. In both cases, the accuracy of the a priori path delay is still insufficient and we have to estimate the residual path delay in the zenith direction from the VLBI data themselves.

During the analysis, atmospheric delays are commonly divided into a hydrostatic and a non-hydrostatic (wet) part. While the hydrostatic part can be modeled with sufficient accuracy of around 1–2 mm if accurate ground pressure measurements are available, the wet part has to be estimated due to its higher variability. Historically, one zenith wet path delay (ZWD) per 1-h observing session was estimated. With the fast slewing VGOS antennas, we can develop a scheduling strategy that would allow us to estimate atmospheric path delay with segments as short as 5 min. To enable a more frequent estimation of ZWD, special emphasis has to be laid on providing observations at different elevation angles within the estimation interval.

Following this idea, a new VLBI observation strategy has been developed for VGOS-INT-S and applied using VieSched++ (Schartner and Böhm 2019). Due to the special geometry of the Mg/Ws baseline with its baseline length of 8418 km, observations at high elevation on one station naturally result in low elevation at the other station as depicted in Fig. 1. Although longer baselines are potentially more sensitive to UT1-UTC, they also have a limited mutually visible sky. For example, the frequently observed KOKEE/WETTZELL baseline has a length of 10358 km, resulting in a maximum observable elevation of only ~ 65°. This can potentially result in a worse determination of the ZWD and thus UT1-UTC.

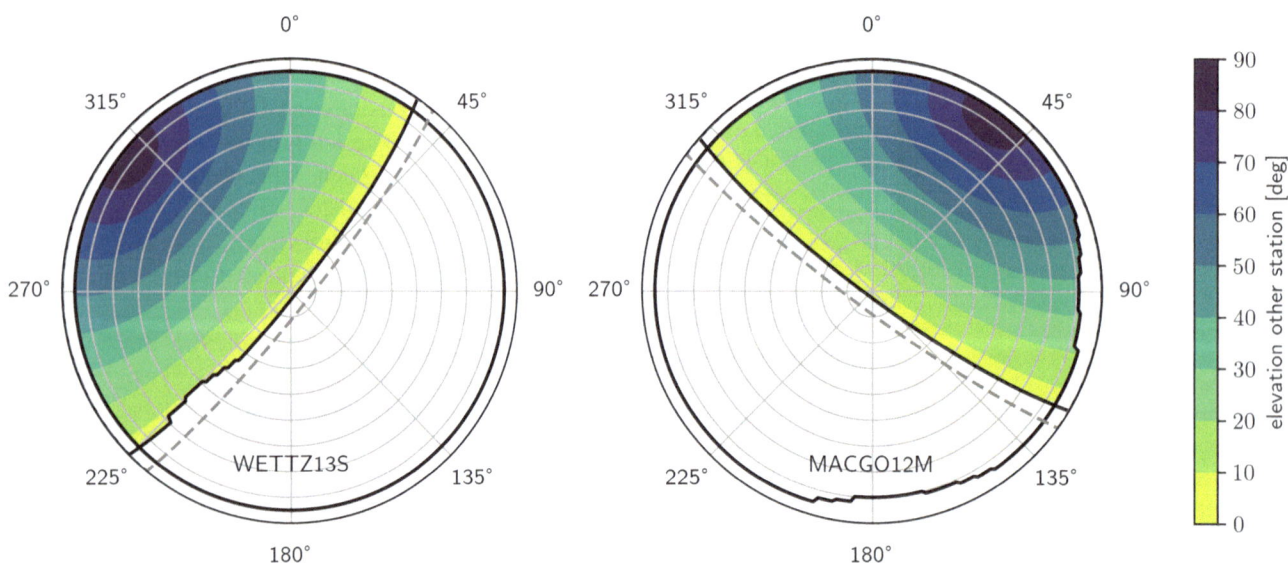

Fig. 1 Mutual visibility color-coded by the elevation of the partner telescope. The black lines represent the station horizon masks while the dashed gray line marks the theoretical horizon

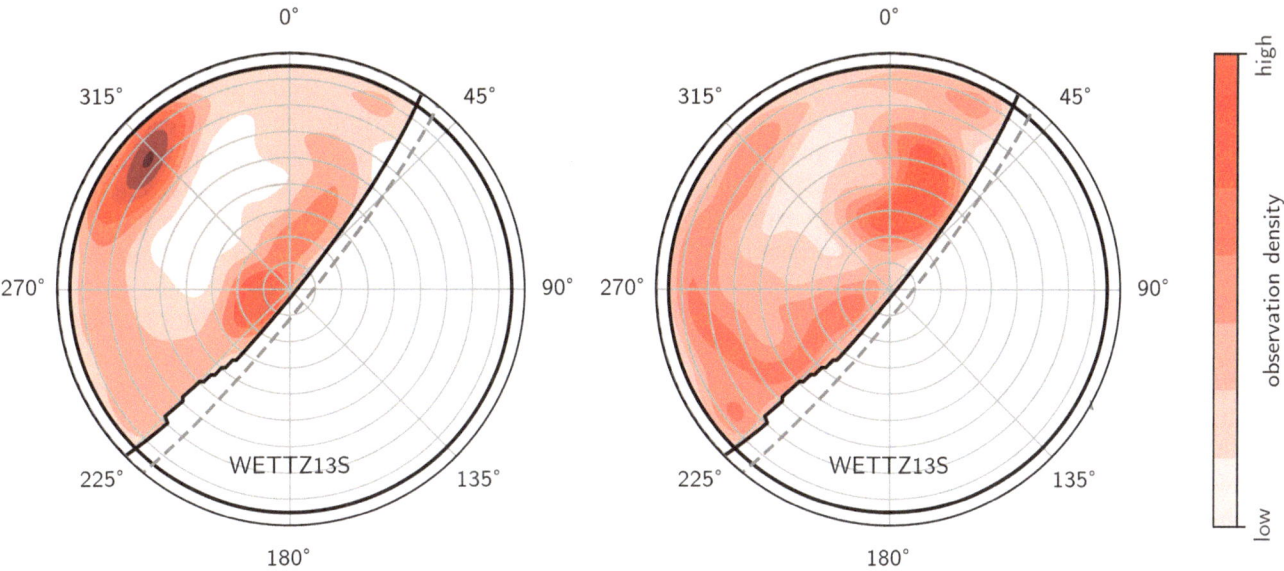

Fig. 2 Scan distribution for station Ws. Left: new observation strategy. Right: old Intensive observing strategy. The darker the color, the higher the number of observations in this area. The black lines represent the station horizon mask while the dashed gray line marks the theoretical horizon

The new observation strategy is based on rapidly alternating between high and low-elevation scans to allow for an improved and potentially higher frequent ZWD determination. Therefore, the following scan sequence is repeated:

- scan with high elevation at Mg (low elevation Wz)
- scan without constraints
- scan with high elevation at Wz (low elevation Mg)
- scan without constraints

Thus, every other scan is especially dedicated to measuring ZWD. The remaining scans are selected in a way to increase the sensitivity towards UT1-UTC, e.g. by observing sources located at the corners of the mutually visible sky (Schartner et al 2021), or by reducing potential systematic errors caused by source-structure effects via observing a high number of different sources.

The effect of this special observing strategy is illustrated in Fig. 2, which depicts the distribution of observations in azimuth and elevation. While the distribution is more balanced using an old observing strategy, two clear clusters are visible with the new observing strategy, one at high elevation and one at low elevation. This confirms that the observing strategy is working as intended.

3 Data

The VGOS-INT-S observing program started on December 7th, 2021 with session S21341. In the years 2021 and 2022, 27 sessions were observed successfully. Visibility data, geodetic databases, and results of the analysis are available at the IVS Data Centers.[1]

Over time, the choice of the SNR target and integration time limit was iteratively adjusted based on station and correlator feedback. The first sessions S21341–S22011 were scheduled conservatively, using a fixed integration time of 30 s independent of the source brightness and thus mimicking the current 24-h operational VGOS (VGOS-OPS) mode to gather some experience on the new baseline. Afterward, the integration time was reduced to increase the number of observations per session, and an SNR-based observing time based on the source brightness and antenna sensitivity was utilized. For S22018–S22053 the minimum integration time was set to 15 s, while it was lowered to 12 s in the remaining sessions while the maximum allowed integration time (except for calibrator scans) was set to 30 s in all sessions. The target SNR per band was set to 15 for all sessions until S22053, while it was lowered to 10 between S22060–S22095, before being increased to 12 again for all sessions after S22109. In practice, the changes were very small and had little effect on the total number of scheduled observations (see Fig. 3), but they helped during the correlation process to recover most observations, especially in cases of reduced antenna sensitivity as discussed later.

To troubleshoot existing hardware-related problems at the stations, two special sessions have been designed. First, S22277 was split into two sections. The first 30 min were observed regularly with an SNR-based integration time and a minimum of 12 s while the second 30 min were scheduled

---

[1]https://ivscc.gsfc.nasa.gov/sessions/.

using a fixed 30-s long integration time. Second, S22284 was scheduled including ONSA13NE (Oe; Sweden) to provide the independent baseline Mg/Oe.

Figure 3 depicts the number of scheduled and successful (defined as analyzed by the NASA analysis center) observations per session. Some sessions suffered from a significant number of non-detections, mostly explained by hardware failures. For S22067, 23 scans were not recorded at station Mg (from 19:45 until 20:00 UTC). For S22074, 12 observations were rejected during analysis due to large residuals. Similarly, 22 observations were rejected for S22214 and 12 observations for S22263. Furthermore, two sessions, namely S22213 and S22215, were scheduled using a different scheduling software instead of VieSched++, explaining the lower number of observations.

A list of the most important technical problems affecting VGOS-INT-S is provided in Table 1. Please note that the dates listed in this table are approximations. In some cases, it is not possible to find out exactly when a problem occurred. For some of the listed problems, it is also unclear how much they affected the performance of VGOS-INT-S. We expect that the LNA failure, first noticed in May 2022, had the most severe effect on the performance of the Intensive sessions due to decreased sensitivity of station WETTZ13S. This corresponds to the decrease of UT1-UTC precision for session S22158 onward as further discussed in Sect. 4.

4   Results

Figure 4 depicts the VGOS-INT-S precision and accuracy, independently derived from the GSFC operational solutions produced with the Solve/νSolve software package. We use two measures of UT1-UTC errors: formal uncertainties derived from the observation SNR following the law of error propagation, called here *precision*, and the differences of the UT1-UTC estimates and the IERS C04 time series, called here *accuracy*. To interpolate the daily IERS C04 UT1-UTC values to the Intensive reference epoch, first, tidal effects with periods < 35 days were subtracted, followed by a Lagrangian interpolation of order four and re-adding the previously subtracted tidal effects. It is visible that the formal errors between S22025 and S22081 are significantly smaller compared to most of the remaining sessions. Within this period, eight sessions have been observed with an average formal error $\sigma_{\text{UT1-UTC}}$ of 3.1 μs and an offset w.r.t. IERS C04 of 1.1 μs. The root mean square error (RMSE) during this period is 31.7 μs. For the remaining sessions, the average formal error is increased to 9.8 μs, the offset w.r.t. IERS C04 is increased to −2.1 μs, and the RMSE is increased to 80.7 μs. The increase in uncertainty might be explained by the technical problems encountered at the stations as listed in Table 1. Especially the LNA failure at station WETTZ13S, noticed on May 2022 (S22158) corresponds well to the decrease in precision depicted in Fig. 4. However, it does not explain

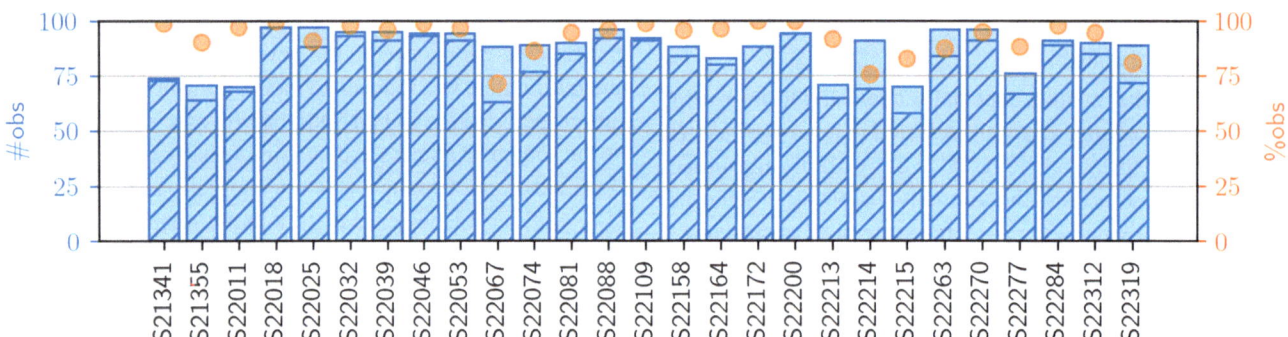

Fig. 3 Number of scheduled observations (blue bars), analyzed observations (blue hatched), and percentage of analyzed observations (orange)

Table 1 List of technical difficulties encountered during VGOS-INT-S sessions

Date		Issue	Effect
2022 Jan–May	Ws	Encoder error	Short downtimes
2022 Feb	Mg	Compressor shutdown	Downtime
2022 May	Mg	Elevation motor coupler failed	Downtime
2022 May–2023 Feb	Ws	LNA failure	Reduced sensitivity
2022 Jun	Mg	Power failure of azimuth motors	Downtime
2022 Jul	Mg	Problems with azimuth motor stop	Downtime
2022 Jul–2023 Feb	Ws	2nd LNA failure	Reduced sensitivity
2022 Aug	Mg	HubPC disk failure	Downtime
2022 Aug–Sep	Mg	Failure of M700 compressor unit	Reduced sensitivity
2022 Jul–Aug	Mg	Phase cal signal failure	Reduced sensitivity
2022 Nov–2023 Feb	Ws	Dewar failure	Downtime
2023 Jan	Mg	Compressor failure	Downtime

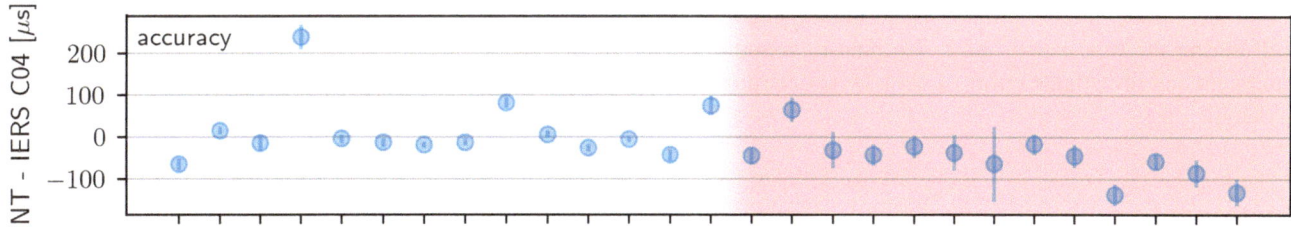

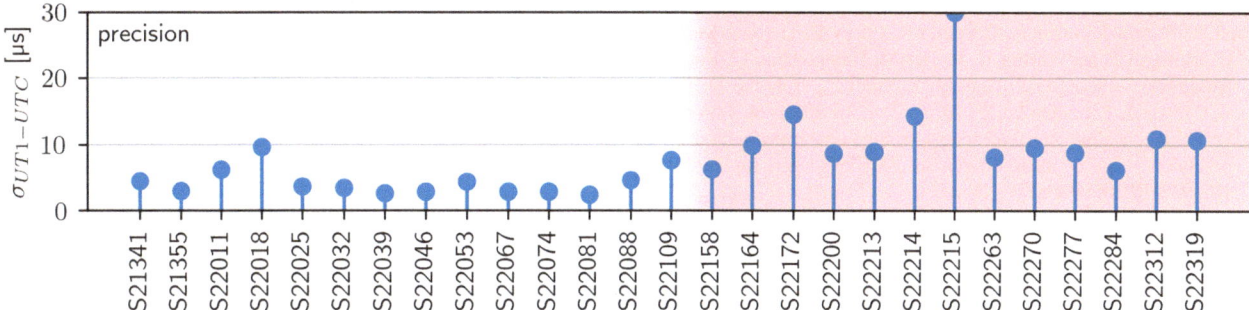

Fig. 4 VGOS-INT-S accuracy (top) and precision (bottom) extracted from the GSFC analysis reports. Accuracy: UT1-UTC estimate w.r.t. IERS C04, the 3σ value is depicted in the error bars. Precision: UT1-UTC formal error σ. The red background depicts sessions affected by decreased sensitivity due to the LNA failure. Note that the exact date of the LNA failure is unknown. It was first noticed in S22158 but might have already occurred before this session

why the previous session (S22109) also suffered decreased precision. This might be a simple coincidence, or it might be that the LNA failure already happened earlier and was only noticed in May 2022.

Still, the very small formal uncertainties and good agreement with IERS C04 between S22025 and S22081 suggest that superior precision in UT1-UTC determination at the Mg/Ws baseline can be achieved. For comparison, during the same time, the VGOS-INT-A sessions at the KOKEE12M/WETTZ13S baseline achieved an average formal error of $4.3\,\mu s$ with an offset w.r.t. IERS C04 of $-8.6\,\mu s$ and a RMSE of $28.7\,\mu s$, although according to simulations, it is expected that based on the baseline geometry alone, VGOS-INT-A should be 40% more sensitive towards UT1-UTC compared to VGOS-INT-S (Schartner et al 2021). The obtained results are also comparable with the VGOS-INT-B sessions, observed between Japan and Sweden, that achieved an RMSE of $23.2\,\mu s$ and a bias of $-3.8\,\mu s$ between December 2019 and February 2020 (Haas et al 2021).

Here, it is to note that the GSFC operational analysis presented above does not yet make use of a more frequent estimation of ZWD enabled by the observation strategy. Instead, ZWD is parameterized as a constant offset only. This highlights that the proposed scheduling strategy provides highly accurate UT1-UTC estimates even when a traditional parameterization is used.

To assess the impact of a reduced ZWD interval enabled by the new scheduling approach, the sessions have been analyzed with 60-min and with 10-min long ZWD intervals using two independent software packages, pSolve and VieVS. The mean of the differences in UT1 between the two pSolve solutions is $0.2\,\mu s$ and the RMSE is $2.0\,\mu s$, which is insignificant. Using VieVS, similar results have been obtained. Compared to IERS C04, the improvement in terms of RMSE based on the new scheduling approach is $0.7\,\mu s$ and, thus, insignificant. Considering hardware failure, an absence of evidence should not be construed as evidence of absence. We need more data with properly working hardware to assess the significance of the impact of scheduling and analysis approaches on UT1 determination. Therefore, the VGOS-INT-S program is continued in 2023 and even extended by 24-h sessions, alternating hourly between the standard and improved scheduling strategy, for improved comparability of the two approaches.

5   Summary

We presented a design of a research and development VLBI observing program for the determination of UT1-UTC at a single baseline between fast slewing radiotelescopes WETTZ13S (Germany) and MACGO12M (Texas, USA).

Although the telescopes suffered a number of technical failures, the results are very encouraging. Despite the shorter baseline length compared to more typical Intensive sessions and the resulting theoretical lower sensitivity towards UT1-

UTC, the VGOS-INT-S sessions performed exceptionally well during the first part of 2022. We plan to continue the campaign and investigate errors of UT1-UTC determination in detail.

References

Haas R, Varenius E, Matsumoto S, Schartner M (2021) Observing UT1-UTC with VGOS. Earth Planets Space 73(1):78. https://doi.org/10.1186/s40623-021-01396-2

Niell A, Barrett J, Burns A, Cappallo R, Corey B, Derome M, Eckert C, Elosegui P, McWhirter R, Poirier M, Rajagopalan G, Rogers A, Ruszczyk C, SooHoo J, Titus M, Whitney A, Behrend D, Bolotin S, Gipson J, Petrachenko B (2018) Demonstration of a broadband very long baseline interferometer system: a new instrument for high-precision space geodesy. Radio Sci 53:1269–1291. https://doi.org/10.1029/2018RS006617

Nothnagel A, Artz T, Behrend D, Malkin Z (2017) International VLBI service for geodesy and astrometry. J Geodesy 91(7):711–721. https://doi.org/10.1007/s00190-016-0950-5

Schartner M (2019) Optimizing geodetic VLBI schedules with VieSched++. Dissertation, Technische Universität Wien. https://doi.org/10.34726/hss.2019.49542

Schartner M, Böhm J (2019) VieSched++: A new VLBI scheduling software for geodesy and astrometry. Publ Astron Soc Pac 131(1002):084501. https://doi.org/10.1088/1538-3873/ab1820

Schartner M, Kern L, Nothnagel A, Böhm J, Soja B (2021) Optimal VLBI baseline geometry for UT1-UTC Intensive observations. J Geodesy 95(7):75. https://doi.org/10.1007/s00190-021-01530-8

Correcting Non-Tidal Surface Loading in GNSS repro3 and Comparison with ITRF2020

Benjamin Männel, Andre Brandt, Susanne Glaser, and Harald Schuh

Abstract

Time-dependent mass variations lead to significant and systematic load-induced deformations of the Earth's crust, impacting space geodetic techniques. Using the ESMGFZ loading models, the impact on the recent IGS reprocessing campaign (repro3) is studied. While non-tidal loading was not corrected in the original repro3, separate solutions were computed by applying the corrections at the solution and the observation level. An initial comparison between the seasonal components in the loading models revealed a good agreement with the periodic functions in the ITRF2020. Based on the considered test period (2012–2016), we found reduced statistical signatures if applying the corrections at the solution level. For the annual amplitudes in the Up direction, an overall reduction of 18% was achieved. Correcting at the observation level provided larger reductions (amplitudes are reduced on average by 42%). Moreover, the consistency of the derived products, i.e., satellite orbits, Earth rotation parameters, and station coordinates, is achieved. Overall, it is recommended to correct non-tidal loading displacements primarily at the observation level. In case of technical restrictions or software limitations, corrections should be applied at the solution level.

Keywords

GNSS · Non-tidal loading · Surface deformation

1 Introduction

Mass re-distribution in atmosphere, oceans, and the terrestrial branch of the global water cycle causes a deformation of the solid Earth and an associated change in the Earth's gravity field, its orientation, and – most important for this study – the geometry of the crust. Surface loading is relevant to reach the accuracy goals of the Global Geodetic Observing System (GGOS) that aim at 1 mm accuracy and 0.1 mm/a stability. In line with the International Earth Rotation and Reference Systems Service (IERS) Conventions 2010 (Petit and Luzum 2010) non-tidal loading was not corrected in the recent GNSS reprocessing campaign of the International GNSS Service (repro3) and the current reference frame realization ITRF2020 (Altamimi et al. 2023). This study aims (1) to compare the ITRF2020 seasonal displacement signals against the loading-predicted surface deformation and (2) to assess the potential impact of associated corrections on the reprocessed GNSS solutions. The main focus of this contribution is, therefore, on the comparison between solutions with non-tidal loading corrections applied at the solution level (abbreviated as SOL) and at the observation level (OBS).

Focusing on large station networks Martens et al. (2020), Mémin et al. (2020), Gobron et al. (2021), Klos et al. (2021) and others investigated the impact of non-tidal loading cor-

B. Männel (✉) · A. Brandt · S. Glaser
Deutsches GeoForschungsZentrum GFZ, Telegrafenberg, Potsdam, Germany
e-mail: benjamin.maennel@gfz-potsdam.de

H. Schuh
Deutsches GeoForschungsZentrum GFZ, Telegrafenberg, Potsdam, Germany

Institute of Geodesy and Geoinformation Science, Technische Universität Berlin, Berlin, Germany

Table 1 Summary of estimation and processing strategy (`repro3`)

A-priori modeling	
Observations	Ionosphere-free linear combination formed by undifferenced GPS observations
Tropospheric correction	Troposphere delays computed with Saastamoinen, mapped with VMF (Böhm et al. 2006)
Ionospheric correction	1st order effect considered with ionosphere-free linear combination, 2nd order correction applied
GNSS phase center	Corrections from dedicated repro3 ANTEX applied (igsR3_2135.atx, http://ftp.aiub.unibe.ch/users/villiger/igsR3_2135.atx)
Gravity potential	GOCO6s up to degree and order 12 (Kvas et al. 2019)
Solid Earth tides	According to IERS 2010 conventions (Petit and Luzum 2010)
Permanent tide	Conventional tide free
Ocean tide model	FES2014b (Lyard et al. 2021)
Ocean loading	Tidal: FES2014b (Lyard et al. 2021)
Atmospheric loading	Tidal: S_1 and S_2 corrections (Ray and Ponte 2003)
High-frequent EOP model	Model of Desai and Sibois (2016)
Mean pole tide	Linear mean pole as adopted by the IERS in 2018
Non-tidal surface loading	Exclusively in OBS processing: ESMGFZ models in CM frame (Dill and Dobslaw 2013)
Parametrization	
Station coordinates	No-net-rotation w.r.t. IGS14 (Rebischung and Schmid 2016)
Troposphere	Zenith wet delays for 0.5 h intervals; two gradient pairs per station and day
GPS orbit modeling	Six initial conditions + nine ECOM2 parameters, pulses at 12 h
Earth rotation	Terrestrial pole coordinates, pole-rates and LOD for 24 h intervals, UT1 tightly constrained to a priori Bulletin A
Receiver clock	Pre-eliminated every epoch, ISB per station for Galileo, per station and satellite for GLONASS
Satellite clocks	Epoch-wise estimated
GNSS ambiguities	Ambiguity fixing for GPS and Galileo
Antenna phase center	Estimated for GPS, GLONASS, and Galileo but tightly constrained to values given in antex

rections applied to previously determined coordinate time series. Overall, the various studies revealed significant RMS reduction and decreased amplitudes on different frequencies especially for the station height coordinates. Incorporating loading corrections directly in the observation modeling was investigated for example by Tregoning and van Dam (2005), Dach et al. (2011), Männel et al. (2019), and Glomsda et al. (2020). The overall advantage of this approach is that corrections are consistently applied to all estimated parameters including Earth rotation parameters and satellite orbits which is of course crucial for reprocessing efforts. A comparison of different loading corrections focusing on reference frames was performed by Glomsda et al. (2022). Furthermore, loading corrections were applied in the DTRF2014 and DTRF2020 realization (Seitz et al. 2022).

The contribution is structured in the following way. After briefly describing the GNSS processing strategy and introducing the ESMGFZ loading models (Sect. 2), we discuss the correction at the solution level (Sect. 3). Section 4 contains the results when applying the corrections at the observation level. Finally, the paper closes with a summary and some conclusions in Sect. 5.

2 GNSS Processing and Loading Corrections

The GNSS processing for this investigation relies on IGS' third reprocessing campaign. As the involved Analysis Cen-

ters decided not to correct for non-tidal loading, the derived GFZ solution is a reference (abbreviated as REF) in this study (Männel et al. 2020, 2021). Table 1 summarizes the applied processing strategy. Compared to previous reprocessings and the operational products, the GPS phase center offsets and the reference frame were adjusted to the published Galileo offsets.[1] While this leads to an independent GNSS-based scale, this does not impact our non-tidal loading investigation. Overall, the GFZ repro3 solution covered 322 stations (on average 185 per day) and 132 satellites, including GPS, GLONASS (from 2012 onwards), and Galileo (from 2014 onwards). According to Rebischung (2021), the daily median formal errors for the station coordinates are 1.0, 1.0, 3.5 mm in North, East, and Up directions. To assess the correction at the observation level, we repeated the repro3, keeping all models but adding the ESMGFZ non-tidal loading corrections (OBS solution). While repro3 was initially performed from 1994 to 2020 (with an extension for 2021–2022) the repeated OBS solution is – related to computational efforts – limited to 2012.0–2016.0. Subsequently, REF and SOL (corrected the derived coordinates) solutions only contain the original repro3 results for these years. Consequently, the investigations in Sects. 3 and 4 are limited to 2012–2016. As the apriori troposphere delays were derived using GPT2 in GFZ's repro3 solution we kept this processing option. For the implications of using a Global

---

[1] For more details see IGSMAIL-8026.

Mapping Function when investigating loading effects, we refer to Steigenberger et al. (2009).

The Earth System Modelling group of Deutsches Geo-ForschungsZentrum (ESMGFZ) in Potsdam (http://isdc.gfz-potsdm.de/esmdata/loading) provides surface loading corrections based on models of the atmosphere, oceans, the terrestrial hydrosphere (Dill and Dobslaw 2013). The fourth model component ensures global mass balance by distributing the excess water mass from atmosphere and terrestrial water storage over the ocean considering loading and self-attraction via the sea level equation. The calculations are performed based on mass distributions provided by the deterministic numerical weather prediction model of the European Centre for Medium-range Weather Forecasts (ECMWF), the Max-Planck-Institute for Meteorology Ocean Model (MPIOM, Jungclaus et al. 2013), and the Land Surface Discharge Model (LSDM, Dill 2008). Corresponding surface deformations in North, East, and Up are provided with a spatial resolution of 0.5° and a temporal sampling of three hours for the atmosphere and ocean and 24 h for the continental hydrosphere and the barystatic sea-level variations. The surface deformations are provided in the center of the Earth's figure (CF) and the center of Earth's mass (CM). As of today, the ESMGFZ non-tidal loading corrections are provided without uncertainties.

An implicit comparison between repro3 results (computed without applying non-tidal loading corrections) and loading models was performed by assessing annual and semi-annual signals derived from the aggregated ESMGFZ loading models (CF frame) against the periodic signals provided along with the ITRF2020 (ITRF2020 2022). For around 80% of the 1344 stations, ITRF2020 reports larger annual amplitudes than predicted by the loading models (Fig. 1). While this is not surprising as the ITRF seasonal coefficients also contain seasonal variations of the observation geometry, systematics from near-field, and thermo-elastic signals, the amplitudes of 52% of the stations agree within 40%. A similar value was reported by Männel et al. (2019) based on a stand-alone PPP solution. Besides the overall agreement, a few stations showed discrepancies. For example, for MAPA (Santana, Brazil) located near the Amazon river, the ESMGFZ overpredicts the annual amplitude by −4 mm in the North and −10 mm in the Up direction (computed as ITRF2020-ESMGFZ). In this case, overprediction may occur as the station is too close to the loading source. The stations in Wuhan, China (WUH2, WUHN) also show larger amplitudes in the up direction for the loading time series (−4.3 and −4.5 mm, respectively), potentially for the same reason. A large horizontal discrepancy of 7.2 mm in the North occurs for UTQI (Barrow, USA), potentially related to the short time series – UTQI was installed in 2017 – and monument-related periodic variations. A more detailed study on the coordinate variability was presented by Boy et al. (2022).

3   Corrections at the Solution Level

As non-tidal loading was not considered in repro3, users might correct the corresponding deformations using the available products, such as ESMGFZ. As shown by Glomsda et al. (2020, 2021) this is possible without introducing inconsistencies at the normal equation level, which is accessible via the provided SINEX files. However, given the required expert knowledge and software capabilities, most users might prefer to simply subtract loading corrections from the extracted coordinate time series, which we call *correction at the solution level.*

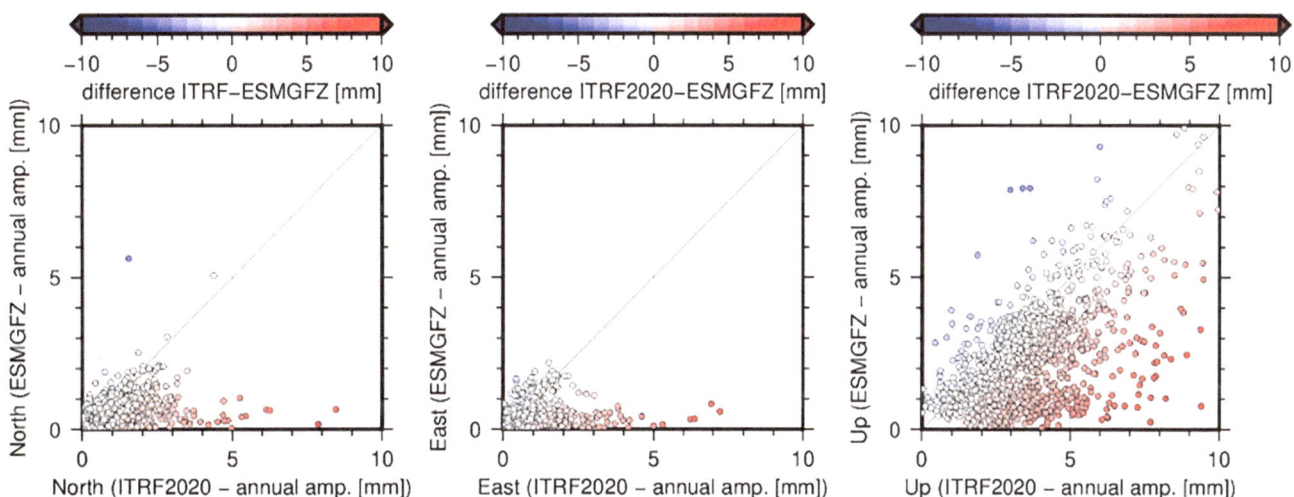

Fig. 1 Comparison of annual amplitudes computed from ESMGFZ loading models and amplitudes provided in the periodic component of ITRF2020

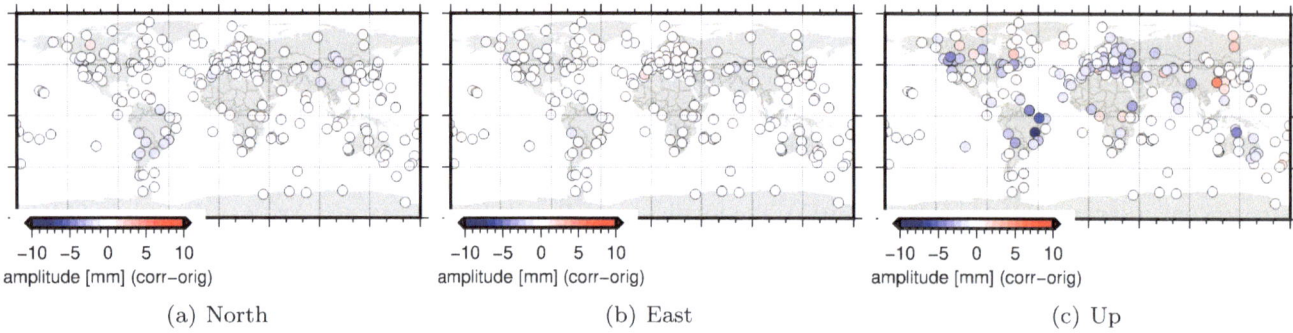

Fig. 2 Correction at the solution level: difference in annual amplitude in (**a**) North, (**b**) East, and (**c**) Up direction. Differences are computed by subtracting the original repro3 solution from a solution where corrections are applied directly to the coordinates

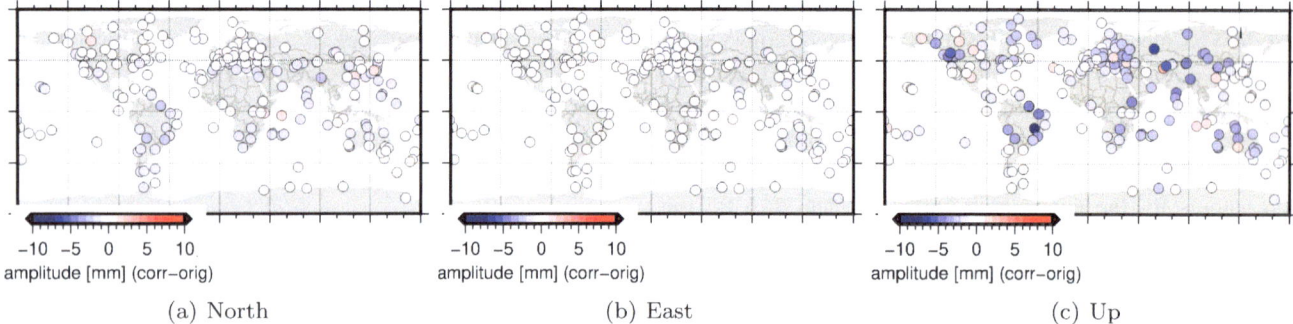

Fig. 3 Correction at the observation level: difference in annual amplitude in (**a**) North, (**b**) East, and (**c**) Up direction. Differences are computed by subtracting the original repro3 solution from the solution where corrections are applied directly at observation level

In the following, the seasonal signals and the coordinate variability (RMS of the time series) are investigated to assess the impact of applying loading corrections. While the first required the estimation of corresponding periodic signals – annual and semi-annual – in the time series assessment the RMS – computed as $\sigma_x = \sqrt{\frac{1}{n-1}\sum_{i=1}^{n}(x_i - \bar{x})^2}$ – was investigated considerung a pure linear station behavior (i.e., using a linear trajectory model). In any case, significant coordinate discontinuities (larger than 2 cm) are considered. With the given dataset, four discontinuities related to antenna replacements (stations: GRAC, JOG2, POLV, SUTM) and one related to earthquakes (SANT, Illapel earthquake 2015) are applied.

Figure 2 shows the difference in annual amplitudes for all stations with more than 600 daily coordinate solutions between 2012.0–2016.0. For the horizontal amplitudes, the effect is small (overall below 0.5 mm), which is not surprising as non-tidal loading creates primary vertical displacements, and horizontal amplitudes are small. The reduction is, on average, from 1.5 to 1.1 mm and from 1.1 to 1.0 mm for North and East, respectively. Nevertheless, amplitudes in North and East are reduced for 217 (82%) and 166 (63%) out of 264 stations. For the vertical amplitudes, an

overall reduction of −0.6 mm from 3.2 to 2.6 mm can be observed; 159 out of 264 stations show smaller amplitudes (Fig. 3). Semi-annual amplitudes, overall smaller in size, are reduced similarly. The mean coordinate variability derived by applying a linear trajectory model to the original repro3 (2012–2016) solution is 2.1, 2.1, and 5.1 mm in North, East, and Up direction. Subtracting non-tidal loading corrections before the time series adjustment leads to slightly reduced RMS values of 1.9, 2.0, and 4.8 mm (Fig. 4). Overall, the variability in North, East, and Up is reduced for 85, 70, and 68% of the stations. Few stations show larger amplitudes and RMS values after applying the corrections at the solution level. For the stations YELL and YEL2 located in Yellowknife, Canada the RMS increased by 1.8 and 2.7 mm. For YELL a similar behavior was reported already by Männel et al. (2019). Comparing the uncorrected coordinate time series and the loading models indicate significant shifts between the model and GNSS for Yellowknife and Wuhan, China (see also Fig. 6). For station YAKT (Yakutsk, Russia), ESMGFZ models and GNSS agrees but the results – RMS increased by 0.3 mm – are biased by unconsidered coordinate variations related to snow and ice on the antenna.

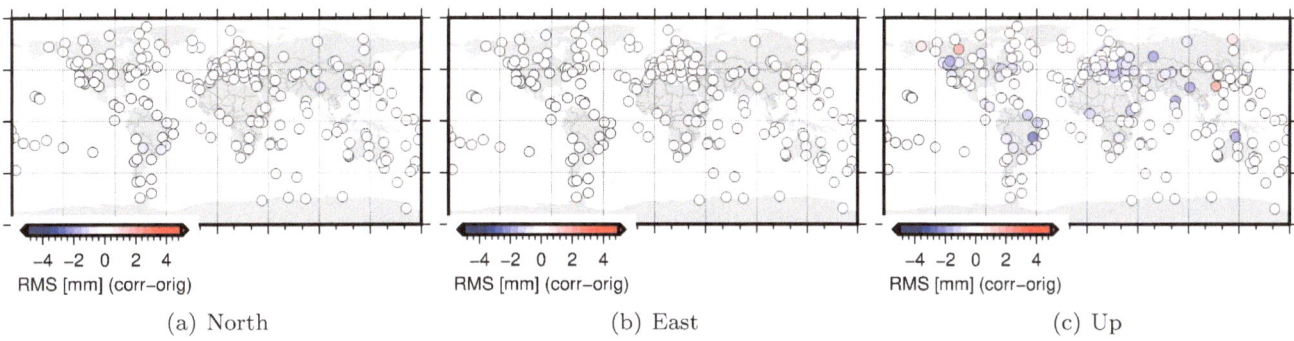

Fig. 4 Correction at the solution level: difference in station coordinate RMS using a linear trajectory model. Differences are computed by subtracting the original repro3 solution from a solution where corrections are applied directly to the coordinates

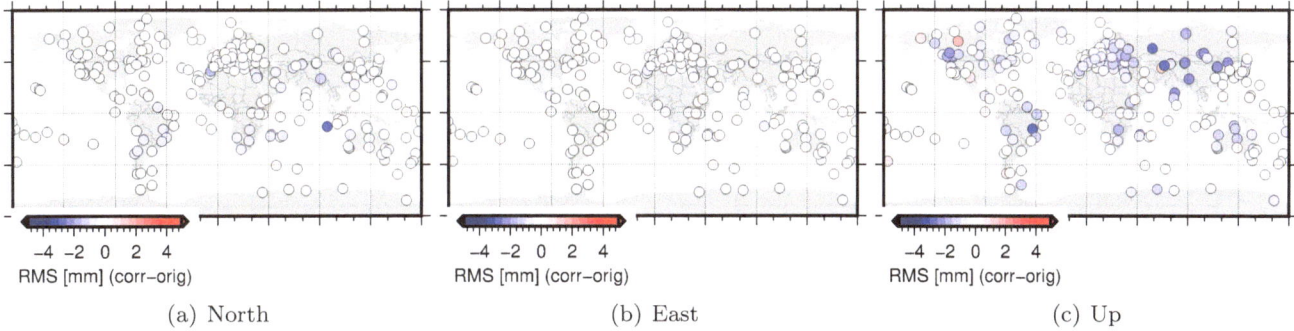

Fig. 5 Correction at the observation level: difference in station coordinate RMS using a linear trajectory model. Differences are computed by subtracting the original repro3 solution from the solution where corrections are applied directly at observation level

4 Corrections at the Observation Level

Figure 3 shows the difference in annual amplitudes for solutions OBS and REF. For the horizontal amplitudes, larger reductions are visible, from 1.5 to 0.8 and from 1.1 to 0.8 mm for North and East, respectively. Annual amplitudes in North and East are reduced for 215 (81%) and 191 (72%) out of 264 stations. An overall reduction of −1.3 mm from 3.2 to 1.9 mm can be observed for the vertical amplitudes. The geographic distribution reveals decreased annual amplitudes, especially for stations in South America (large signals in terrestrial water storage) and central Asia (significant atmospheric loading). A small fraction of the considered stations (5%) show larger amplitudes if correcting for surface loading. Similar to the SOL results, semi-annual amplitudes are reduced but are overall small in size. The mean coordinate variability derived by applying a linear trajectory model to the original repro3 (2012–2016) solution is 2.1, 2.1, and 5.1 mm in North, East, and Up direction. Correcting for non-tidal loading corrections at the observation level leads to slightly reduced RMS values of 1.8, 2.0, and 4.5 mm (Fig. 5). Overall, the variability in North, East, and Up is reduced for 90, 80, and 84% of the stations. Around 3% of the stations show RMS values increased by more than 0.5 mm; among them stations at Yellowknife (Canada). Compared

with the SOL results, the stations in Wuhan show significant improvement.

Figure 6 shows an individual comparison between GNSS and loading models and between uncorrected and corrected height time series for selected stations discussed already above. BRAZ (Brasila, Brazil) is strongly affected by hydrological loading with the models predicting more than 20 mm peak-to-peak variations. The GNSS time series represent that displacement quite well; a correlation factor of 0.81 is derived between the model and GNSS time series (smoothed with a monthly moving average[2]). Consequently, the corrected time series shows a significant reduction in annual amplitude (from 10.9 mm to 3.2 mm) and RMS (9.7 mm to 6.7 mm). Applying the correction at the observation level has no additional impact on the amplitude but reduces the RMS by additional 0.4 mm. For the stations in Jiufeng (JFNG) and Wuhan (WUH2) located within a distance of around 14 km the models also predict peak-to-peak variations of 20 mm driven by the hydrological loading. This strong signal is dominated by Yangtze and Han rivers and by the several hundred lakes within the Jianghan Plain. Despite data gaps in the GNSS, time series for JFNG, WUH2, and the co-located station WUHN agree, especially in 2015. However, the GNSS-based motion patterns differ significantly

---

[2]The correlation is 0.62 if using the pure time series.

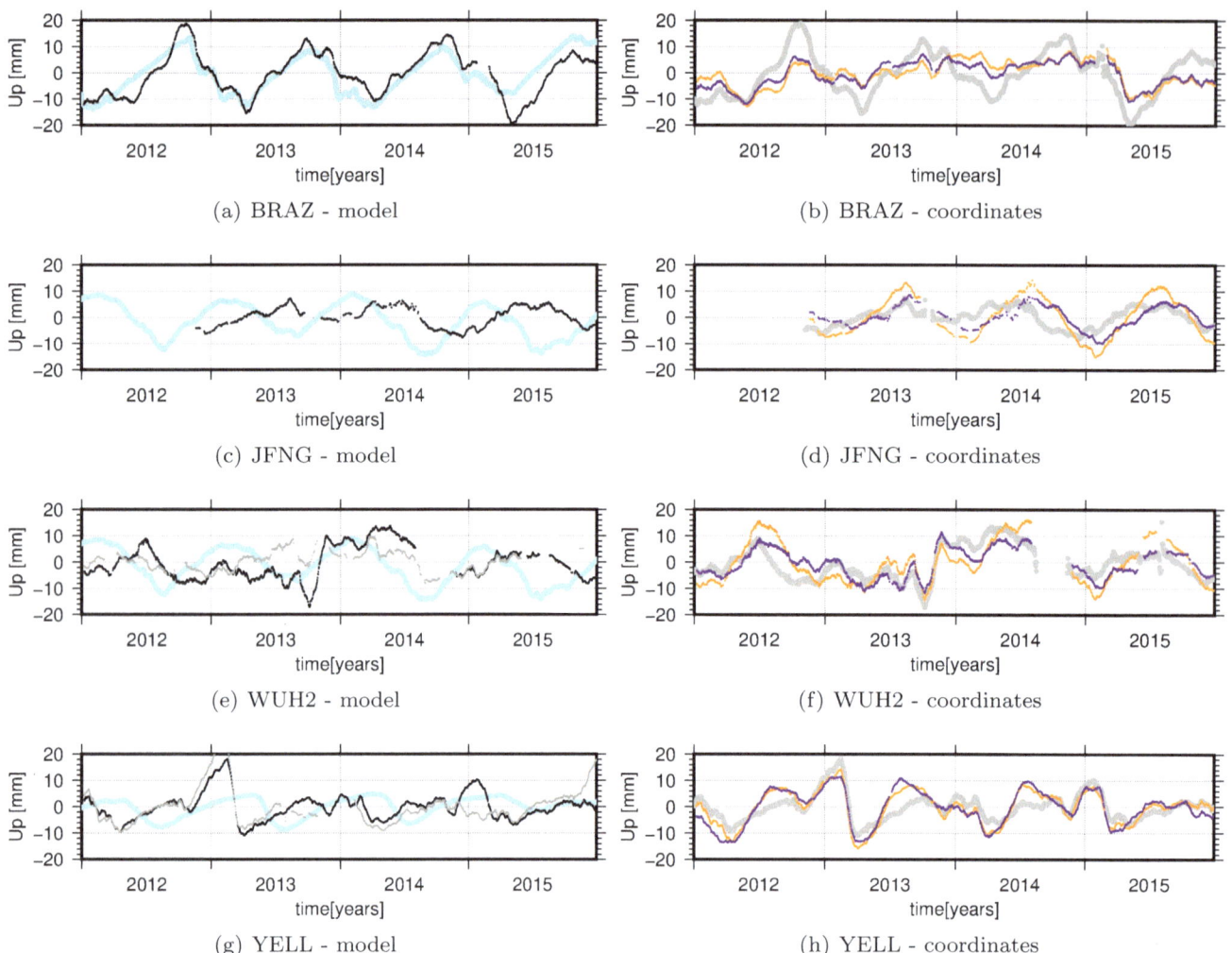

Fig. 6 Comparison of the model-based displacements (*blue*) and the derived GNSS time series (*black*) for stations BRAZ (Brasilia, Brazil), JFNG (Jiufeng, China), WUH2 (Wuhan, China), and YELL (Yellowknife, Canada) in figures (**a, c, e, g**); co-located stations WUHN and YEL2 are added to figure (**e**) and (**g**) (*grey*). Comparison of time series, original (*grey*), loading corrected at solution level (*orange*) and at observation level (*purple*) in figures (**b, d, f, h**). Results based on linear trajectory model, moving average for GNSS time series applied

in phase; a shift of around 100 days (loading is ahead) is determined for JFNG and WUH2 with a correlation of 0.7 and 0.5 respectively. If applying the correction at the solution level, annual amplitudes increase from around 3–4 mm to 8–10 mm. Correcting at the observation level performs better for these stations leading to reduced amplitudes in Wuhan and only slight increases in Jiufeng. The stations YELL and YEL2 located close to Canada's Great Slave Lake, also show large differences between GNSS results and model-based displacement series (GNSS is ahead by around 90 days). Applying loading correction leads to significantly increased amplitudes independent of the level at which the correction is applied.

5 Summary and Conclusions

Time-dependent mass variations in the atmosphere, oceans, and hydrosphere lead to significant deformations of the Earth's surface. Associated displacement corrections are provided in dedicated non-tidal loading models. An initial comparison between seasonal signals determined from the ESMGFZ models and ITRF2020's periodic functions revealed a good agreement with amplitude differences below 10 mm. Overall, the ITRF2020 amplitudes are larger than the loading amplitudes for around 80% of the stations while the amplitudes agree within 40% for half of the 1344 stations.

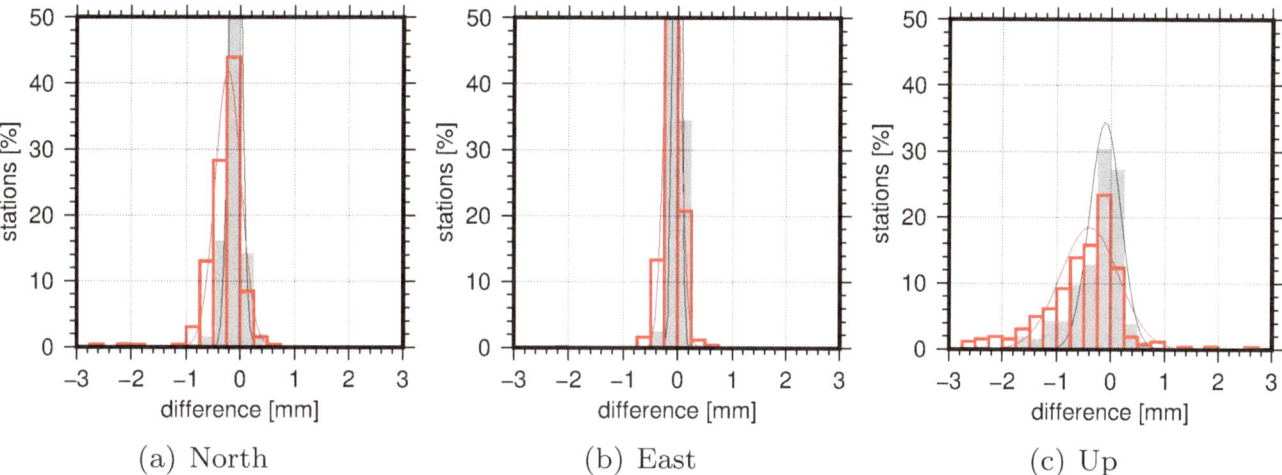

(a) North   (b) East   (c) Up

Fig. 7 Histogram of the RMS differences between original GNSS time series and solution derived with applying the loading corrections at the solution level (*solid, grey*), and at the observation level (*red outline*). The equivalent normal distribution was added

While the beneficial impact of applying loading corrections on GNSS-based time series has been shown in different studies, our focus is on the differences between correcting at the solution and the observation level. For the vertical amplitudes a significant reduction by 18% was found at the solution level, while a decrease by 42% was achieved at the observation level. Figure 7 compares the resulting RMS differences for both solutions with respect to the original repro3 solution. While the negative values indicate the overall improvement, the additional improvement related to the observation level is visible as the difference between the grey (SOL results) and red (OBS results) bars. Overall, applying non-tidal loading corrections at the solution level decreased the height RMS by 5% while corrections at the observation level lead to a 10% reduction.

Finally, we recommend to apply the corrections at the observation level whenever possible. This is also beneficial for the consistency between the simultaneously estimated station coordinates, satellite orbits, and Earth rotation parameters.

Acknowledgements The authors want to thank IGS (Johnston et al. 2017) for making the GNSS data publicly available. We would also like to thank two anonymous reviewers for their assistance in evaluating this paper and their helpful recommendations.

Author Contribution B.M. and S.G. defined the study. A. B. and B.M. processed the GNSS data and prepared the different solutions. All authors contributed to the analysis, interpretation, and discussion of the results. B.M. prepared the manuscript with major contributions from S.G. and H.S. All authors read and approved the final manuscript.

Data Availability Statement All GNSS data are available at IGS data centers, the non-tidal loading corrections are available at http://rz-vm115.gfz-potsdam.de:8080/repository The GFZ repro3 products are available via Männel et al. (2021), while the SOL and OBS results can be made available upon individual request.

References

Altamimi Z, Rebischung P, Collilieux X, Métivier L, Chanard K (2023) Itrf2020: an augmented reference frame refining the modeling of nonlinear station motions. J Geodesy 97(5):47. https://doi.org/10.1007/s00190-023-01738-w

Böhm J, Werl B, Schuh H (2006) Troposphere mapping functions for GPS and VLBI from European Centre for medium-range weather forecasts operational analysis data. J Geophy Res 111(B2):B02,406. https://doi.org/10.1029/2005JB003629

Boy JP, Rebischung P, Altamimi Z (2022) Comparison of ITRF2020 residual displacements with environmental loading models. https://www.refag2022.org/wp-content/uploads/2022/11/2022REFAG_BOY_ID37_lr.pdf, REFAG 2022, Thessalonik, Greece

Dach R, Böhm J, Lutz S, Steigenberger P, Beutler G (2011) Evaluation of the impact of atmospheric pressure loading modeling on GNSS data analysis. J Geod 85(2):75–91. https://doi.org/10.1007/s00190-010-0417-z

Desai SD, Sibois AE (2016) Evaluating predicted diurnal and semidiurnal tidal variations in polar motion with gps-based observations. J Geophys Res Solid Earth 121(7):5237–5256. https://doi.org/10.1002/2016JB013125

Dill R (2008) Hydrological model LSDM for operational Earth rotation and gravity field variations. Scientific technical report STR, vol 369. GFZ

Dill R, Dobslaw H (2013) Numerical simulations of global-scale high-resolution hydrological crustal deformations. J Geophys Res 118(9):5008–5017. https://doi.org/10.1002/jgrb.50353

Glomsda M, Bloßfeld M, Seitz M, Seitz F (2020) Benefits of non-tidal loading applied at distinct levels in VLBI analysis. J Geodesy 94(9):90. https://doi.org/10.1007/s00190-020-01418-z

Glomsda M, Bloßfeld M, Seitz M, Seitz F (2021) Correcting for site displacements at different levels of the Gauss-Markov model - A case study for geodetic VLBI. Adv Space Res 68(4):1645–1662. https://doi.org/10.1016/j.asr.2021.04.006

Glomsda M, Bloßfeld M, Seitz M, Angermann D, Seitz F (2022) Comparison of non-tidal loading data for application in a secular terrestrial reference frame. Earth Planets Space 74(1):87. https://doi.org/10.1186/s40623-022-01634-1

Gobron K, Rebischung P, Van Camp M, Demoulin A, de Viron O (2021) Influence of aperiodic non-tidal atmospheric and oceanic loading deformations on the stochastic properties of global GNSS vertical land motion time series. J Geophys Res Solid Earth 126(9):e2021JB022,370. https://doi.org/10.1029/2021JB022370. https://agupubs.onlinelibrary.wiley.com/doi/abs/10.1029/2021JB022370, e2021JB022370 2021JB022370

ITRF2020 (2022) ITRF2020 solution. https://itrf.ign.fr/en/solutions/ITRF2020, online; accessed 13-February-2023

Johnston G, Riddell A, Hausler G (2017) The international GNSS service. Springer International Publishing, Cham, Switzerland, pp 967–982. https://doi.org/10.1007/978-3-319-42928-1

Jungclaus JH, Fischer N, Haak H, Lohmann K, Marotzke J, Matei D, Mikolajewicz U, Notz D, von Storch JS (2013) Characteristics of the ocean simulations in the Max Planck Institute Ocean Model (MPIOM) the ocean component of the MPI-Earth system model. J Adv Model Earth Sy 5(2):422–446. https://doi.org/10.1002/jame.20023

Klos A, Dobslaw H, Dill R, Bogusz J (2021) Identifying the sensitivity of GPS to non-tidal loadings at various time resolutions: examining vertical displacements from continental Eurasia. GPS Solut 25(3):89. https://doi.org/10.1007/s10291-021-01135-w

Kvas A, Mayer-Gürr T, Krauss S, Brockmann JM, Schubert T, Schuh WD, Pail R, Gruber T, Jäggi A, Meyer U (2019) The satellite-only gravity field model goco06s. https://doi.org/10.5880/ICGEM.2019.002. http://dataservices.gfz-potsdam.de/icgem/showshort.php?id=escidoc:4081892

Lyard FH, Allain DJ, Cancet M, Carrere L, Picot N (2021) FES2014 global ocean tide atlas: design and performance. Ocean Sci 17(3):615–649. https://doi.org/10.5194/os-17-615-2021

Männel B, Dobslaw H, Dill R, Glaser S, Balidakis K, Thomas M, Schuh H (2019) Correcting surface loading at the observation level: impact on global GNSS and VLBI station networks. J Geod 93:2003–2017. https://doi.org/10.1007/s00190-019-01298-y

Männel B, Brandt A, Bradke M, Sakic P, Brack A, Nischan T (2020) Status of IGS reprocessing activities at GFZ. Springer, Berlin, Heidelberg, pp 1–7. https://doi.org/10.1007/1345_2020_98

Männel B, Brandt A, Bradke M, Sakic P, Brack A, Nischan T (2021) GFZ repro3 product series for the International GNSS Service (IGS). https://doi.org/10.5880/GFZ.1.1.2021.001. GFZ Data Services

Martens HR, Argus DF, Norberg C, Blewitt G, Herring TA, Moore AW, Hammond WC, Kreemer C (2020) Atmospheric pressure loading in GPS positions: dependency on GPS processing methods and effect on assessment of seasonal deformation in the contiguous USA and Alaska. J Geodesy 94(12):115. https://doi.org/10.1007/s00190-020-01445-w

Mémin A, Boy JP, Santamaría-Gómez A (2020) Correcting GPS measurements for non-tidal loading. GPS Solut 24(2):45. https://doi.org/10.1007/s10291-020-0959-3

Petit G, Luzum B (2010) IERS Conventions (2010). IERS Technical Note 36. Verlag des Bundesamts für Kartographie und Geodäsie, Frankfurt am Main, iSBN 3-89888-989-6

Ray R, Ponte R (2003) Barometric tides from ECMWF operational analyses. Ann Geophys 21(8):1897–1910

Rebischung P (2021) Terrestrial frame solutions from the third igs reprocessing: the igs contribution to itrf2020. In: Tour de lÍGS. https://www.igs.org/tour-de-ligs-repro3/

Rebischung P, Schmid R (2016) Igs14/igs14.atx: a new framework for the igs products. In: AGU Fall Meeting, San Francisco, CA. https://mediatum.ub.tum.de/doc/1341338/le.pdf

Seitz M, Bloßfeld M, Angermann D, Seitz F (2022) DTRF2014: DGFI-TUM's ITRS realization 2014. Adv Space Res 69(6):2391–2420. https://doi.org/10.1016/j.asr.2021.12.037

Steigenberger P, Böhm J, Tesmer V (2009) Comparison of GMF/GPT with VMF1/ECMWF and implications for atmospheric loading. J Geod 83(10):943. https://doi.org/10.1007/s00190-009-0311-8

Tregoning P, van Dam T (2005) Effects of atmospheric pressure loading and seven-parameter transformations on estimates of geocenter motion and station heights from space geodetic observations. J Geophys Res 110(B3):n/a–n/a. https://doi.org/10.1029/2004JB003334, b03408

Upgrading the Metsähovi Geodetic Research Station

Markku Poutanen, Mirjam Bilker-Koivula, Joona Eskelinen, Ulla Kallio,
Niko Kareinen, Hannu Koivula, Sonja Lahtinen, Jyri Näränen,
Jouni Peltoniemi, Arttu Raja-Halli, Paavo Rouhiainen, and Nataliya Zubko

Abstract

Metsähovi Geodetic Research Station (MGRS) of the National Land Survey of Finland, has undergone a major upgrade. The first observations at MGRS were started in 1978. A decade-long reform began in 2012, during which all major systems were renewed. This included Global Navigation Satellite System (GNSS) station, Satellite Laser Ranging (SLR) system, and a dedicated geodetic Very Long Baseline Interferometer (VLBI) system. Furthermore, the absolute gravimeter (AG) was upgraded, the superconducting gravimeter (SG) was renewed, and the station infrastructure was completely refurbished. When completed, MGRS will be one of the northernmost stations in the core network of the Global Geodetic Observing System (GGOS) of the International Association of Geodesy (IAG). MGRS has a full suite of co-located major geodetic instrumentation, and local geodetic networks and facilities to connect various observing techniques (local ties). Together, the core stations form the solid backbone for maintaining the International Terrestrial Reference Frame (ITRF) and monitoring the orientation of the Earth in space and producing information for computing precise satellite orbits, including GNSS. The stability of the stations and their long and consistent series of measurements is paramount both for global and regional networks. We present recent developments at MGRS and introduce the instrumentation that already contributes and will contribute in the future to various IAG services.

Keywords

Core stations · Geodetic observations · GGOS · Reference frames

1   Introduction and History

The Metsähovi Geodetic Research Station (MGRS) at the village Kylmälä in the Kirkkonummi municipality, Finland, was established in 1975. In the same area, there are also the Metsähovi Radio Observatory of Aalto University and small optical telescopes of the University of Helsinki. MGRS is managed by the Finnish Geospatial Research Institute (FGI) of the National Land Survey (NLS) (until 2015 the Finnish Geodetic Institute, FGI). It is one of the northernmost geodetic stations in the core network of the Global Geodetic Observing System (GGOS) of the International Association of Geodesy (IAG), located 60° north.

The first geodetic measurements at MGRS, as part of the global International Satellite Laser Ranging Service (ILRS) network, started in 1978 with an in-house made Satellite Laser Ranging (SLR) system consisting of a 63 cm telescope and a Ruby laser capable of transmitting four pulses per minute. In 1993 the system was replaced by a 100 cm telescope and a 1 Hz-capable Nd:YAG laser. In 2005 it had to be abandoned because the maintenance of the laser, in particular, became impossible and the telescope was unsuitable for modern kHz or faster operation. Due to the lack of funding,

M. Poutanen (✉) · M. Bilker-Koivula · J. Eskelinen · U. Kallio ·
N. Kareinen · H. Koivula · S. Lahtinen · J. Näränen · J. Peltoniemi ·
A. Raja-Halli · P. Rouhiainen · N. Zubko
Finnish Geospatial Research Institute FGI, National Land Survey of
Finland, Espoo, Finland
e-mail: markku.poutanen@nls.fi

© The Author(s) 2023
J. T. Freymueller, L. Sánchez (eds.), *Gravity, Positioning and Reference Frames*,
International Association of Geodesy Symposia 156, https://doi.org/10.1007/1345_2023_203

the acquisition of a modern kHz-capable SLR system could only be started in 2012. For further details of the SLR history, see Raja-Halli et al. (2019).

The first permanent GPS station, now identified as METS00FIN, was established in 1992. First, data were sent to the Cooperative International GPS Network (CIGNET) and since 1994 to the International GNSS Service (IGS), and from 1995 also to the EUREF Permanent GNSS Network (EPN). Because the antenna is placed on a 25 m tall mast, the thermal expansion is eliminated with an invar wire-based compensator (Koivula et al. 1998). Additionally, a NASA/UNAVCO GNSS receiver is sharing the antenna of METS00FIN.

MGRS has hosted and provided local support to the Détermination d'Orbite et Radiopositionnement Intégré par Satellite (DORIS) radio beacon since 1990 (Koivula et al. 1998). The DORIS station is located about three kilometres from the Metsähovi main station to minimise radio frequency interference. It is operated and maintained by the French Space Agency (Centre national d'études spatiales, CNES). The DORIS system has been upgraded four times since the initial installation. METG00FIN GNSS station is located at the DORIS station and it is part of the Réseau GNSS pour l'IGS et la NAvigation (REGINA) infrastructure maintained by CNES and Institut national de l'information géographique et forestière (IGN). METG00FIN submits data also to EPN and IGS networks.

FGI procured its first absolute gravimeter (AG), JILAg-5, in 1986. In 2003 the JILAg-5 was replaced by the FG5-221, which was subsequently updated to FG5X-221 in 2014. In Finland, the FG5X-221 is used to, e.g., monitor the gravity change at MGRS and other locations due to postglacial land uplift, and to maintain the national gravity network. It is also the national standard for free-fall acceleration of gravity. FGI AG has been used in several places abroad, from Svalbard to Antarctica, and many AGs have visited the MGRS (Arnautov et al. 1982; Gitlein 2009; Bilker-Koivula et al. 2021).

In 1994 a dedicated gravity building was built at MGRS (Fig. 1). The building has two laboratory rooms, one for absolute gravimeters and one for superconductive gravimeters. In 1994 FGI procured a superconductive gravimeter (SG) GWR T020, which was continuously operated until 2016. It was then replaced with more modern GWR iOSG-022 and iGRAV-013 superconductive gravimeters (Virtanen and Raja-Halli 2018).

The geodetic Very Long Baseline Interferometry (VLBI) observations have been made since 2004 with the 13.7 m radio telescope of the Aalto University Metsähovi Radio Observatory in the same area (Poutanen and Koivula 2007). An S/X receiver and Mk5A data acquisition system had been acquired for the purpose. Later the system was upgraded to Mk5B and DBBC2. Other equipment, like the hydrogen maser for timing, were already available. Observing time for a few geodetic VLBI sessions per year was purchased from

Fig. 1 Aerial photo of the MGRS. The gravimetric laboratory is at the left and the old main building is in the centre with the new VGOS telescope in the background. The SLR observatory is on the right and behind it is the new main building

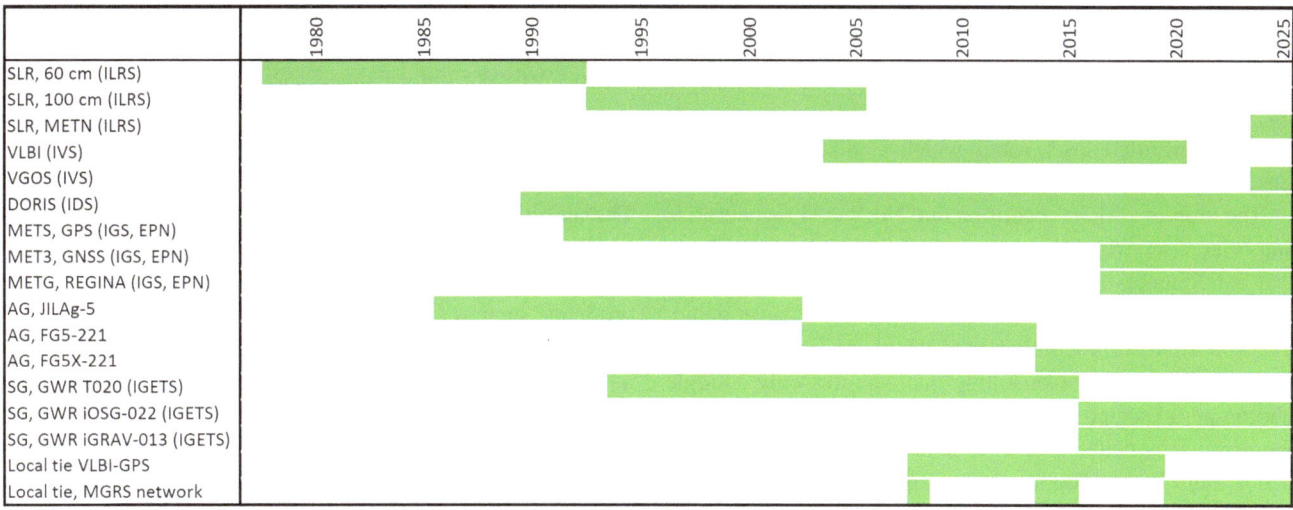

Fig. 2 MGRS equipment timeline and contribution to the IAG services. 2025 timeline is an estimation

Aalto University until 2021. Observations will continue with the new dedicated geodetic VLBI system at MGRS after the commissioning phase.

Techniques and facilities for GPS local tie measurements between the METS00FIN and the legacy telescope, simultaneously with the VLBI sessions, were developed (Kallio and Poutanen 2012). Observations started in 2008 and continued until 2015. The validation of the GPS local tie system was made with the automatic robot tachymeter monitoring during two VLBI sessions in the EMRP SIB60 project in 2015. The local tie vector from METS00FIN to legacy telescope MET-SAHOV in the ITRF2020 solution is based on the terrestrial measurements within the same project (Pollinger et al. 2015; Jokela et al. 2016; Kallio et al. 2016). Terrestrial techniques were further used to measure the consistent set of local tie vectors between all space geodetic instruments at MGRS.

All instruments and reference benchmarks and pillars are established on solid granite bedrock, except the DORIS/REGINA station. Even though the implication is that a minimal danger for movements of the benchmarks or instruments exists, regular control of the benchmarks (local tie measurements) is necessary and performed.

The reference points of the national height (N2000) and gravity (FOGN) systems are also located at Metsähovi. MGRS is also an International Height Reference Frame (IHRF) core site.

At the beginning of the millennium, the equipment of MGRS was partly outdated and did not meet the new requirements set for IAG's services. The modernization of MGRS started in 2012 with special funding from the Ministry of Agriculture and Forestry and later on also by the internal funding of the NLS. The modernisation includes the renewal or updating of all major measurement systems, as well as the construction of a new VLBI Global Observing System

(VGOS) system, SLR system and improvement of the station infrastructure and working facilities (Fig. 1). The MGRS data contribute through IAG services to various global geodetic and Earth-exploring purposes (Fig. 2). From 2023, the newly established Geodetic Infrastructure Unit of the NLS will be responsible for the daily operation and maintenance of instruments and facilities and the Ministry of Agriculture and Forestry has allocated special funding for the operation of MGRS.

2 Renewal of Major Instruments

2.1 SLR

The SLR technique is used to determine satellite orbits, and the observations from the northern location of MGRS are important for it. The design and integration of the cost-effective and modern third-generation SLR system was done in-house because there were no commercial solutions available for procuring a complete turnkey system. The system was designed to fulfil data quality and quantity requirements for a modern SLR system with the ILRS network (ILRS 2023). Most of the major subsystems, such as the observatory dome (Baader Planetarium GmbH), the telescope (Cybioms Corp.), and the operating system (DiGOS GmbH) were procured from commercial companies. A modern laser, built by High Q Laser Production GmbH, had been procured earlier for the modernisation attempt of the second-generation SLR system around the time the system failed and it was decided to move the laser into the third-generation system as legacy instrumentation.

The MGRS SLR system is bi-static, i.e. it has a twin telescope mounted on a shared gimbal. A ten cm refractor with a Coudé focus is used to transmit the laser pulses and a 50 cm reflector with the receiver system at the Cassegrain focus is used to capture the return pulses. Coudé focus allows for the laser to be situated downstairs in a separate and climate-controlled laser room which is a requirement for stable laser operation.

The laser is a 2 kHz repetition rate Nd:YVO4 regenerative amplifier laser which has a native wavelength of 1062 nm that is frequency-doubled to 532 nm for SLR operation. The laser pulse length is approximately 12 ps. The receiver detector is a time-walk compensated single-photon avalanche diode built by PESO Consulting that captures only a single photon per expected return pulse. The quantum efficiency of the receiver is better than 40% and it has a time resolution better than 20 ps. The system is triggered with an FPGA-based range-gate generator, built by DiGOS GmbH that can handle all foreseeable operating scenarios. Timing is based on an Eventech A033-ET event timer with a non-linearity better than 2 ps. The time and frequency base is provided by a GNSS-controlled oven-controlled crystal oscillator (Meinberg GmbH) with short-term stability of 5 ps. The system is operated via a dedicated operation control system written in C++ and real-time operations are handled via a dedicated Linux Ubuntu operating system with a low latency kernel.

Expected system performance, based on radar link budget calculations (Degnan 1993), system component specifications and experiences of SLR stations with similar instrumentation, has the normal point accuracy of 1 mm at a distance of 300 km and 1 cm at a distance of 20,000 km. The system is capable of daytime observations up to GNSS orbits. This is especially important considering that statistically most clear skies occur mainly between March and September when there is almost no dark time at the MGRS latitudes. Based on the observations made with a local cloud sensor (Boltwood CloudSensor II) we have estimated that Metsähovi SLR will have suitable weather for observations approximately 30–40% of the time (Del Pino et al. 2017).

A new observatory building was inaugurated in 2014 and all of the major system components were in place in late 2015. However, there have been unforeseen difficulties with the commissioning of the telescope subcomponent by the manufacturer. We anticipate test observations in 2024.

2.2 VGOS

Geodetic VLBI is the only technique capable of observing all Earth Orientation Parameters (EOP), especially UT1-UTC, and therefore invaluable e.g. for GNSS. Development towards the VGOS concept (Petrachenko et al. 2009) requires dedicated, relatively small (12 m) and fast slewing radio telescopes, using broadband receivers. To meet the requirements, the capability of the legacy telescope is not sufficient.

The VGOS performances should respond to the high demands for accuracy, reliability, and timeliness of the global reference frame. This in turn requires an order-of-magnitude improvement in the geodetic VLBI measurement accuracy. The ambitious goals of VGOS are 1 mm accuracy for positioning, 0.1 mm/yr for velocity, continuous operation, and extremely fast processing of the data to obtain the geodetic products, such as EOPs, in near real-time.

The new MGRS VGOS telescope has been installed and commissioned during 2018–2020. The telescope has been designed by MT Mechatronics GmbH. The technical characteristics of the telescope meet the VLBI2010 antenna specifications described in (Petrachenko et al. 2009).

The telescope dish is mounted on a steel pedestal embedded in a large, heavily reinforced concrete block which is firmly attached to the solid bedrock. To ensure the stability of the telescope reference point (intersection of the azimuth and elevation axis of the telescope), the pedestal was additionally layered with extra insulation and a shell at the end of 2022. Twelve temperature sensors have been installed inside the pedestal to monitor its temperature.

The telescope has a ring-focus antenna with a 13.2 m diameter main reflector. It is a high-speed slewing antenna with $12°/s$ in azimuth and $6°/s$ in elevation axes, the acceleration on both axes is $2.5°/s^2$. The surface accuracy of the main and secondary reflectors is better than 0.3 mm RMS and 0.1 mm RMS, which has a good margin to work within the VGOS frequencies and above.

The telescope is equipped with a broadband receiver manufactured by the IGN-Yebes (Spain) technology development centre in October 2019. The receiver has a quad-ridge feed horn (QRFH), designed to measure both linear polarisations at a frequency range of 2.1–14.1 GHz. The signal from the receiver is passed through the filtering and pre-amplifier modules, and each polarization component is divided into low (2.1–5.6 GHz) and high (3.6–14.1 GHz) frequency bands. Initial RFI-background evaluation has been carried out. Substantial interference sources were identified, especially in the 2–3 GHz range, thus limiting the use of the lower band. The vertical and horizontal polarisation components of the signal are distributed into 4 (5) channels by the filter bank module and further transferred over a fibre link to the backend located in the instrumentation room of the station's main building. At the backend, the signal is digitised with a Digital Baseband Converter DBBC3, produced by Hat-Lab (Tuccari et al. 2014). A FlexBuff system will be used for signal recording. The first light with the receiving system was obtained at the end of 2019. The finalisation of the VGOS signal chain is in progress with an expected test observation period of the whole system beginning in 2024.

2.3 GNSS

A new GNSS station, MET300FIN, was established as part of the renewal of MGRS, which was done in parallel with the modernisation of the Finnish permanent GNSS network FinnRef. All the new FinnRef stations (47 stations in total) have the same design and the same type of instruments. The choke ring antenna has been installed on the top of a steel grid mast, which has been attached to the bedrock whenever possible. The steel grid masts were preferred to concrete pillars because they were much easier to install in remote locations. We had experience with both concrete and steel grid masts, but we did not notice any difference. The setup at MGRS and other FinnRef stations allows operating other receivers parallel to the official stations. The technical details of the instrumentation are available in the site logs, e.g. by EPOS, EPN, or IGS websites (https://gnss-metadata.eu/, https://epncb.eu/, https://igs.org/). The MET300FIN was included in the EPN and IGS networks in 2017.

The MET300FIN station has achieved the same high quality as the METS00FIN. The RMS of the position time series are approximately 1.0 and 3.2 mm in north/east and up, respectively, without any strong seasonal signals. The receivers of METS00FIN and MET300FIN were both upgraded from Javad TRE_G3TH DELTA to Javad TRE_3 DELTA in spring 2018 to enable new tracking features. However, the change caused a small but significant jump in the time series as shown in Fig. 3. The same discontinuity

was observed in the METS00FIN time series. Lahtinen (2022) computed a zero baseline between parallel-tracked data of the two receiver types and found a difference of 0.7 mm in the east component that corresponds to the detected change in the position time series. However, as the effect is on a one mm level, it may not appear in the analysis with a higher noise level.

2.4 DORIS

With the global coverage of DORIS stations, the International DORIS Service (IDS) provides the geodetic community with various space geodetic products that are independent of, e.g., Global Navigation Satellite Systems (GNSS). DORIS is thus considered one of the major geodetic techniques of a geodetic core station. The latest (4th) generation DORIS system was installed at MGRS in 2021. The infrastructure for the system has also been updated within the overall MGRS upgrade, with a new instrument shelter installed in 2022.

The coordinates between the main station instrument reference points and the DORIS beacon are regularly verified that they are not moving locally relative to each other. The height of the REGINA point (originally a mobile VLBI monument from 1988) has been levelled relative to a nearby bedrock point. During the first decade, there has been a small subsidence, which has been stabilised (Fig. 4). Locally

Fig. 3 Detrended position time series of MET300FIN computed by the NKG GNSS Analysis Centre (Lahtinen et al. 2018). The vertical dashed line shows the epoch of the receiver change at the station

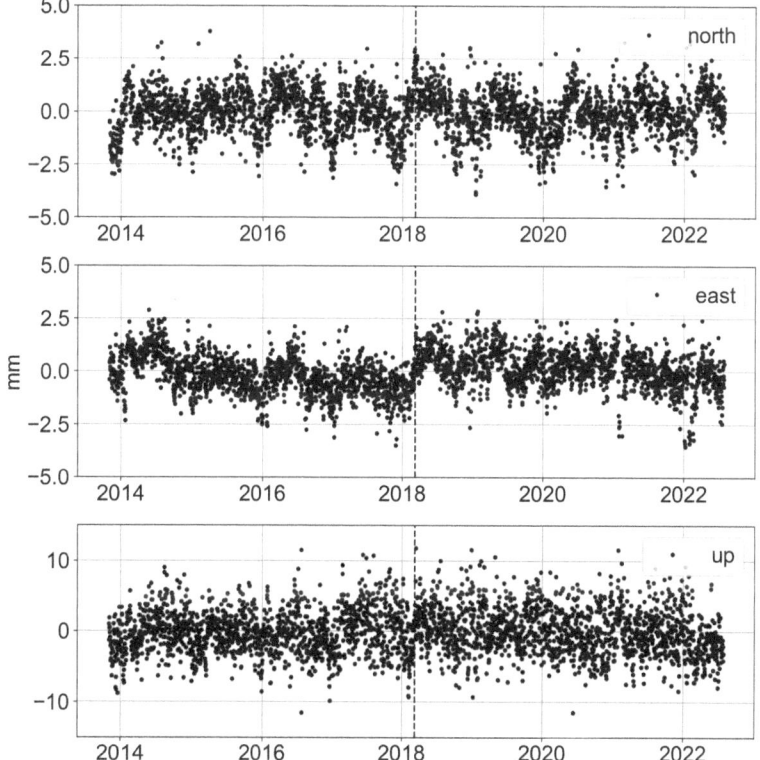

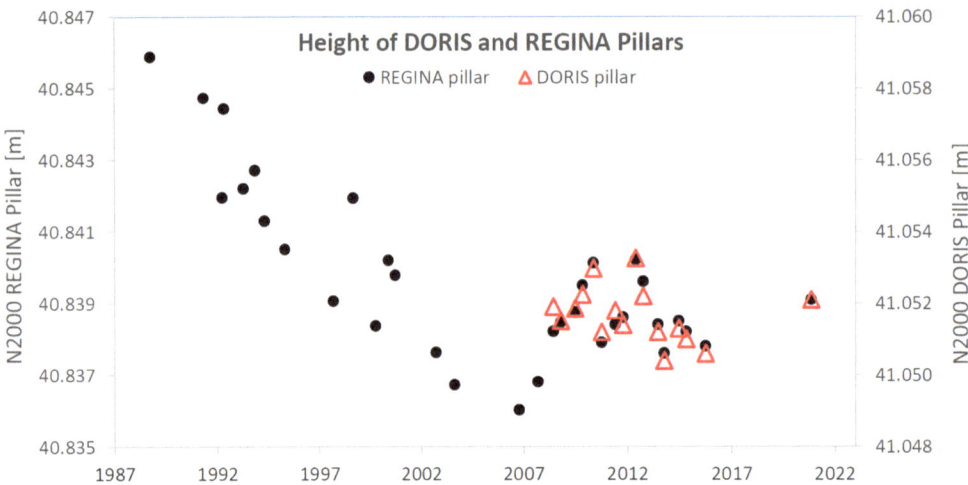

Fig. 4 Height of DORIS and REGINA pillars in the national N2000 height system. REGINA antenna mast is placed on the old pillar which was originally built for 1988 Mobile VLBI. The time series to the current DORIS pillar is shorter

the height difference between the DORIS beacon and the REGINA pillar has been monitored regularly. There are no relative height changes between these two instruments. Because the heights are relative to a bedrock benchmark, the postglacial uplift, about 4 mm/yr, is not visible in the time series.

As the REGINA station METG00FIN is not established on bedrock, we prefer not to constrain its velocity to the METS00FIN and MET300FIN station velocities in the reference frame-related works.

2.5   Gravity Instruments

Absolute (AG) and superconducting (SG) gravimeters are prerequisites for creating and maintaining national gravity systems and for researching temporal and spatial variations of gravity. They also contribute to global and regional gravity networks. The absolute gravimeter FG5X-221 is also the Finnish national standard for the free-fall acceleration of gravity.

The FG5X-221 absolute gravimeter provides measurements traceable to the SI. Additionally, the superconducting gravimeters monitor the gravity changes in time. Due to the metrological traceability and detailed time series of the gravity change, the laboratory is suitable for absolute gravimeter comparisons. The MGRS contributes also to the realisation of the International Terrestrial Gravity Reference System, ITGRS (Wziontek et al. 2021).

Both gravity laboratories at MGRS have multiple concrete pillars connected directly to the bedrock and mechanically separated from the building to minimize disturbance to the instruments. Thermal stability better than 0.2 °C can be achieved in the laboratories during measurements. Bilateral

AG comparisons have regularly taken place on two pillars in the AG laboratory. These include comparisons with the FG5-233 (Lantmäteriet), FG5-220 (IfE, Univ. Hannover), FG5-301 (BKG), FG5-101 (BKG) and the GBL-P-1 (TsNI-IGAiK). The FG5X-221 routinely participates in international comparisons of absolute gravimeters (e.g. Wu et al. 2020).

The new SG's, GWR iOSG-022 and iGRAV-013, replacing the first superconducting gravimeter (GWR SG-T020) are located in the same laboratory room on separate concrete pillars (Virtanen and Raja-Halli 2018). The iGrav is a transportable model and may be taken out of the laboratory for off-site measurements.

SGs are the most sensitive relative gravity instruments with an accuracy in the order of 10^{-11} m/s^2. One of the largest un-modelled gravity signals comes through the direct attraction from the changes in groundwater level as well as changes in the local hydrology (e.g., soil moisture, surface water, snow, Mäkinen et al. 2014). To understand the water-related gravimetric signals, a suite of environmental sensors has been installed at MGRS. These include boreholes with pressure gauges, soil moisture sensors, a gamma ray spectrometer for snow water content evaluation, and meteorological sensors including a snow depth sensor. Data from SGs is available through the International Geodynamics and Earth Tide Service (IGETS) database. The high-quality SG time series of MGRS is one of the longest in the world and has been used in many studies of e.g., hydrology (Boy and Hinderer 2006) solid Earth and ocean tides (Boy and Lyard 2008), and Earth's free core nutation (Rosat and Lambert 2009).

The scale and drift of the SGs are routinely determined using measurements from the FG5X-221. In turn, the time series of the SGs are used to correct absolute gravity mea-

Fig. 5 Absolute and superconducting time series in Metsähovi. Both time series have been corrected for polar motion, tides and standard air pressure. A linear trend was removed from the superconductive gravimeter time series

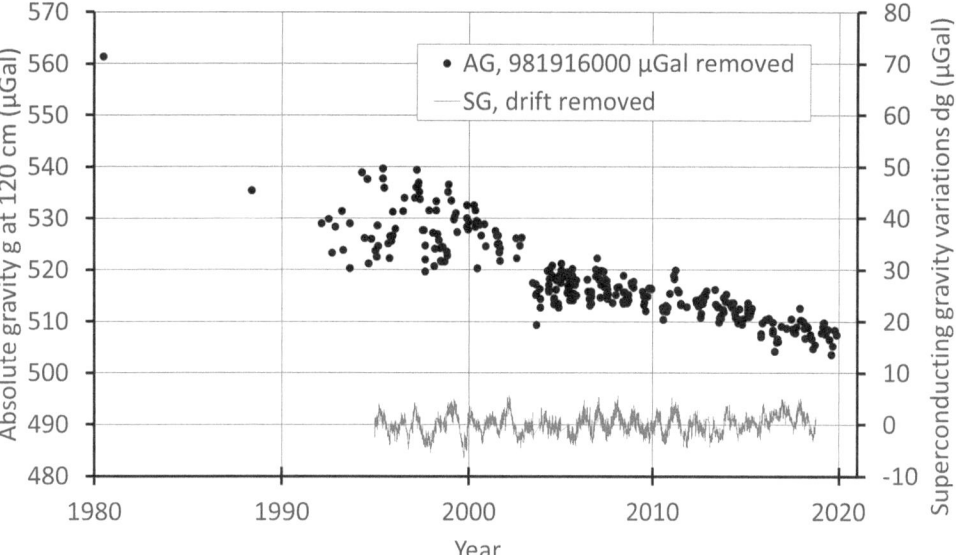

surements for temporal gravity changes and possibly detect malfunctioning of the absolute gravimeter (Fig. 5).

2.6 Local Tie

The local tie is said to be the fifth technique in the ITRF combination. It connects the systems of different techniques, and without the local ties common use of techniques would not be possible.

Two different techniques are used for local ties at MGRS, based on kinematic GNSS or terrestrial tachymetric monitoring measurements. Both techniques use an indirect method (Dawson et al. 2007) to determine the reference point. The GNSS antennas or the targets/prisms rotate along with the radio telescope and their positions are measured in hundreds of telescope orientations. By using the angle positions and the adjusted coordinates of the targets, the coordinates of the reference point and telescope axis orientation and offset can be estimated (Kallio and Poutanen 2012).

Terrestrial monitoring measurements are connected to the local survey network, which is aligned to the ITRF by constraining the horizontal rotation of the local GNSS network to zero. The vertical axes of the tachymeter station points are aligned with the ellipsoidal normal at the tachymeter station points by converting the tachymeter angle observations with deflection of vertical corrections to each direction separately. Components of the deflection of vertical are calculated using the gravimetric geoid (Saari and Bilker-Koivula 2018). No further transformations are needed because the 3D adjustment is performed in the global reference frame. The scale of the network is traceable to the definition of the metre through the calibration of the tachymeter in the Nummela standard

baseline (Jokela 2014). The Metsähovi model (Kallio and Poutanen 2012) is used in reference point estimation.

The GNSS-based local ties do not need a local survey network. The post-processed kinematic solutions are calculated for the vectors between two rotating GNSS antennas attached at the side of the radio telescope dish and the permanent GNSS point. The GNSS antennas have individual absolute calibration tables and their phase centre and phase variation are corrected for every single phase observation in each antenna position in every observation epoch. Also, the influence of the difference in tropospheric delay due to the height differences is corrected before the final trajectory calculation by applying the troposphere model GMT3 for each phase observation in each epoch (Landskron and Böhm 2018).

The GPS-based local tie measurements were continued since 2008 until 2015 simultaneously with VLBI sessions between the METS00FIN GPS point and the legacy telescope (Jokela et al. 2016), while the terrestrial monitoring were conducted during two VLBI sessions in 2015 using eight Leica GPR1 prisms, which were installed onto the counterweights of the legacy telescope (Kallio et al. 2016).

The MGRS local tie network was expanded during the EMPIR GeoMetre project 2019–2022 (Kallio et al. 2022; Pollinger et al. 2022) to also include the GGOS telescope, the new SLR telescope, and the MET300FIN GNSS station (Fig. 6). In 2020, the new VGOS telescope was equipped with several Bohnenstingel spherical prisms on the counterweight of the telescope, and the GNSS antennas were installed onto the dish edges of the VGOS telescope. The terrestrial and GPS-based monitoring measurements to the new VGOS telescope were started in 2020 and repeated in 2021. The terrestrial and GPS-based local tie results were compared – the coordinates

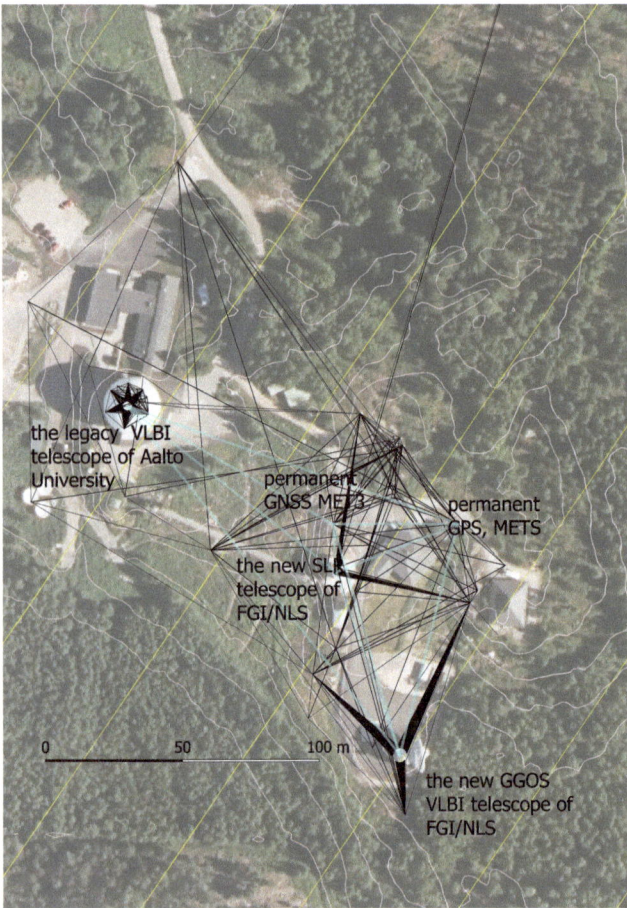

Fig. 6 The local survey and monitoring networks at MGRS (black). The yellow contour lines show the geoid slope with a contour interval of 2 mm. The local ties, vectors between the space geodetic instruments, are connected to the local survey network through monitoring networks

of reference points in the simultaneous GPS and tachymeter monitoring measurements, performed in the summer of 2021, differed by 0.9200, 0.3100, and 0.1200 mm in North, East and Up directions, respectively (Kallio et al. 2023). Furthermore, in 2022, the terrestrial indirect method was applied to the SLR telescope with two spherical prisms which rotated with the SLR telescope (Raja-Halli et al. 2022).

3   Infrastructure

Refurbishing and reconstruction of the MGRS infrastructure and working facilities improve or enable new functionality. A new satellite laser ranging observatory was built for the SLR system in 2014. The new VGOS system required additional infrastructure work to host the new VGOS telescope. A new main building was constructed with especially the needs of VGOS in mind (Fig. 1). It has sufficient lab space to allow maintenance work of the VGOS receiver as well as a temperature-controlled server/electronics room for the

backend. The whole building is made as radio frequency interference (RFI) free as possible via the use of special RFI shield mesh in the walls, using as RF silent building electronics as possible. The RFI shield is installed on all outer walls, floor, and ceiling of the building to prevent RFI towards the telescope. Some rooms, such as the server room where the VLBI backend will be located, have also shielding in the internal walls.

The new VGOS system will increase the data production volume of MGRS by several orders of magnitude. The data production of the system is more than 16 Gb/s while on target. As the data needs to be correlated with data from other systems, it must be transferred to correlation centres. To allow efficient data transfer, the MGRS internet connection has been upgraded to allow transfers of up to 100 Gb/s via the Finnish University and Research Network FUNET. MGRS has also been connected via the FUNET optical cable to the Finnish realisation of UTC (UTC MIKE), maintained by the Finnish metrological institute VTT MIKES. The MGRS time and frequency base can be compared to the UTC MIKE via commercial White Rabbit time and frequency transfer technology as well as dedicated in-house built frequency-only transfer technology. The latter is currently under development but preliminary results indicate better performance than White Rabbit. The ultimate aim would be to use UTC MIKE as the time and frequency base for many of the measurements at MGRS providing a direct link between the geodetic measurements and the SI unit of one second. However, currently, the feasibility of the link for such operational use is still under investigation and the link is used for research and technology development purposes only (Calvés et al. 2018).

MGRS also has a GPS-based Network Time Protocol (NTP) server that provides a time signal to the local station network allowing the measurements to be synchronised to the same time base. VLBI especially requires a very stable frequency source. For the commissioning phase, we are using reference frequency from the nearby hydrogen masers of Aalto University Metsähovi Radio Observatory. The cable length between the masers and the backend is approximately 100 m. For operational use and to provide the best possible data, the H-maser needs to be closer. To this end, we are in the process of procuring an H-maser that will be located in the same room as the VLBI backend.

MGRS has since 2013 hosted a triangular trihedral 1.5 m corner reflector (CR) for interferometric synthetic aperture radar (InSAR) satellites, owned by the German Aerospace Agency DLR (Gisinger et al. 2022). The CR is used for radiometric and geometric calibration of InSAR data of satellites such as TerraSAR-X and Sentinel-1. InSAR is also being intensively studied as a newly emerging technique for geodesy as it provides continuous data on heights and their changes over large geographical areas. MGRS is an ideal site as a testbed for such CRs as the bedrock provides a stable foundation and local surveys of the CR reference point position can be directly linked to the International Terrestrial Reference Frame through the IGS station(s) on-site.

Almost all measurements at MGRS require some kind of weather or environment data either directly in the processing and analysis or as metadata to be included as added information. With the station upgrade it became possible to create a more easily managed, maintained and calibrated central solution for providing the weather data. Additionally, there are detectors for soil moisture, groundwater and snow depth measurements. Calibration of sensors is done according to the manufacturer's specifications. It is critical to use properly calibrated sensors as, e.g., false pressure readings would distort the SLR and gravity results in a way that might not be immediately obvious in the data analysis.

4   Conclusions and Discussion

During the last decade, most of the MGRS activities were concentrated on technical issues and construction work of new instrumentation. When the installation and commissioning of equipment are completed, the main attention will be shifted back to research.

MGRS measurements contribute to the global and regional geodetic reference frames and global geodesy as shown in Fig. 2. As an example, the GNSS receivers produce data for IGS and EPN networks, and together with other EPN stations provide the primary link of the national reference frame to global and European reference frames.

The long and continuous GNSS time series at MGRS with only a few equipment changes has made the station very valuable for reference frame realisations. However, it is noteworthy that REGINA station METG00FIN does not co-locate at the Metsähovi main station, and its station velocity should not be constrained to METS/MET300FIN in the reference frame related works.

Long gravity time series of AG and SG are equally valuable for regional and global research. The status as the National Standards Laboratory of the free fall acceleration enables participation in metrological research projects also in the future and mutual comparison of gravimeters at MGRS.

MGRS is also one of the northernmost GGOS Core stations, and therefore data are valuable for Arctic region research. Especially, SLR is crucial for improving the orbit information of both GNSS and Earth exploring satellites in the Arctic where climate change-induced phenomena, like ice sheet decay, are pronounced.

MGRS is Finland's contribution to the UN General Assembly's 2015 resolution 69/266 A "Global geodetic reference frame for sustainable development". As a small country, Finland has limited resources for global geodetic research, but MGRS with its upgraded instrumentation and its long history will maintain the internationally recognized status of Finnish geodesy. Recent international evaluation of the FGI drew special attention to the Metsähovi Geodetic Research Station, which is "unique, at a national and international level in terms of its research infrastructure" (Hämäläinen et al. 2023).

References

Arnautov GP, Kalish YeN, Kiviniemi H, Stus YuF, Tarasiuk VG, Scheglov SN (1982) Determination of absolute gravity values in Finland using laser ballistic gravimeter. Publications of the Finnish Geodetic Institute, 97, Helsinki, 1982, p 18

Bilker-Koivula M, Mäkinen J, Ruotsalainen H, Näränen J, Saari T (2021) Forty-three years of absolute gravity observations of the Fennoscandian postglacial rebound in Finland. J Geodesy 95(2):1–18

Boy JP, Hinderer J (2006) Study of the seasonal gravity signal in superconducting gravimeter data. J Geodyn 41:227–233

Boy JP, Lyard F (2008) High-frequency non-tidal ocean loading effects on surface gravity measurements. Geophys J Int 175(1):35–45

Calvés GM, Wallin A, Näränen J, Fordell T (2018) Initial results from the MIKES-Metsahovi time and frequency link for the VGOS radio telescope. In: 10th IVS General Meeting, IVS 2018

Dawson J, Sarti P, Johnston GM, Vittuari L (2007) Indirect approach to invariant point determination for SLR and VLBI systems: an assessment. J Geodesy 81(6–8):433–441

Degnan JJ (1993) Millimeter accuracy satellite laser ranging: a review. In: Smith DE, Turcotte DL (eds) Contributions of space geodesy to geodynamics: technology. https://doi.org/10.1029/GD025p0133

Del Pino J, Raja-Halli A, Salmins K, Näränen J (2017) Sky clarity comparison between Riga and Metsahovi SLR stations. Presented at the 2017 ILRS Technical Workshop, Riga, Latvia, October 02-05, 2017. https://cddis.nasa.gov/2017_Technical_Workshop/docs/papers/session3/ilrsTW2017_s3_paper_delPino.pdf

Gisinger C, Libert L, Marinkovic P, Krieger L, Larsen Y, Valentino A et al (2022) The extended timing annotation dataset for Sentinel-1 – product description and first evaluation results. IEEE Trans Geosci Remote Sens 60:1–22

Gitlein O (2009) Absolutgravimetrische Bestimmung der Fennoskandischen Landhebung mit dem FG5-220. Wissenschaftliche arbeiten der fachrichtung geodäsie und geoinformatik der Leibniz universität Hannover, Nr. 281, p 175. ISSN 0174-1454

Hämäläinen K, Fritsch D, Jensen A, Verhagen S (2023) International evaluation of the Finnish Geospatial Research Institute, FGI. Publications of the Ministry of Agriculture and Forestry 2023:5. ISBN 978-952-366-564-4, https://urn.fi/URN:ISBN:978-952-366-564-4

ILRS (2023) System performance standards. https://ilrs.gsfc.nasa.gov/network/system_performance/index.html

Jokela J (2014) Length in geodesy – on metrological traceability of a geospatial measurand. Doctoral thesis, Aalto University. School of Engineering. http://urn.fi/URN:ISBN:978-951-711-310-6

Jokela J, Kallio U, Koivula H, Lahtinen S, Poutanen M (2016) FGI's contribution in the JRP SIB60 "metrology for long distance surveying". In: Proceedings of the 3rd Joint International Symposium on Deformation Monitoring (JISDM). https://doi.org/10.1007/978-3-642-20338-1-8

Kallio U, Poutanen M (2012) Can we really promise a mm accuracy for the local ties on a Geo-VLBI Antenna. In: Geodesy for Planet Earth. Springer Science Business Media, pp 35–42. https://doi.org/10.1007/978-3-642-20338-1-5

Kallio U, Lösler M, Bergstrand S, Haas R, Eschelbach C (2016) Automated simultaneous local ties with GNSS and robot tachymeter. In: Behrend D, Baver KD, Armstrong KL (eds) IVS 2016 General Meeting Proceedings "New Horizons with VGOS". NASA/CP-2016-219016

Kallio U, Klügel T, Marila S, Mähler S, Poutanen M, Saari T, Schüler T, Suurmäki H (2022) Datum problem handling in local tie surveys at Wettzell and Metsähovi. In: International Association of Geodesy Symposia. Springer, Berlin, Heidelberg, pp 1–11. https://doi.org/10.1007/1345_2022_155

Kallio U, Eskelinen J, Jokela J, Koivula H, Marila S, Näränen J, Poutanen M, Raja-Halli A, Rouhiainen P, Suurmäki H (2023) Validation of GNSS-based reference point monitoring of the VGOS VLBI telescope at Metsähovi. In: 5th Joint International Symposium on Deformation Monitoring (JISDM), 20–22 June 2022, Valencia, Spain. https://doi.org/10.4995/JISDM2022.2022.13691

Koivula H, Ollikainen M, Poutanen M (1998) The Finnish Permanent GPS Network - FinnRef. In: Proceedings of the 13th General meeting of the Nordic Geodetic Commission, Gävle, Sweden, 25–29 May, 1998

Lahtinen S (2022) Reference frame densifications for Nordic and Baltic countries - from local analysis to common and consistent GNSS solutions. Aalto University publication series DOCTORAL THESES, 54/2022, FGI publications no 167

Lahtinen S, Häkli P, Jivall L, Kempe C, Kollo K, Kosenko K, Pihlak P, Prizginiene D, Tangen O, Weber M, Paršeliūnas E, Baniulis R, Galinauskas K (2018) First results of the Nordic and Baltic GNSS Analysis Centre. J Geodetic Sci 8(1):34–42. https://doi.org/10.1515/jogs-2018-0005

Landskron D, Böhm J (2018) VMF3/GPT3: refined discrete and empirical troposphere mapping functions. J Geodesy 92(4):349–360

Mäkinen J, Hokkanen T, Virtanen H, Raja-Halli A, Mäkinen RP (2014) Local hydrological effects on gravity at Metsähovi, Finland: implications for comparing observations by the superconducting gravimeter with global hydrological models and with GRACE. In: Marti U (ed) Gravity, geoid and height systems. International Association of Geodesy Symposia, vol 141. Springer, Cham. https://doi.org/10.1007/978-3-319-10837-7_35

Petrachenko B, Niell A, Behrend D, Corey B, Böhm J, Charlot P, Collioud A, Gipson J, Haas R, Hobiger T, Koyama Y, MacMillan D, Malkin Z, Nilsson T, Pany A, Tuccari G, Whitney A, Wresnik J (2009) Design aspects of the VLBI2010 system – progress report of the IVS VLBI2010 Committee. NASA/TM-2009-214180

Pollinger F, Bauch A, Meiners-Hagen K, Astrua M, Zucco M, Bergstrand S, Görres B, Kuhlmann H, Jokela J, Kallio U, Koivula H, Poutanen M, Neyezhmakov P, Kupko V, Merimaa M, Niemeier W, Saraiva F, Schön S, van den Berg SA, Wallerand JP (2015) Metrology for long distance surveying: a joint attempt to improve traceability of long distance measurements. In: Rizos C, Willis P (eds) IAG 150 years. International Association of Geodesy Symposia, vol 143. Springer, Cham, pp 651–656. https://doi.org/10.1007/1345_2015_154

Pollinger F, Courde C, Eschelbach C, García-Asenjo L, Guillory J, Hedekvist PO, Kallio U, Klügel T, Neyezhmakov P, Pesce D, Pisani M, Seppä J, Underwood R, Wezka K, Wiśniewski M (2022) Large-scale dimensional metrology for geodesy – first results from the European GeoMetre Project. In: International Association of Geodesy Symposia. Springer, Berlin, Heidelberg. https://doi.org/10.1007/1345_2022_168

Poutanen M, Koivula H (eds) (2007) Geodetic operations in Finland. Finnish Geodetic Institute, Kirkkonummi, p 39

Raja-Halli A, Näränen J, Poutanen M (2019) Metsähovi research station – four decades of SLR observations 1978 – 2005 & 2016->. 19th International Workshop on Laser Ranging. Celebrating 50 years of SLR: remembering the past and planning for the future. https://cddis.nasa.gov/lw19/Anniversary/Metsahovi_RajaHalli_history.pdf

Raja-Halli A, Kallio U, Näränen J, Riikonen E (2022) Determination of the reference point of Metsähovi SLR telescope, S-06 P-05, presented at the 22nd International Workshop on Laser Ranging, Guadalajara, Spain, November 07-11, 2022

Rosat S, Lambert SB (2009) Free core nutation resonance parameters from VLBI and superconducting gravimeter data. Astron Astrophys 503(1):287–291

Saari T, Bilker-Koivula M (2018) Applying the GOCEbased GGMs for the quasi-geoid modelling of Finland. J Appl Geodesy 12(1):15–27. https://doi.org/10.1515/jag-2017-0020

Tuccari G, Walter A, Buttaccio S, Casey S, Felke A, Lindqvist M, Wunderlich M (2014) DBBC3: an EVN and VGOS all-inclusive VLBI system. In: Baver KD, Behrend D (eds) IVS 2014 General Meeting Proceedings "VGOS: The New VLBI Network". Science Press, Beijing, pp 86–90. ISBN: 978-7-03-042974-2

Virtanen H, Raja-Halli A (2018) Parallel observations with three superconducting gravity sensors during 2014–2015 at Metsähovi Geodetic Research Station, Finland. Pure Appl Geophys 175:1669–1681. https://doi.org/10.1007/s00024-017-1719-3

Wu S, Feng J, Li C, Su D, Wang Q, Hu R et al (2020) The results of CCM. G-K2. 2017 key comparison. Metrologia 57(1A):07002

Wziontek H, Bonvalot S, Falk R, Gabalda G, Mäkinen J, Pálinkáš V, Rülke A, Vitushkin (2021) Status of the international gravity reference system and frame. J Geod 95:7. https://doi.org/10.1007/s00190-020-01438-9

Assessing the Potential of VLBI Transmitters on Next Generation GNSS Satellites for Geodetic Products

Shrishail Raut, Susanne Glaser, Nijat Mammadaliyev, Patrick Schreiner, Karl Hans Neumayer, and Harald Schuh

Abstract

The next-generation Global Navigation Satellite Systems (NextGNSS) satellites are planned to be equipped with inter-satellite links and ultra-stable clocks as well as a dedicated Very Long Baseline Interferometry (VLBI) transmitter. This will enable the VLBI network to observe the satellites along with extra-galactic radio sources. The study aims to evaluate the potential benefits by placing VLBI transmitters on NextGNSS satellites. This will empower the NextGNSS to determine UT1-UTC, which is otherwise impossible directly. Furthermore, VLBI observations of satellites would allow for independent validation of satellite orbit determination. In this study, we investigate geodetic parameters such as station positions and Earth Rotation Parameters (ERPs) and the impact of different network geometry on these parameters. Based on the initial findings, it appears that using satellites and quasars in VLBI can define a datum with No-Net Rotation (NNR) without the need for No-Net Translation (NNT) conditions. When both NNR and NNT are imposed, the Helmert transformation parameters are smaller compared to when only NNR is imposed. This can be improved by optimizing the network geometry. Furthermore, the study's findings indicate that VLBI observations can determine the satellite's orbit with cm-level accuracy. The performance of the ERPs is better in a uniformly distributed network especially when only NNR condition is imposed.

Keywords

ERPs · GGOS · Next-generation GNSS · Precise orbit determination · TRF · VLBI transmitter

1 Introduction

A global geodetic endeavor is to improve the space geodetic techniques contributing to the global Terrestrial Reference Frames (TRFs). The Global Geodetic Observing System (GGOS) sets requirements on TRFs of 1 mm accuracy and 0.1 mm/year stability and scale accuracy to 0.1 ppb and 0.01 ppb/yr long-term stability (Gross et al. 2009). However, these requirements are not fulfilled yet. To achieve this goal, there are proposed plans e.g., for next-generation GNSS (NextGNSS) constellations like the Kepler system proposed by the German Aerospace Center[1][2] (DLR). The Kepler constellation features in particular new optical sensors and precise inter-satellite observations via inter-satellite optical links allowing for perfect time synchronization (Giorgi et al.

S. Raut (✉) · N. Mammadaliyev · H. Schuh
GFZ German Research Centre for Geosciences, Space Geodetic Techniques, Potsdam, Germany

Technische Universität Berlin, Chair Satellite Geodesy, Berlin, Germany
e-mail: raut@gfz-potsdam.de

S. Glaser · P. Schreiner · K. H. Neumayer
GFZ German Research Centre for Geosciences, Space Geodetic Techniques, Potsdam, Germany

[1]https://www.dlr.de/kn/desktopdefault.aspx/tabid-17411/.
[2]https://www.kepler.global/conf/.

© The Author(s) 2023
J. T. Freymueller, L. Sánchez (eds.), *Gravity, Positioning and Reference Frames*,
International Association of Geodesy Symposia 156, https://doi.org/10.1007/1345_2023_217

2019; Glaser et al. 2020; Michalak et al. 2021). There are also plans to incorporate new observation types provided by Very Long Baseline Interferometry (VLBI) transmitters onboard the NextGNSS, which will be investigated in this work.

VLBI as one of the main space geodetic techniques observe extra-galactic radio sources (mostly quasars) and can determine the full set of Earth Orientation Parameters (EOP), i.e., Polar Motion (PM), UT1-UTC, and Celestial Pole Offsets. In contrast, GNSS cannot determine UT1-UTC in an absolute sense and can only determine the negative time-derivative of UT1-UTC, i.e., Length of Day (LOD). Placing a VLBI transmitter on a GNSS satellite can enable the VLBI stations to observe GNSS satellites besides the quasars possibly allowing to transfer UT1-UTC (Sert et al. 2022).

Studies have been performed to assess the benefits of placing a VLBI transmitter on the satellite (e.g., Plank 2013, Männel 2016, Anderson et al. 2018, Wolf et al. 2022). Plank et al. (2017) and Tornatore et al. (2014) focused on the direct observations of the GNSS signals with a VLBI network. Whereas, Hellerschmied et al. (2018) concentrated on observations to a dedicated VLBI transmitter on board a satellite. McCallum et al. (2016), Jaradat et al. (2021) discussed the technical aspects and challenges of signal generation on a Galileo satellite for VLBI observations. Mammadaliyev et al. (2021) simulated VLBI observations to a VLBI transmitter placed on a Low Earth Orbit (LEO) satellite in addition to the quasar observations. The future ESA mission GENESIS was approved recently, which aims to install a VLBI transmitter onboard a satellite for the co-location in space of all four space geodetic techniques (Delva et al. 2023).

In this work, we investigated the potential of a VLBI transmitter on one Galileo-like Medium Earth Orbit (MEO) satellite (semi-major axis: 29600 km) on geodetic products such as station coordinates, ERPs, and satellite orbits by examining different station networks. We will perform our analysis using dynamic Precise Orbit Determination (POD) with the EPOS-OC software (Zhu et al. 2004), which is also capable of simulating and processing all four main space geodetic techniques.

2   Scheduling

In VLBI, the observation plans (also referred to as schedules) are required as the initial step for conducting real and simulated observations, which are generated by dedicated software. Except for VieSched++ (Schartner and Böhm 2019), most of them were developed for classical geodetic VLBI purposes i.e. scheduling only the quasar observations and

usually did not support scheduling of the VLBI to satellite observations. We adopted a strategy that was followed in Mammadaliyev et al. (2021) for scheduling observations to the MEO satellite and quasars. To generate a 'quasar-only schedule', the satellite observations were excluded from the 'satellite + quasar' schedule. This ensures that the number of quasar observations remains the same across different scenarios (Table 1).

We followed a source-based strategy for the source selection so as to obtain a uniform distribution in the sky. A total of 64 sources were selected by the scheduler, i.e., one source for every 64 sky segments of the equal area as done in Sun (2013). We created two sets of schedules based on two different station networks to assess the performance of station networks on the estimated parameters. The first network consists of 13 stations from the R1 VLBI sessions organized by International VLBI Service for Geodesy and Astrometry (Nothnagel et al. 2017), referred to as 'Network A'. As network A has only three stations in the southern hemisphere, therefore for the second network, we added three stations in the southern hemisphere in addition to network A stations to improve the global distribution of the network geometry referred to as 'Network B'. Figure 1 shows the participating stations of both networks. The priority is given to quasar observation over satellite observation as the primary objective of VLBI is to obtain UT1-UTC which can only be determined from quasar observations. Therefore, we scheduled in such a way that we have approximately 6500 quasar observations and approximately 1000 satellite observations in one day from network A (see Fig. 2). Similarly, there are approximately 9000 quasar observations and 1300 satellite observations from network B.

3   Estimation of Parameters

This study consists of three main scenarios and the acronyms assigned to the different scenarios as given in Table 1 will be used hereafter. In the first scenario, i.e. VLBI observations to quasars (VoQ), we imposed datum with No-Net Rotation (NNR) and No-Net Translation (NNT) conditions which are standard for VLBI to get a minimum constraint solution. The NNR and NNT conditions were imposed in the second scenario, i.e. VLBI observations to quasar + satellite ($VoQS_{RT}$). However, this results in an over-constrained solution as satellite observations are basically sensitive to the geocenter, and imposing NNT would not be necessary (e.g., Glaser et al. 2015). Therefore, we only imposed NNR in the $VoQS_R$ scenario to achieve a minimum constraint solution.

Table 1 Scenario description and their parameterization

Scenarios		Estimated parameters			Datum conditions	
Acronym	Description	Station coordinates	Full set of ERPs	Orbit	NNR	NNT
VoQ	VLBI observations to Quasars	✔	✔	N.A.	✔	✔
VoQS$_{RT}$	VLBI observations to Quasars + Satellite	✔	✔	✔	✔	✔
VoQS$_R$	VLBI observations to Quasars + Satellite	✔	✔	✔	✔	✖

Fig. 1 Station observation networks. The maroon triangle represents network A and the three additional stations are represented by a maroon inverted triangle. These additional stations together with network A comprise network B. The green dot represents the stations that are used for the datum realization

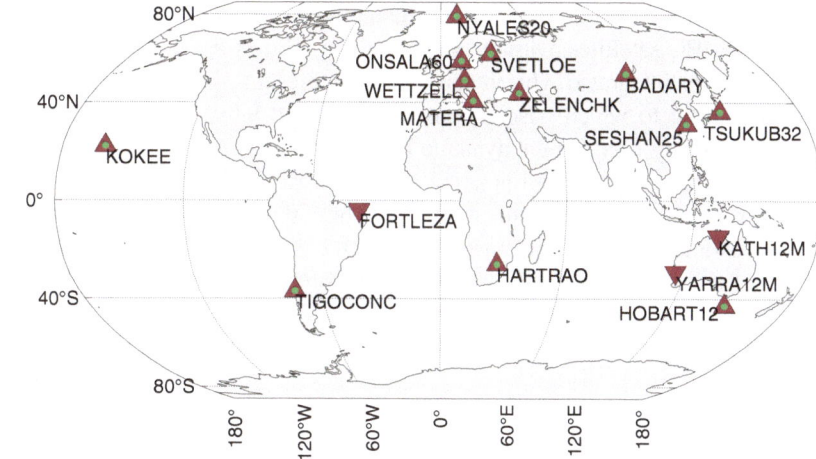

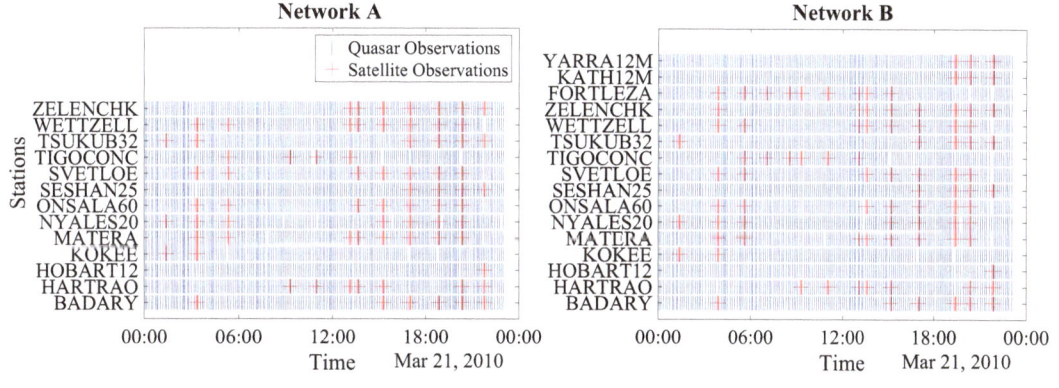

Fig. 2 Satellite and quasar observations as seen by the stations in one day (other days show a similar pattern)

3.1 Simulation

The simulation of VLBI observations uses the VLBI delay model recommended by IERS Conventions 2010 (Petit and Luzum 2010) for quasar observations, and the model for Earth satellites by Klioner (1991) for satellite observations. We assume perfect models in the simulations as the best-case scenario; this study did not include systematic errors in orbit modeling. The VLBI module of EPOS-OC software reads the observation epoch, baseline, and target source from the schedule as input to compute the group delays and keeps them for further processing. White noise for stations of 30 picoseconds was added to all group delays. We did not introduce tropospheric and clock modeling errors in this study,

which is planned for future work. The simulated group delay observations (white noise added) were subsequently used to generate datum-free Normal Equations Systems (NEQs) for each day (one arc). This is followed for all 10 days, and further, these 10 generated datum-free NEQs were stacked together to obtain one stacked datum-free NEQ.

3.2 Recovery (Precise Orbit Determination)

To determine the satellite's orbit with VLBI we use dynamic Precise Orbit Determination (POD) with EPOS-OC. This approach uses models of the forces, i.e., gravitational and non-gravitational, to calculate the sum of the acceleration forces acting on the satellite. There were several models used in our study, for example, EIGEN-6C (Shako et al. 2014) as Earth's gravity potential, and the Solar Radiation Pressure (SRP) is taken care of by the empirical CODE (Center for Orbit Determination in Europe) orbit model (ECOM) in the reduced version (Beutler et al. 1994; Springer et al. 1999), to mention a few. We did not consider accelerations due to Earth's albedo and atmospheric drag as their effects are quite small for MEO satellites. During simulation, the orbit of the satellite is first integrated. This integrated orbit represents the reference orbit to which observations are simulated which are subsequently used for the dynamic POD in the recovery run. The dynamic orbit determination process in EPOS-OC is done by least squares adjustment. In both VoQS scenarios, we also estimated satellite parameters such as the initial state vector of the daily arcs in the form of position and velocity, and reduced ECOM parameters.

3.3 TRF Solution

To obtain a TRF solution, we impose datum on the stacked NEQ (explained in Sect. 3.1) with NNR and/or NNT of 1 mm to the 13 common stations in both networks, i.e., all stations in network A and this stacked NEQ is inverted for estimating the final solution. The parameters estimated for the scenarios are illustrated in Table 1. We estimated one set of station coordinates from 10 days, daily polar motion (PX, and PY), and daily UT1-UTC for all the scenarios. We did not estimate source positions and kept them fixed to their a priori values.

4   Result and Discussion

4.1 Satellite's Orbit

We computed the differences in orbital components for various scenarios (i.e., along-track, cross-track, and radial components) w.r.t. the reference orbit of the simulation (see Fig. 3) and computed the root mean square (RMS) for the 10 days (see Fig. 4). For network A the RMS values in the $VoQS_{RT}$ case are 1 cm in radial, 0.8 cm in cross-track, and 1.5 cm in the along-track direction. The values in the $VoQS_R$ case, whereby no NNT condition was applied, are remarkably larger with values of 3.5 cm, 2.7 cm, and 4.5 cm respectively. Significantly better values could be achieved with network B. The RMS values for the $VoQS_{RT}$ scenario are 0.6 cm (radial), 0.5 cm (cross-track), and 0.8 cm (along-track). In the case without NNR, i.e. $VoQS_R$, with network B we get again slightly larger deviations. With values of 1.9 cm, 1.5 cm, and 2 cm, however, significantly lower than for network A. We can deduce from the results that the

addition of three stations in the southern hemisphere, i.e., network B, improved the orbital components in all three directions by up to 50 % across both VoQS scenarios. The network geometry plays a more vital role in NNR-only than in both NNR and NNT scenarios. For these $VoQS_R$ scenarios (without NNT), we see improvements in radial, cross-track, and along-track directions of approximately 48 %, 42 %, and 54 % for network B compared to network A, respectively.

4.2 Station Coordinates

The formal errors of estimated station coordinate averaged over x, y, z for VoQ and $VoQS_{RT}$ scenarios in 3D are around 2 mm for both networks (see Fig. 5). The results from these two scenarios have similar performances. Now we omit the NNT condition since this information should be given by the satellite observations. We obtain the results of the $VoQS_R$ scenario, where the formal errors for network A increase strongly up to 8 mm. Whereas, for the same scenario for network B, the formal errors are up to only 2.5 mm. In $VoQS_R$ scenario, the addition of three stations resulted in a reduction of formal errors in the 3D coordinate by 68 %.

The scenarios observed from network B have approximately 40 % more observations compared to network A due to three additional stations. To quantify the significance of the improvements in the estimated parameters from both networks, we computed the expected improvement due to different Degrees of Freedom (DoF). The anticipated improvement in the parameters from network B is approximate 15 % w.r.t. network A. So the improvements mentioned earlier can be considered significant as they are more than the expected improvement, i.e., 15 %.

4.3 Helmert Parameters

The Helmert transformation parameters between estimated station positions of VoQ, $VoQS_{RT}$, and $VoQS_R$ scenarios w.r.t. their a priori, and corresponding standard deviations (see Fig. 6) were computed. This is performed for the stations that participated in the datum definition (see Fig. 1). The Helmert transformation is used to see differences between the networks expressed by three translations, three rotations, and one scale factor. The translational parameters T_x, T_y, T_z and rotational parameter R_x for $VoQS_R$ scenario from network A are approximately up to ± 5 mm and from network B it reduces up to ± 1 mm. Nevertheless, we can say that it is possible to realize datum on mm-level without imposing NNT condition. Furthermore, with better network geometry, the Helmert parameters do improve considerably w.r.t. a priori TRF.

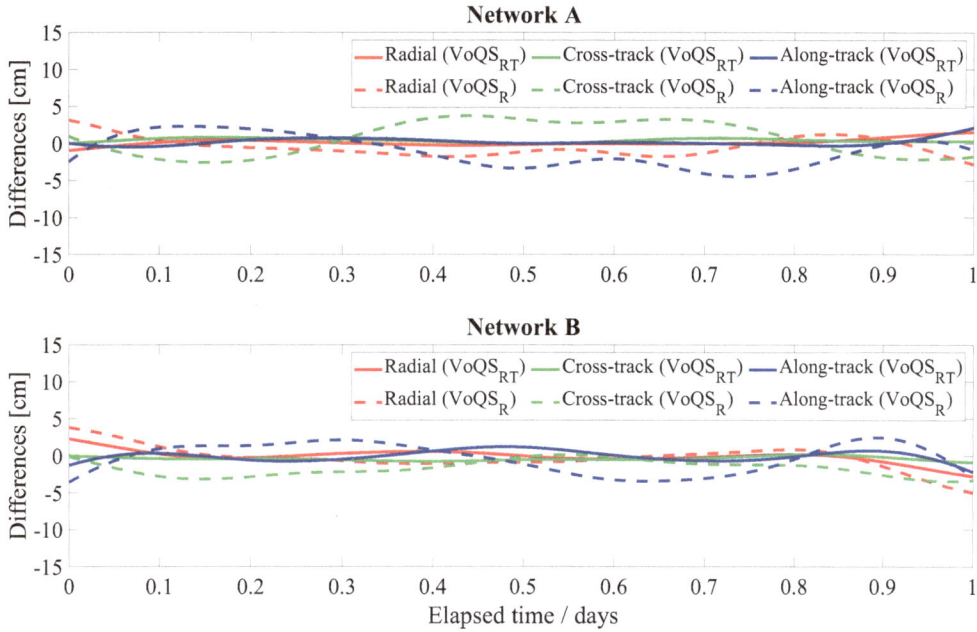

Fig. 3 Time series of the differences in orbital components w.r.t. their reference orbit for both networks (for one day)

Fig. 4 RMS values of differences in orbital components w.r.t. their reference orbit for both networks (for 10 days)

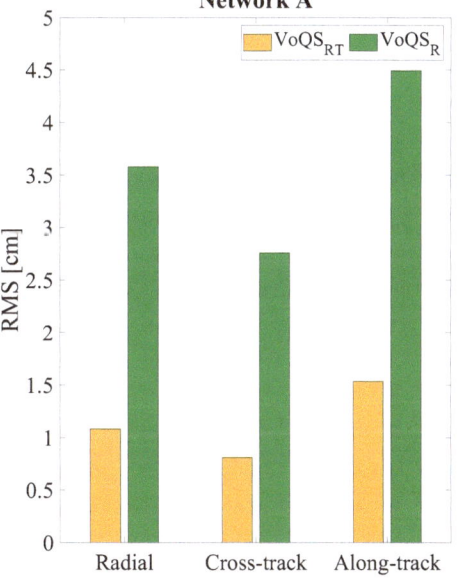

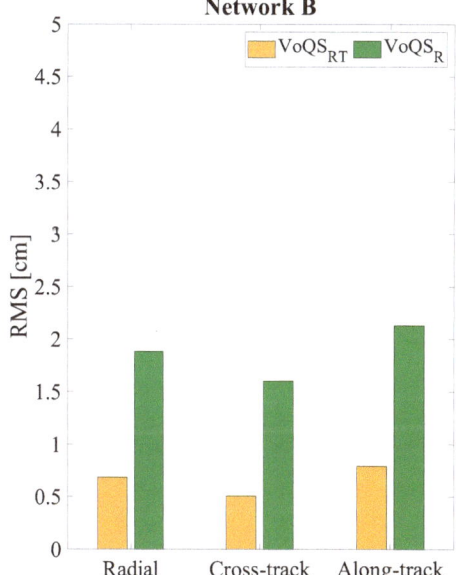

4.4 Earth Rotation Parameters

The mean formal errors in PM and UT1-UTC for both networks and all scenarios are shown in Fig. 7. For VoQ and $VoQS_{RT}$ scenarios from network A, the formal errors in ERPs are up to 40 μas, which corresponds to about 1.3 mm, and from network B, it is up to approximately 30 μas. In $VoQS_R$ scenario, we observed relatively high formal errors in PM of up to 115 μas, specifically PY with network A. Whereas, for the same scenario for network B, the formal errors were up to 65 μas. By improving the global distribution of the station network, we observe

a significant reduction w.r.t. expected improvement for the $VoQS_R$ scenario in formal errors of PX, PY, and UT1-UTC by 38 %, 40 %, and 33 %, respectively.

5 Conclusions and Outlook

We performed simulations of a VLBI transmitter on one next-generation GNSS satellite and by performing POD in addition to quasar observation for a period of 10 days. In this study, two different VLBI station networks were considered. The stations in network B consist of network A plus three

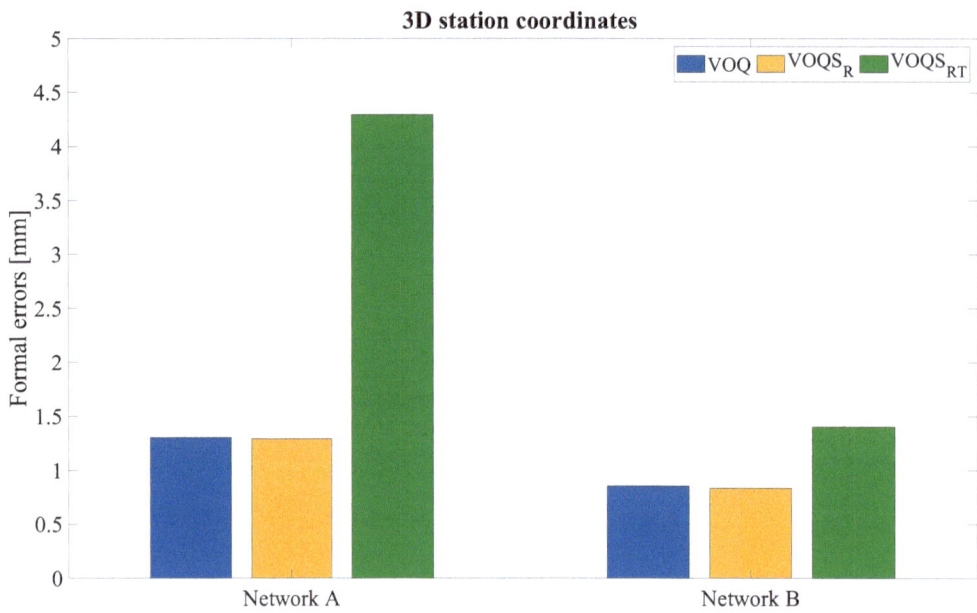

Fig. 5 Mean of formal errors in 3D station coordinates averaged over x, y, z for both networks and all scenarios

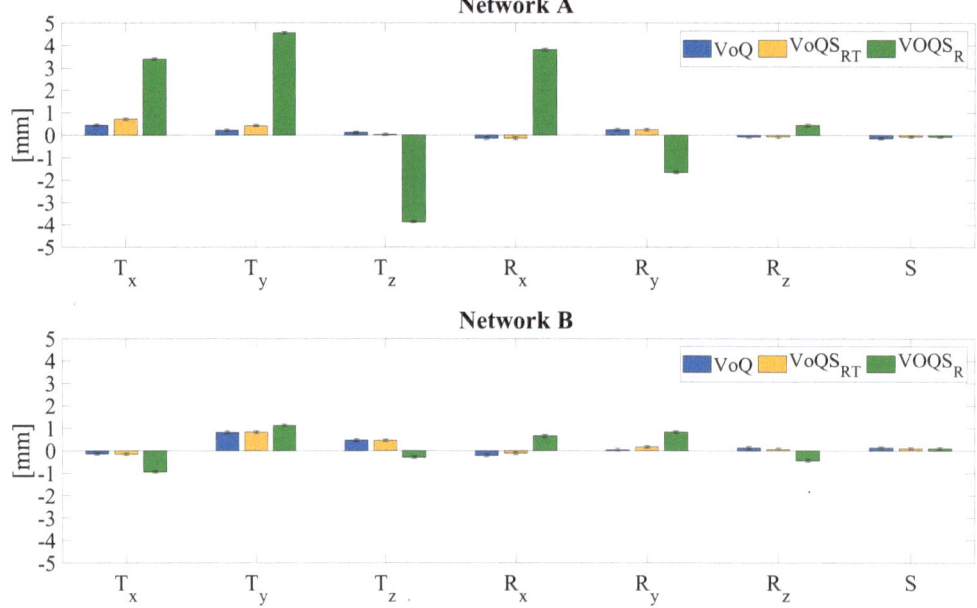

Fig. 6 Helmert transformation parameters (three translation T_x, T_y, T_z, three rotations R_x, R_y, R_z and Scale S) of the estimated station positions (datum only) w.r.t. their a priori for both networks and all scenarios, along with their corresponding standard deviation (shown as error bars)

stations in the southern hemisphere. In total, we formulated the study for three scenarios, i.e., VoQ scenario (quasar-only), VoQS$_{RT}$ scenario (quasar + one satellite, NNR/NNT), and VoQS$_R$ scenario (quasar + one satellite, NNR only). The expected improvement in formal errors of the estimated parameters due to more observations in network B w.r.t. network A is around 15 %. If the actual improvement is more than the expected value, it means that the parameters have improved due to the better network geometry and not because of more observations. We noticed that network

geometry plays a vital role in estimating satellite orbits, station positions, and ERPs in this case.

By using simulated VLBI observations for POD, we recovered the orbit of the MEO satellite with cm-level accuracy. To quantify the effect of network geometry for the VoQS$_R$ scenario on the results, a relatively poor network geometry as network A resulted in orbit recovery up to 5 cm (RMS value). The addition of three stations in the southern hemisphere, i.e., network B, significantly improved orbit recovery up to 60 %. If the network geometry for the

Fig. 7 Mean of formal errors in ERPs for both networks and all scenarios

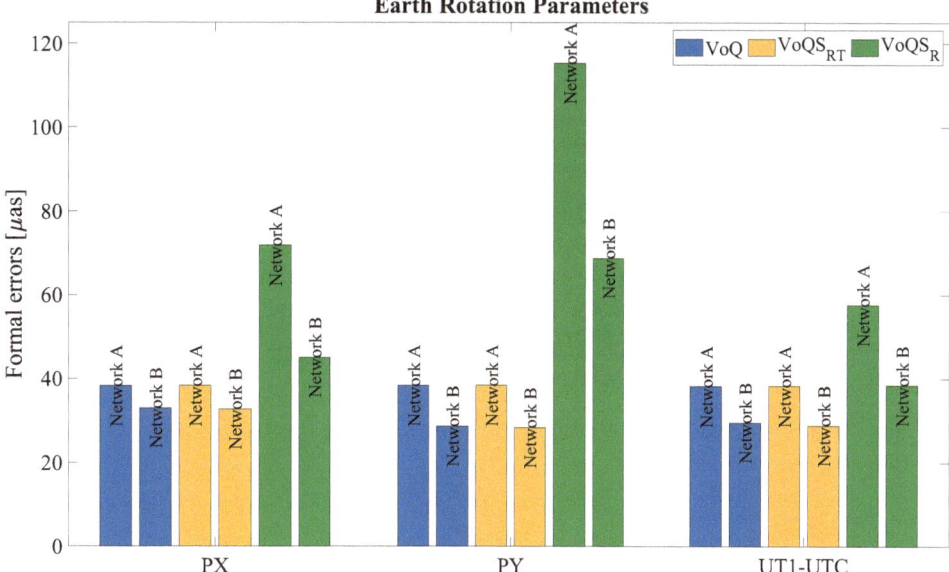

VoQS$_R$ scenario is good, one gets near the results of the VoQS$_{RT}$ without having to impose NNT anymore. These are promising findings for upcoming satellite missions, which may be equipped with a VLBI transmitter.

The expected improvement due to an increase in observations because of the addition of one MEO satellite in the quasar + satellite (VoQS) scenario w.r.t. quasar-only (VoQ) scenario is 6 %. The station positions and ERP estimated from VoQS scenario do not lead to a considerable difference compared to VoQ scenario, as it is less than the expected improvement of 6 %. Upon further examination of different stations, it appears that some stations have got better, which is around the expected improvement, while others have not shown any improvement at all. This could be due to differences in satellite observations among the stations. It can be stated that the inclusion of satellite observations does not appear to have any adverse impact on the estimated parameters.

We observed the additional benefits of the addition of a MEO satellite such as, when VLBI observations are extended to satellite observations, it allows us to realize the origin of the reference frame, thus imposing that the NNT condition is not necessary. Based on the estimated Helmert transformation parameters, we realized datum on mm-level by imposing NNR condition only. The VoQS$_R$ scenario results in translations (T_x, T_y, T_z) and rotation (R_x) of up to ±5 mm for network A. However, better network geometry, i.e. network B leads to significant improvement resulting in Helmert parameters around 1 mm. The mean formal errors in station positions for VoQS$_R$ scenario from network A were approximate up to 6 mm. Here again, the better network geometry, i.e. network B leads to a reduction by up to 75 %. The ERPs formal errors show an improvement up to 40 % estimated from a better network geometry i.e., network

B w.r.t network A. By looking at the results for station positions and ERPs, we can say that in the VoQS$_R$ scenario, the improvements due to three additional stations are more significant w.r.t. the expected improvement than the VoQS$_{RT}$ scenario.

In the following steps, we plan to combine GNSS and VLBI into one common space-tie satellite to investigate how well the datum information can be transferred from the GNSS network to the VLBI network. In addition, we would like to extend our study to next-generation GNSS satellites with optical links and future observation types. Furthermore, the impact of the VGOS stations and a lower noise level could also be studied.

Acknowledgements This work has been supported by the Deutsche Forschungsgemeinschaft (DFG) under Grant number GL 1028/1-1 (NextGNSS4GGOS – Next generation GNSS constellations for GGOS-compliant geodetic solutions). We want to express our gratitude to the reviewers for their valuable comments and suggestions that have contributed to enhancing the quality of the manuscript.

Conflict of Interest The authors declare that they have no conflict of interest.

References

Anderson JM, Beyerle G, Glaser S, Liu L, Männel B, Nilsson T, Heinkelmann R, Schuh H (2018) Simulations of VLBI observations of a geodetic satellite providing co-location in space. J Geodesy 92(9):1023–1046. https://doi.org/10.1007/s00190-018-1115-5

Beutler G, Brockmann E, Gurtner W, Hugentobler U, Mervart L, Rothacher M, Verdun A (1994) Extended orbit modeling techniques at the CODE processing center of the International GPS service for geodynamics (IGS): theory and initial results. Manuscripta Geodaetica 19(6):367–386

Delva P, Altamimi Z, Blazquez A, Blossfeld M, Böhm J, Bonnefond P, Boy JP, Bruinsma S, Bury G, Chatzinikos M, et al (2023) GENESIS: co-location of geodetic techniques in space. Earth Planets Space 75(1):5. https://doi.org/10.1186/s40623-022-01752-w

Giorgi G, Schmidt TD, Trainotti C, Mata-Calvo R, Fuchs C, Hoque MM, Berdermann J, Furthner J, Günther C, Schuldt T, et al (2019) Advanced technologies for satellite navigation and geodesy. Adv Space Res 64(6):1256–1273. https://doi.org/10.1016/j.asr.2019.06.010

Glaser S, Fritsche M, Sośnica K, Rodríguez-Solano CJ, Wang K, Dach R, Hugentobler U, Rothacher M, Dietrich R (2015) A consistent combination of GNSS and SLR with minimum constraints. J Geodesy 89(12):1165–1180. https://doi.org/10.1007/s00190-015-0842-0

Glaser S, Michalak G, König R, Neumayer KH, Männel B, Schuh H (2020) Reference system origin and scale realization within the future GNSS constellation "Kepler". J Geodesy 94(117). https://doi.org/10.1007/s00190-020-01441-0

Gross R, Beutler G, Plag HP (2009) Global geodetic observing system: meeting the requirements of a global society on a changing planet in 2020, chap Integrated scientific and societal user requirements and functional specifications for the GGOS. Springer, Berlin, Heidelberg, pp 209–224. https://doi.org/10.1007/978-3-642-02687-4_7

Hellerschmied A, McCallum L, McCallum J, Sun J, Böhm J, Cao J (2018) Observing APOD with the AuScope VLBI array. Sensors 18(5):1587. https://doi.org/10.3390/s18051587

Jaradat A, Jaron F, Gruber J, Nothnagel A (2021) Considerations of VLBI transmitters on Galileo satellites. Adv Space Res 68(3):1281–1300. https://doi.org/10.1016/j.asr.2021.04.048

Klioner SA (1991) General relativistic model of VLBI observables. In: Proc. AGU Chapman Conf. on Geodetic VLBI: Monitoring Global Change, Carter, WE (ed.). NOAA Technical Report, 137, pp 188–202

Mammadaliyev N, Schreiner PA, Glaser S, Neumayer K, König R, Heinkelmann R, Schuh H (2021) Precise orbit determination with VLBI to satellites: a simulation study. In: 25th Working Meeting of the European VLBI Group for Geodesy and Astrometry (EVGA)

Männel B (2016) Co-location of geodetic observation techniques in space, Geodätischgeophysikalische Arbeiten in der Schweiz, vol 97. Schweizerische Geodätische Kommission. https://www.sgc.ethz.ch/sgc-volumes/sgk-97.pdf

McCallum J, Plank L, Hellerschmied A, Böhm J, Lovell J (2016) Technical challenges in VLBI observations of GNSS sources. In: Proceedings of the first international workshop on VLBI observations of near-field targets, pp 1864–1113

Michalak G, Glaser S, Neumayer K, König R (2021) Precise orbit and Earth parameter determination supported by LEO satellites, inter-satellite links and synchronized clocks of a future GNSS. Adv Space Res 68(12):4753–4782. https://doi.org/10.1016/j.asr.2021.03.008. https://www.sciencedirect.com/science/article/pii/S0273117721002003, scientific and Fundamental Aspects of GNSS - Part 2

Nothnagel A, Artz T, Behrend D, Malkin Z (2017) International VLBI service for geodesy and astrometry: delivering high-quality products and embarking on observations of the next generation. J Geodesy 91(7):711–721. https://doi.org/10.1007/s00190-016-0950-5

Petit G, Luzum B (2010) IERS conventions (2010) Technical Note No. 36. Frankfurt am Main: Verlag des Bundesamts für Kartographie und Geodäsie pp 179–182. https://www.iers.org/IERS/EN/Publications/

Plank L (2013) VLBI satellite tracking for the realization of frame ties. PhD thesis, TU Vienna. https://resolver.obvsg.at/urn:nbn:at:at-ubtuw:3-415

Plank L, Hellerschmied A, McCallum J, Böhm J, Lovell J (2017) VLBI observations of GNSS-satellites: from scheduling to analysis. J Geodesy 91(7):867–880. https://doi.org/10.1007/s00190-016-0992-8

Schartner M, Böhm J (2019) VieSched++: a new VLBI scheduling software for geodesy and astrometry. Publ Astron Soc Pacific 131(1002):084501. https://doi.org/10.1088/1538-3873/ab1820

Sert H, Hugentobler U, Karatekin O, Dehant V (2022) Potential of UT1ÚUTC transfer to the Galileo constellation using onboard VLBI transmitters. J Geodesy 96(10):1–13. https://doi.org/10.1007/s00190-022-01675-0

Shako R, Förste C, Abrikosov O, Bruinsma S, Marty JC, Lemoine JM, Flechtner F, Neumayer H, Dahle C (2014) Observation of the System Earth from Space - CHAMP, GRACE, GOCE and future missions: GEOTECHNOLOGIEN Science Report No. 20, chap EIGEN-6C: a high-resolution global gravity combination model including GOCE data. Springer, Berlin, Heidelberg, pp 155–161. https://doi.org/10.1007/978-3-642-32135-1_20

Springer TA, Beutler G, Rothacher M (1999) A new solar radiation pressure model for GPS satellites. GPS Solut 2(3):50–62. https://doi.org/10.1007/PL00012757

Sun J (2013) VLBI scheduling strategies with respect to VLBI2010, vol 92. Dep. of Geodesy and Geoinformation of the Vienna Univ. of Technology

Tornatore V, Haas R, Casey S, Duev D, Pogrebenko S, Calvés GM (2014) Direct VLBI observations of global navigation satellite system signals. In: Earth on the edge: science for a sustainable planet. Springer, pp 247–252. https://doi.org/10.1007/978-3-642-37222-3_32

Wolf H, Böhm J, Schartner M, Hugentobler U, Soja B, Nothnagel A (2022) Dilution of precision (DOP) factors for evaluating observations to Galileo satellites with VLBI. Springer, Berlin, Heidelberg, pp 1–8. https://doi.org/10.1007/13452022_165

Zhu S, Reigber C, König R (2004) Integrated adjustment of CHAMP, GRACE, and GPS data. J Geodesy 78(1–2):103–108. https://doi.org/10.1007/s00190-004-0379-0

Potential of Lunar Laser Ranging for the Determination of Earth Orientation Parameters

Liliane Biskupek, Vishwa Vijay Singh, Jürgen Müller, and Mingyue Zhang

Abstract

The distance between the observatories on the Earth and the retro-reflectors on the Moon has been regularly measured with Lunar Laser Ranging (LLR) since 1970. In recent years, LLR observations have been carried out at infrared wavelength (OCA, WLRS), resulting in a better distribution of LLR normal points over the lunar orbit and retro-reflectors with a higher accuracy, also leading to a higher number of LLR observations in total. By analysing LLR data, Earth Orientation Parameters (EOPs) can be determined along with other parameters of the Earth-Moon system. Focusing on ΔUT1 and terrestrial pole coordinates the accuracies have improved significantly compared to the previous results. In the past, the reported uncertainties of the estimated parameters were published as three times the formal error from the least-squares adjustment to account for small random and systematic errors in the LLR analysis. To investigate if such a scaling factor is still needed, a sensitivity analysis was performed. The current best accuracies are $12.36\,\mu s$ for ΔUT1, $0.47\,mas$ for x_p and $0.59\,mas$ for y_p. Also the determined corrections to the long-periodic nutation coefficients of the MHB2000 model are now significantly smaller with higher accuracies, i.e., accuracies better than $0.18\,mas$ are obtained.

Keywords

Earth rotation phase · Lunar Laser Ranging · Nutation · Terrestrial pole coordinates

1 Introduction

Lunar Laser Ranging (LLR) has been measuring the distance between the Earth and the Moon with laser pulses for more than 53 years. Currently four observatories perform regular measurements: The Côte d'Azur Observatory, France (OCA), the Apache Point Observatory Lunar Laser ranging Operation, USA (APOLLO), the Matera Laser Ranging Observatory, Italy (MLRO) and the Geodetic Observatory Wettzell, Germany (WLRS). In the past also the McDonald Laser Ranging Station, USA (MLRS) and the Lure Obser-

vatory on Maui/Hawaii, USA (LURE) contributed to the measurements. On the Moon there are five retro-reflectors where laser pulses from the observatories are reflected back to Earth. The measurement of round trip travel times with short laser pulses over 5 min to 15 min is used to calculate a so-called normal point (NP) (Michelsen 2010) which is the observable in the LLR analysis. With the analysis of the LLR data, contributions to terrestrial, lunar and celestial reference frames (Müller et al. 2009a; Hofmann et al. 2018; Pavlov 2019) as well as the understanding of the lunar interior (Williams et al. 2013; Pavlov et al. 2016) are possible. One major task of LLR is to test the validity of General Relativity in the solar system. Test quantities include, e.g., the equivalence principle, temporal variation of the gravitational constant G, Yukawa term, metric parameters, and geodetic precession (Williams et al. 2012; Viswanathan et al. 2018; Hofmann and Müller 2018; Zhang et al. 2020; Biskupek

L. Biskupek (✉) · V. V. Singh · J. Müller · M. Zhang
Institute of Geodesy (IfE), Leibniz University Hannover, Hannover, Germany
e-mail: biskupek@ife.uni-hannover.de; singh@ife.uni-hannover.de; mueller@ife.uni-hannover.de; zhang@ife.uni-hannover.de

© The Author(s) 2024

J. T. Freymueller, L. Sánchez (eds.), *Gravity, Positioning and Reference Frames*, International Association of Geodesy Symposia 156, https://doi.org/10.1007/1345_2024_238

2021). Furthermore, the determination of Earth Orientation Parameters (EOPs) is also possible from LLR data. It includes parameters like terrestrial pole coordinates and the Earth rotation phase ΔUT1 (Singh et al. 2022a; Biskupek et al. 2022), the celestial pole coordinates (Zerhouni and Capitaine 2009; Cheng et al. 2019) and coefficients for precession and nutation (Hofmann et al. 2018; Biskupek et al. 2012). As special case the Universal Time at a specific location ΔUT0 can be determined. ΔUT0 and the coefficients of the nutation series are of particular interest, since these parameters are otherwise only determined from Very Long Baseline Interferometry (VLBI).

2   Data and Analysis

The current LLR dataset includes 30172 NPs over the time span April 1970 to April 2022. Starting 2015, many NPs have been measured with laser pulses at infra-red (IR) wavelength, enabling distance measurements near new and full Moon for OCA and WLRS (Chabé et al. 2020; Eckl et al. 2019). This leads to a better coverage of the lunar orbit over the synodic month, i.e. the time span in which Sun, Earth, and Moon return to a similar constellation again. With a better coverage of the lunar orbit, it is possible to estimate various parameters of the Earth-Moon system with higher accuracy and reduced internal correlation. This benefit, together with a higher number of NPs per night, gives the motivation for the determination of EOPs from LLR.

The parameter estimation with the LUNAR analysis software consists of several parts. One part is the calculation of the ephemeris of the eight planets, Sun, Moon, Pluto and asteroids (Ceres, Vesta, and Pallas) as well as the orientation of the core and mantle of the Moon. The needed initial positions and velocities are taken from the DE440 ephemeris (Park 2021). The calculation of the Moon's rotation is carried out simultaneously with the ephemeris calculation. Another part is the calculation of the Earth-Moon distance. The rotation of the Earth is described by two series of EOPs. For the time span 04.1970 to 01.01.1983 the Kalman Earth Orientation Filter (KEOF) COMB2019 series (Ratcliff and Gross 2020) is used and from 02.01.1983 on the IERS EOP C04 series (Bizouard et al. 2019). The difference between these series is the input data, only the COMB series includes LLR data. For this reason, the series fits the LLR analysis better in the initial phase of the observations. From the 1980s on, the differences between the series are small (only a few mas and ms) and the IERS series is used for its shorter latency. All other models in the LLR analysis follow the recommendations of the IERS Conventions 2010 (Petit and Luzum 2010). The last part of the analysis is the parameter estimation itself with the calculation of the residuals between the observed NPs and calculated Earth-Moon distance in a least-squares adjustment (LSA). The NPs are treated as uncorrelated for the stochastic model of the LSA and are weighted according to their accuracies.

3   Determination of Earth Orientation Parameters

The terrestrial pole coordinates, x_p and y_p, describe the change of the rotation axis with respect to the Earth's surface. The Earth rotation phase ΔUT1 and the Length of Day (LOD) refer to the rotation of the Earth about its axis. All these parameters are summarised as Earth Rotation Parameters (ERPs). Together with the celestial pole offsets, as corrections to the conventional precession–nutation model, they define the EOPs.

For the analysis of LLR data, the Barycentric Celestial Reference System (BCRS) is used as the inertial system. The coordinates of the observatories and retro-reflectors are given in their respective body-fixed reference systems, like the International Terrestrial Reference System (ITRS) for the Earth and the Principle Axis System (PAS) for the Moon, and are transformed during the analysis into the inertial system. For the Earth, the transformation from ITRS to the Geocentric Celestial Reference System (GCRS) is given by

$$\mathbf{r}_{GCRS} = \mathbf{Q}(dt)\,\mathbf{R}(dt)\,\mathbf{W}(dt)\,\mathbf{r}_{ITRS} . \qquad (1)$$

Here $\mathbf{W}(dt)$ includes the terrestrial pole coordinates x_p and y_p. The Earth rotation phase ΔUT1 is part of $\mathbf{R}(dt)$. Finally $\mathbf{Q}(dt)$, represented here according to the Fukushima–Williams parametrisation via precession and nutation (Fukushima 2003; Williams 1994), contains the coefficients of the nutation series. As the rotation matrix (Eq. 1) is included in the LLR analysis model, the various parameters of the formula can be estimated directly in the least-squares adjustment of the LLR data together with the other parameters of the Earth-Moon system. A more detailed description of the EOP determination from LLR data is given in Biskupek (2015), Hofmann et al. (2018), Singh et al. (2022a), Biskupek et al. (2022).

Another approach to determine ERPs from LLR data is given by Dickey et al. (1985), Müller (1991) and Pavlov (2019), where the ERPs are determined from the post-fit residuals of the least-squares adjustment of LLR data. In this way the variation of longitude ΔUT0 can be determined (Chapront-Touzé et al. 2000) by

$$\Delta\text{UT0} = \Delta\text{UT1} + \frac{(x_p \sin(\lambda) + y_p \cos(\lambda))\tan(\phi)}{15 \times 1.002737909} , \qquad (2)$$

as combination of ΔUT1 and the terrestrial pole coordinates x_p, y_p, with the observatories longitude λ and latitude ϕ. The variation of latitude VOL is given by

$$\text{VOL} = x_p \cos\lambda - y_p \sin\lambda . \tag{3}$$

The disadvantage of this approach is that the correlations between the ERPs and the other parameters of the Earth-Moon system can not be investigated compared to the approach via Eq. (1). However, to better assess the results of the two approaches, they will be compared in a future study.

3.1 Earth Rotation Parameters

In the LLR analysis, different cases for the ERP estimation can be set up, e.g., by selecting certain time spans of data, specific nights on which a minimum number of NPs is available, or selecting NPs from specific observatories. Previous investigations (Singh et al. 2022a; Biskupek et al. 2022) show that the accuracy of the determined ERPs has greatly improved from 2000. Nevertheless, the reported uncertainties of the estimated parameters from the LLR analysis were normally published as three times the formal error from the LSA (3σ). In the past, it was assumed that some small random and systematic errors remained in the LLR modelling and analysis, and affect the determined parameters. To give more realistic uncertainties for the determined parameters and to also consider possible shortcomings in the analysis a scaling factor for the formal errors of the least-squares adjustment was used (Müller 1991; Biskupek 2015; Hofmann et al. 2018; Singh et al. 2022b). Systematic errors include, e.g., the uneven distribution of NPs during the synodic month and the constellation of Earth and Moon when observing an LLR NP, because of the inaccuracy of atmospheric delay models for low altitude observations. Further error sources are the imperfection of lunar ephemeris and rotation, e.g., because of simplified modelling of the asteroids, tidal deformations affecting the gravitational potential of Earth and Moon, modelling of the lunar core and unstable delay offsets for the calibration. These errors are different for each observation. Random errors result from the general measurement accuracy of LLR. They are different for each night and depend on the observatory. To assess whether such a scaling factor is necessary when estimating ERPs from LLR, a sensitivity analysis was carried out by creating variations in the fitted and fixed parameters to obtain multiple solutions. The fitted ERPs from different calculations were then compared to each other. Four cases were run for the different calculations:

1. Case 1.1: Initial values of all parameters (including the velocities of the LLR observatories) from a standard

solution of LUNAR. All standard parameters along with the ERPs for selected nights were fitted.
2. Case 1.2: Similar to case 1.1, except only ERPs on selected nights were fitted and the standard parameters were kept fixed.
3. Case 2.1: Initial values of all parameters from a solution of LUNAR which was obtained by fixing the velocities of the LLR observatories to ITRF2020 values. All standard parameters except the velocities of the LLR observatories were fitted, along with the ERPs for selected nights.
4. Case 2.2: Similar to case 2.1, except only ERPs on selected nights were fitted and the standard parameters (including the velocities of the LLR observatories) were kept fixed.

For the sensitivity analysis, case 2.1 was selected as standard case against which the results of the other cases are compared. This case was taken because the specific LLR network of the observatories, determined from the LLR analysis, is stabilised by the fixed ITRF velocities.

The ERPs for the sensitivity analysis were determined from the NPs of all LLR observatories. The minimum number of NPs was 15 which results in 491 nights in the time span 04.1984–03.2022. Each component of the ERPs, that is x_p, y_p, and ΔUT1, was determined in a separate adjustment procedure. The IERS C04 series was used as the a-priori ERP. Its values have been fixed for the nights not considered in the fit, which helps to keep the LLR internal network closer to the ITRF. Studies with more LLR data sets are discussed in Singh (2023).

The results of the sensitivity analysis are given in Table 1. All values in the table are the WRMS, weighted according to the number of NPs per night. Column three of Table 1 shows the formal errors of the LSA, without a scaling factor (1σ (2.1)). This means, from the individual 491 formal errors for the determined ERP component, the WRMS is calculated, weighted according to the number of NPs per night. For column four, the mean of the formal errors of the cases 1.1, 1.2 and 2.2 is calculated for each night. This mean value is

Table 1 The values in each column are the WRMS values, weighted according to the number of NPs per night. The last three columns show the formal errors of the LSA (1σ values) for the standard case 2.1, the difference of standard case 2.1 to the mean of the other cases (MC) and the standard deviation (Std. Dev.) of the cases 1.1, 1.2 and 2.2. For each estimated ERP, the results are split into two time spans, before and after 2000

ERP component	Time span	1σ (2.1)	SV(2.1)–MC	Std. Dev.
x_p [mas]	<2000.0	5.22	0.45	0.38
	>2000.0	0.47	0.10	0.10
y_p [mas]	<2000.0	3.84	0.74	0.59
	>2000.0	0.59	0.11	0.09
ΔUT1 [μs]	<2000.0	39.30	38.61	46.97
	>2000.0	6.18	5.52	4.77

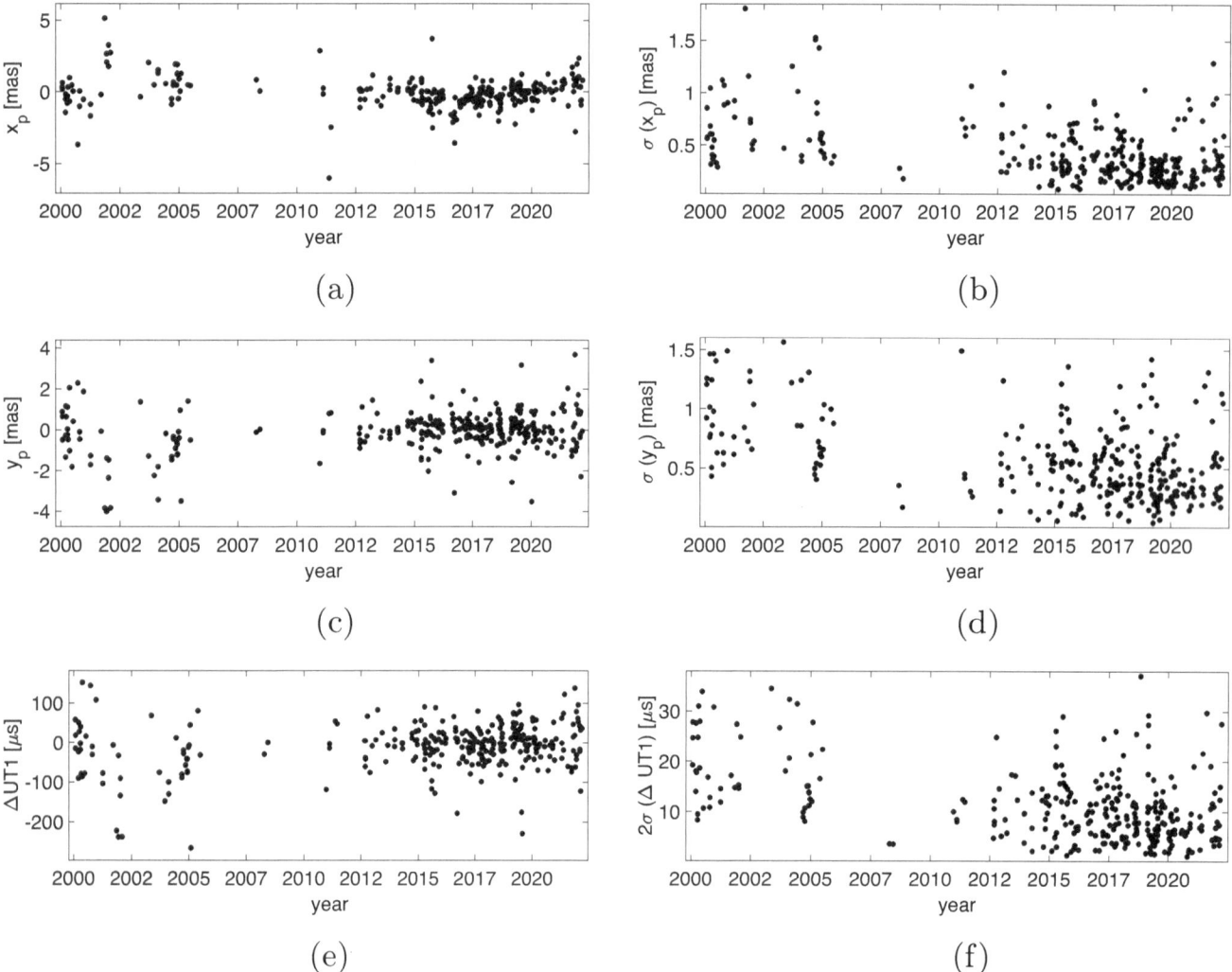

Fig. 1 Results for the determination of ERPs from the NPs of all observatories. The left plots show the differences to the a-priori IERS C04 series and the right plots the uncertainties. For the pole coordinates the uncertainties are given as the formal errors of the LSA, for ΔUT1 as two times the formal errors. (**a**) x_p differences. (**b**) $\sigma(x_p)$ uncertainties. (**c**) y_p differences. (**d**) $\sigma(y_p)$ uncertainties. (**e**) ΔUT1 differences. (**f**) $2\sigma(\Delta$UT1) uncertainties

subtracted from case 2.1, and from all that differences, the WRMS is calculated and given in column four. Finally, for the result in column five, the standard deviation of the three cases 1.1, 1.2 and 2.2 (Std. Dev.) is calculated for each night, and from these values the WRMS is given in column five. For each estimated ERP, the results are split into two time spans, before and after 2000, as the results become much better after 2000 (Singh et al. 2022a).

For the two pole coordinates the results are very similar. The formal errors from the LSA are bigger than the differences between the standard case and the other cases, and also bigger than the standard deviation of the three cases. This is true for both time spans, before and after 2000. The result shows that a scaling factor for the formal errors of the adjustment is not necessary for the pole coordinates with the current version of the analysis model. The results for the Earth rotation phase ΔUT1 differ from those of the

pole coordinates. For the time span before 2000, only the difference between the standard case and the other cases is smaller than the formal errors of the adjustment. The standard deviations of the three cases are bigger than the formal errors of the adjustment. To ensure that the formal errors of the adjustment are bigger than the other values and represent a realistic uncertainty, at least a scaling factor of two is required here. Further investigations with more subsets show that a scaling factor of three is needed (Singh 2023). Also for the time span after 2000, the formal errors of the adjustment are only slightly bigger than the difference of the standard case to the other cases and the standard deviations of the other cases. So, a scaling factor of two should be used for ΔUT1.

Figure 1 shows the results from the sensitivity analysis for case 2.1 as differences to the a-priori IERS C04 series and the uncertainties. The individual sub-figures show two aspects: (1) The large amount of data after 2015, because of the IR measurements provided by OCA, and (2) the improved uncertainties in the calculated ERP values, from 2015 onwards. The improved uncertainties are due to the high accuracy of the data. For the pole coordinates, the differences to the a-priori IERS C04 EOP series vary in the range of -6.0 mas to 5.1 mas for x_p and -4 mas to 3.7 mas for y_p. The uncertainties (1σ) vary between 0.1 mas to 1.8 mas for x_p and 0.0 mas to 1.6 mas for y_p. For ΔUT1, the differences to the a-priori IERS C04 EOP series vary in the range of $-266.2\,\mu$s to $151.4\,\mu$s. The uncertainties (2σ) vary between $0.1\,\mu$s to 37.1μs.

From the sensitivity analysis the best current uncertainty for the pole coordinates (1σ) is 0.47 mas for x_p and 0.59 mas for y_p. Using the Earth radius at the equator as 6378 km, 1 mas corresponds to 3 cm spatial resolution on Earth's surface. Thus, the uncertainty for the pole coordinates result in 1.41 cm and 1.77 cm spatial resolution, respectively. The current best uncertainty for ΔUT1 is $12.36\,\mu$s (2σ). Using the Earth radius, $10\,\mu$s corresponds to 4.6 mm on the Earth's surface and lead to a spatial resolution of 5.69 mm for ΔUT1. As the ERP components were determined in separate adjustments the given uncertainties might be too optimistic because correlations between UT1 and the pole coordinates are not taken into account. In an earlier study by Singh et al. (2022a), the simultaneous determination of the two pole coordinates x_p and y_p has already been investigated. The uncertainties for the pole coordinates from a simultaneous determination were only slightly higher than from a separate determination with 15 NPs per night in the least-squares adjustment.

Compared to other space geodetic techniques like VLBI, Global Navigation Satellite System (GNSS) and Satellite Laser Ranging (SLR) the results from LLR are still worse. The uncertainties for ΔUT1 from VLBI are about $3\,\mu$s to $5\,\mu$s from 24h sessions and $15\,\mu$s to $20\,\mu$s from intensive sessions. For x_p, y_p from VLBI, the uncertainties are about $50\,\mu$as to $80\,\mu$as (Schuh and Behrend 2012; Raut et al. 2022), about $10\,\mu$as to $30\,\mu$as from SLR (Sciarretta et al. 2010) and about $5\,\mu$as to $20\,\mu$as from GNSS (Capitaine 2017; Zajdel et al. 2020). Nevertheless, when the LLR results become better in the future, a possible contribution as an independent technique could be considered.

3.2 Nutation

As mentioned in Eq. (1), all EOPs are needed for the transformation of station coordinates from the ITRS into the inertial system. The IAU 2000 nutation model is described in the IERS Conventions 2010 (Petit and Luzum 2010) as a series

for nutation in longitude $\Delta\psi$ and obliquity $\Delta\epsilon$, referred to the mean ecliptic of date:

$$\Delta\psi = \sum_{i=1}^{n} (A_i + A_i't)\,\sin(ARG) + (A_i'' + A_i'''t)\,\cos(ARG)$$
(4)

$$\Delta\epsilon = \sum_{i=1}^{n} (B_i + B_i't)\,\cos(ARG) + (B_i'' + B_i'''t)\,\sin(ARG)$$
(5)

with $ARG = \sum_j^5 N_j F_j$, N_j: multipliers, F_j: Delaunay parameters and time t measured in Julian centuries from epoch J2000. n defines the number of terms the model is composed of, 678 lunisolar and 687 planetary terms with *in-phase* (first part of the sum in Eqs. (4) and (5)) and *out-of-phase* (second part of the sum) coefficients. This series is based on the REN2000 nutation solution (Souchay et al. 1999) for the rigid Earth, which is convolved to the nutation model MHB2000 for the non-rigid Earth by the transfer function from Mathews et al. (2002). This model is used as a-priori nutation model in the LLR analysis, where the non-time-dependent coefficients can be determined along with other parameters of the Earth-Moon system.

Nutation by definition refers to a dynamic reference system. In order to minimise possible small systematic deviations in orientation between the kinematic realisation of the inertial system based on VLBI and the inertial system which is dynamically realised by the LLR-based ephemeris computation, an additional perturbation rotation matrix is defined as

$$\mathbf{S}(dt) = \begin{bmatrix} 1 & \Theta_z & -\Theta_y \\ -\Theta_z & 1 & \Theta_x \\ \Theta_y & -\Theta_x & 1 \end{bmatrix}$$
(6)

where Θ_x is used to adjust the ecliptic angle, Θ_y allows an adjustment of the GCRS equator, Θ_z an adjustment between the vernal equinox and the origin along the equator. The perturbation rotations are modelled as time-dependent quantities with $\Theta = \Theta_0 + \dot{\Theta}\Delta t$. A similar approach to fitting the reference systems is also described in Hilton and Hohenkerk (2004), Yagudina (2009), Zerhouni and Capitaine (2009), Williams et al. (2013). In a first step of the analysis, the angles Θ_x and Θ_y were determined with the fixed nutation model to minimise the deviations between the reference systems. In a second step, the components of Eq. (6) were fixed to determine the coefficients of the nutation series.

Table 2 gives the values from the LLR analysis for the periods with the largest contribution to the nutation angles. These periods are: 18.6-year, 182.62-day, 13.66-day, 9.3-year, and 365.26-day, sorted in order of their largest contribution. The values are given as differences to the a-priori model with uncertainties as three times the formal errors from the LSA. The current results are compared to

Table 2 Results from the determination of nutation coefficients as differences to the main periods of the MHB2000 model. Results are compared between a determination in 2018 (Hofmann et al. 2018) and the current results of 2023. All values are given in [mas]

Period	Results 2018	Results 2023
$A_{18.6y}$	1.42±0.18	0.79±0.05
$B_{18.6y}$	−0.18±0.08	−0.20±0.03
$A''_{18.6y}$	−0.68±0.12	0.61±0.05
$B''_{18.6y}$	−0.06±0.07	0.05±0.02
$A_{9.3y}$	−1.12±0.12	0.23±0.03
$B_{9.3y}$	−0.27±0.05	−0.04±0.01
$A''_{9.3y}$	−1.55±0.12	0.12±0.03
$B''_{9.3y}$	0.17±0.05	0.07±0.01
$A_{365.3d}$	1.05±0.07	0.30±0.06
$B_{365.3d}$	−0.51±0.03	−0.04±0.03
$A''_{365.3d}$	0.65±0.05	0.28±0.03
$B''_{365.3d}$	0.04±0.02	0.08±0.02
$A_{182.6d}$	0.51±0.02	0.17±0.05
$B_{182.6d}$	−0.06±0.01	0.19±0.02
$A''_{182.6d}$	−0.57±0.02	0.73±0.05
$B''_{182.6d}$	−0.07±0.01	0.04±0.02
$A_{13.6d}$	1.49±0.07	0.10±0.03
$B_{13.6d}$	−0.65±0.03	0.01±0.02
$A''_{13.6d}$	−1.42±0.10	−0.10±0.03
$B''_{13.6d}$	0.27±0.04	−0.11±0.01

those from Hofmann et al. (2018), where a shorter time span of NPs was used, in particular fewer NPs measured in IR were used. Looking at the differences to the a-priori model, the 2022 results are smaller than the 2018 results in most cases, and the uncertainties have improved by a factor of two. The largest improvement is for the 13.66-day period, where the benefit from IR OCA data and the associated more homogeneous observation of the lunar orbit is clearly visible. The uncertainties are still the formal errors with a scaling factor of three to be comparable with the 2018 results. In future, a sensitivity analysis similar to the ERPs will also be carried out for the determination of the nutation coefficients from the LLR analysis in order to assess the need for such a scaling factor.

4 Conclusions

A 52-year LLR data set has been analysed to determine EOPs. For the determination of the terrestrial pole coordinates and the Earth rotation phase a sensitivity analysis was performed in order to assess the need for a scaling factor of the formal errors from the LSA. Different cases and time periods were investigated. For the terrestrial pole coordinates, a scaling factor is not needed. However, for the Earth rotation phase, a scaling factor of two (after 2000) seems to be reasonable. The current best results are 0.47 mas for x_p, 0.59 mas for y_p and 12.36 μs with a scaling factor of two

for ΔUT1. Nevertheless, the LLR uncertainties might be too optimistic because correlations between UT1 and the polar coordinates are not taken into account when determining the ERPs components separately. Therefore as next step, UT1 and the pole coordinates will be determined together and analysed to find the best strategy for ERP determination from LLR data. It will also be further investigated, which parameters of the Earth-Moon system should be determined together with the ERPs. This will lead to a more realistic estimation of their uncertainties.

Compared to results for the nutation coefficients from the year 2018, the current differences to the a-priori MBH2000 model are smaller in most cases, and the uncertainties have improved by a factor of two. Here, the high number of IR NPs and the more homogeneous tracking of the lunar orbit are beneficial, especially for the 13.66-day nutation period.

With more IR data from the observatories OCA and WLRS, it is expected that the parameters of the LSA and also the EOPs can be further improved. Compared to other space geodetic techniques, the results from LLR still lag behind. However, the results are still important as LLR is the only technique other than VLBI which can provide ΔUT1 and nutation values with some good accuracy, and therefore can be used to verify the VLBI results. In future a combined analysis of LLR and VLBI data for the EOPs determination is planned.

Acknowledgements We acknowledge with gratitude, that more than 53 years of processed LLR data have been obtained under the efforts of the personnel at McDonald Laser Ranging Station, USA, Côte d'Azur Observatory, France, Lure Observatory on Maui/Hawaii, USA, Apache Point Observatory Lunar Laser ranging Operation, USA, Matera Laser Ranging Observatory, Italy and Geodetic Observatory Wettzell, Germany. We also acknowledge with thanks the funding by the Deutsche Forschungsgemeinschaft (DFG, German Research Foundation) under Germany's Excellence Strategy (EXC-2123 QuantumFrontiers - Project-ID 390837967).

Declarations

- Availability of data: LLR data is collected, archived, and distributed under the auspices of the International Laser Ranging Service (ILRS) (Pearlman et al. 2019). All LLR NPs used for these studies are available from the Crustal Dynamics Data Information System (CDDIS) at NASA's Archive for Space Geodesy Data, USA, (Noll 2010) at the website.[1] The KEOF COMB2019 EOP time series is available at the website[2] and the IERS C04 EOP time series is available at the website.[3]

---

[1] https://cddis.nasa.gov/Data_and_Derived_Products/SLR/Lunar_laser_ranging_data.html.

[2] https://keof.jpl.nasa.gov/combinations/latest/.

[3] https://www.iers.org/IERS/EN/DataProducts/EarthOrientationData/eop.html.

References

Biskupek L (2015) Bestimmung der Erdorientierung mit Lunar Laser Ranging. PhD thesis, Leibniz Universität Hannover. https://doi.org/10.15488/4721

Biskupek L, Müller J, Hofmann F (2012) Determination of nutation coefficients from lunar laser ranging. In: Kenyon S, Pacino MC, Marti U, et al (eds) Geodesy for Planet Earth, International Association of Geodesy Symposia, vol 136. Springer, Berlin, Heidelberg, pp 521–525. https://doi.org/10.1007/978-3-642-20338-1_63

Biskupek L, Müller J, Torre JM (2021) Benefit of new high-precision LLR data for the determination of relativistic parameters. Universe 7(2). https://doi.org/10.3390/universe7020034

Biskupek L, Singh VV, Müller J (2022) Estimation of earth rotation parameter UT1 from lunar laser ranging observations. Springer, Berlin, Heidelberg, pp 1–7. International Association of Geodesy Symposia. https://doi.org/10.1007/1345_2022_178

Bizouard C, Lambert S, Gattano C, et al (2019) The IERS EOP 14C04 solution for Earth orientation parameters consistent with ITRF 2014. J Geodesy 93(5):621–633. https://doi.org/10.1007/s00190-018-1186-3

Capitaine N (2017) The determination of earth orientation by VLBI and GNSS: principles and results. In: Arias EF, Combrinck L, Gabor P, et al (eds) The science of time 2016. Springer International Publishing, Cham, pp 167–196. https://doi.org/10.1007/978-3-319-59909-0_23

Chabé J, Courde C, Torre JM, et al (2020) Recent progress in lunar laser ranging at grasse laser ranging station. Earth Space Sci 7(3):e2019EA000,785. https://doi.org/10.1029/2019EA000785

Chapront-Touzé M, Chapront J, Francou G (2000) Determination of UT0 with LLR observations. In: Proceedings of the Journées 1999 "Motion of Celestial Bodies, Astrometry and Astronomical Reference Frames", pp 217–220

Cheng YT, Liu JC, Zhu Z (2019) Analyses of celestial pole offsets with VLBI, LLR, and optical observations. Astron Astrophys 627:A81. https://doi.org/10.1051/0004-6361/201834785

Dickey JO, Newhall XX, Williams JG (1985) Earth orientation from lunar laser ranging and an error analysis of polar motion services. J Geophys Res 90(B11):9353–9362

Eckl JJ, Schreiber KU, Schüler T (2019) Lunar laser ranging utilizing a highly efficient solid-state detector in the near-IR. In: Prochazka I, Sobolewski R, James RB, et al (eds) Quantum Optics and Photon Counting 2019, International Society for Optics and Photonics, vol 11027. SPIE, p 1102708. https://doi.org/10.1117/12.2521133

Fukushima T (2003) A new precession formula. Astron J 126(1):494–534. http://stacks.iop.org/1538-3881/126/i=1/a=494

Hilton JL, Hohenkerk CY (2004) Rotation matrix from the mean dynamical equator and equinox at J2000.0 to the ICRS. Astron Astrophys 413(2):765–770. https://doi.org/10.1051/0004-6361:20031552

Hofmann F, Müller J (2018) Relativistic tests with lunar laser ranging. Classic Quant Gravity 35(3):035,015. https://doi.org/10.1088/1361-6382/aa8f7a

Hofmann F, Biskupek L, Müller J (2018) Contributions to reference systems from Lunar Laser Ranging using the IfE analysis model. J Geodesy 92(9):975–987. https://doi.org/10.1007/s00190-018-1109-3

Mathews PM, Herring TA, Buffett BA (2002) Modeling of nutation and precession: New nutation series for nonrigid Earth and insights into the Earth's interior. J Geophys Res 107(B4):ETG 3-1–ETG 3-30. https://doi.org/https://doi.org/10.1029/2001JB000390

Michelsen EL (2010) Normal point generation and first photon bias correction in APOLLO Lunar Laser Ranging. PhD thesis, University of California, San Diego

Müller J (1991) Analyse von Lasermessungen zum Mond im Rahmen einer post-Newton'schen Theorie. PhD thesis, Technische Univer-

sität München, Deutsche Geodätische Kommission bei der Bayerischen Akademie der Wissenschaften, Reihe C, Nr. 383

Müller J, Biskupek L, Oberst J, et al (2009) Contribution of lunar laser ranging to realise geodetic reference systems. In: Drewes H (ed) Geodetic Reference Frames, International Association of Geodesy Symposia, vol 134. Springer, Berlin, Heidelberg, pp 55–59. https://doi.org/10.1007/978-3-642-00860-3_8

Noll CE (2010) The crustal dynamics data information system: A resource to support scientific analysis using space geodesy. Adv Space Res 45(12):1421–1440. https://doi.org/10.1016/j.asr.2010.01.018

Park RS, Folkner WM, Williams JG, et al (2021) The JPL planetary and lunar ephemerides DE440 and DE441. Astron J 161(3):105. https://doi.org/10.3847/1538-3881/abd414

Pavlov D (2019) Role of lunar laser ranging in realization of terrestrial, lunar, and ephemeris reference frames. J Geodesy 94(1):5. https://doi.org/10.1007/s00190-019-01333-y

Pavlov DA, Williams JG, Suvorkin VV (2016) Determining parameters of Moon's orbital and rotational motion from LLR observations using GRAIL and IERS-recommended models. Celest Mech Dyn Astron 126(1):61–88. https://doi.org/10.1007/s10569-016-9712-1

Pearlman MR, Noll CE, Pavlis EC, et al (2019) The ILRS: approaching 20 years and planning for the future. J Geodesy 93(11):2161–2180. https://doi.org/10.1007/s00190-019-01241-1

Petit G, Luzum B (eds) (2010) IERS Conventions 2010. No. 36 in IERS Technical Note, Verlag des Bundesamtes für Kartographie und Geodäsie, Frankfurt am Main

Ratcliff JT, Gross RS (2020) Combinations of earth orientationmeasurements: SPACE2019, COMB2019,and POLE2019. Tech. Rep. JPL Publication 20-3

Raut S, Heinkelmann R, Modiri S, et al (2022) Inter-comparison of UT1-UTC from 24-hour, intensives, and VGOS sessions during CONT17. Sensors 22(7). https://doi.org/10.3390/s22072740

Schuh H, Behrend D (2012) VLBI: A fascinating technique for geodesy and astrometry. J Geodynam 61:68–80. https://doi.org/https://doi.org/10.1016/j.jog.2012.07.007

Sciarretta C, Luceri V, Pavlis EC, et al (2010) The ILRS EOP time series. Artif Satell 45:41–48. https://doi.org/10.2478/v10018-010-0004-9

Singh VV (2023) Lunar laser ranging - improved modelling and parameter estimation. PhD thesis, Leibniz Universität Hannover. https://doi.org/10.15488/14298

Singh VV, Biskupek L, Müller J, et al (2022a) Earth rotation parameter estimation from LLR. Adv Space Res 70(8):2383–2398. https://doi.org/10.1016/j.asr.2022.07.038

Singh VV, Biskupek L, Müller J, et al (2022b) Estimation of lunar ephemeris from lunar laser ranging. EGU abstract. https://doi.org/10.5194/egusphere-egu22-2815, EGU General Assembly 2022, Vienna, Austria, 23–27 May 2022, EGU22-2815

Souchay J, Loysel B, Kinoshita H, et al (1999) Corrections and new developments in rigid earth nutation theory - III. Final tables "REN-2000" including crossed-nutation and spin-orbit coupling effects. Astron Astrophys Suppl Ser 135:111–131. https://doi.org/10.1051/aas:1999446

Viswanathan V, Fienga A, Minazzoli O, et al (2018) The new lunar ephemeris INPOP17a and its application to fundamental physics. Month Not Roy Astrono Soc 476(2):1877–1888. https://doi.org/10.1093/mnras/sty096

Williams JG (1994) Contributions to the Earth's obliquity rate, precession, and nutation. Astron J 108(2):711–724. https://doi.org/10.1086/117108

Williams JG, Turyshev SG, Boggs DH (2012) Lunar laser ranging tests of the equivalence principle. Class Quant Grav 29(18):184,004. https://doi.org/10.1088/0264-9381/29/18/184004

Williams JG, Boggs DH, Folkner WM (2013) DE430 lunar orbit, physical librations, and surface coordinates. JPL Interoffice Memorandum IOM 335-JW,DB,WF-20080314-001, Jet Propulsion Laboratory, California Institute of Technology, Passadena, California

Yagudina EI (2009) Lunar numerical theory EPM2008 from analysis of LLR data. In: Soffel M, Capitaine N (eds) Astrometry, Geodynamics and Astronomical Reference Systems, Lohrmann-Observatorium and Observatoire de Paris, Proceedings of the Journées 2008 Systèmes de Référence Spatio-Temporels, pp 61–64

Zajdel R, Sośnica K, Bury G, et al (2020) System-specific systematic errors in earth rotation parameters derived from GPS, GLONASS,

and Galileo. GPS Sol 24(3):74. https://doi.org/10.1007/s10291-020-00989-w

Zerhouni W, Capitaine N (2009) Celestial pole offsets from lunar laser ranging and comparison with VLBI. Astron Astrophys 507(3):1687–1695. https://doi.org/10.1051/0004-6361/200912644

Zhang M, Müller J, Biskupek L (2020) Test of the equivalence principle for galaxy's dark matter by lunar laser ranging. Celest Mech Dyn Astron 132(4):25. https://doi.org/10.1007/s10569-020-09964-6

List of Reviewers

Alexander Kehm
Alexander Neidhardt
Alexandre Couhert
Anna Klos
Anna Riddell
Daniel Willi
Daniele Borio
Dariusz Tomaszewski
Derek VanWestrum
Dmitry Pavlov
Dominik Próchniewicz
Erricos C. Pavlis
Frédéric Jaron
Gabriel Guimaraes
Georgios S. Vergos
Grzegorz Klopotek
Hana Krasna
Hartmut Wziontek
Hussein A. Abd-Elmotaal
Iván Dario Herrera Pinzón
Jacek Paziewski
Jamie McCallum
Janusz Bogusz
Jean-Paul Boy
Jeff Kanney
Jianghui Geng
Johannes Böhm
Jonas Ågren
José Rodréguez
Jungang Wang
Karine Le Bail

Kevin Ahlgren
Kyriakos Balidakis
Laura Sánchez
Lennard Huisman
Leonid Petrov Marcin Rajner
Matthias Schartner
Mehdi Eshagh
Mohammad Ali Goudarzi
Paul Groves
Pawel Wielgosz
Periklis-Konstantinos Diamantidis
Petr Stepanek
Phillip McFarland
Przemyslaw Dykowski
Rebekka Steffen
Riccardo Barzaghi
Richard Stanaway
Ryan Hardy
Ryan Hippenstiel
Rüdiger Haas
Safoora Zaminpardaz
Salim Masoumi
Sebastien Lambert
Thalia Nikolaidou
Tobias Nilsson
Toshimichi Otsubo
Tzvetan Simeonov
Ulrich Schreiber
Virendra M. Tiwari
Víctor Puente
Zohreh Adavi

J. T. Freymueller, L. Sánchez (eds.), *Gravity, Positioning and Reference Frames*,
International Association of Geodesy Symposia 156, https://doi.org/10.1007/978-3-031-63855-8

Author Index

A
Abbondanza, C., 139
Ågren, J., 3–8
Alfredsson, A., 3–8
Aljebreen, S., 63–68
Al-Jubreen, S., 21–26
Alonso, I., 53, 54
Al-Qahtani, A., 21–26, 63–68
Al-Shahrani, S., 21–26, 63–68
Alshawaf, F., 110
Altamimi, Z., 119, 139, 147, 148, 167–172, 185, 189, 209
Alvarez-Calderón, A., 153–163
Alves Costa, S.M., 153–163
Ampatzidis, D., 139–145
Andersen, O.B., 13, 23, 63
Anderson, J.M., 228
Antokoletz, E.D., 153–163
Arnautov, G.P., 218
Artz, T., 183
Ayan, M., 22

B
Bai, D., 88
Balasubramanian, G., 88
Balidakis, K., 109–115, 129–136
Banville, S., 73
Barnéoud, J., 189–193
Barzaghi, R., 40, 63, 141
Becker, D., 45, 46
Beer, S., 102
Behrend, D., 120, 239
Beilin, J., 189–193
Bender, M., 80
Bennitt, G., 109
Beutler, G., 230
Bevis, M., 109
Bierman, G.J., 135
Bilker-Koivula, M., 14, 217–225
Biondi, R., 79–85
Bishop, C.M., 96
Biskupck, L., 235–240
Bizouard, C., 185, 236
Bjerhammar, A., 54
Blewitt, G., 140, 168
Blitzkow, D., 153–163
Bloßfeld, M., 183, 185
Bock, O., 110
Böder, V., 102
Boehm, J., 75
Böhm, J., 121, 196, 197, 204, 210, 223, 228
Bolotin, Y.V., 46

Bond, J., 176
Borio, D., 93–95, 97
Bos, M.S., 13
Boy, J.-P., 12, 15, 16, 130, 167–172, 211, 222
Brack, A., 73–78
Brandt, A., 209–215
Bregni, S., 95
Brenot, H., 79–85
Breva, Y., 101–107
Bruinsma, S., 4, 22
Brunini, C., 154
Bruyninx, C., 175–176

C
Caizzone, S., 102
Calvés, C.M., 224
Capitaine, N., 236, 239
Carrère, L., 169
Chabé, J., 189, 236
Champollion, C., 79
Chanard, K., 167–172
Chapront-Touzé, M., 236
Charlot, P., 120, 185, 195, 198
Chatzinikos, M., 142
Cheng, M.K., 57
Cheng, Y., 13
Cheng, Y.T., 236, 239
Cina, A., 176
Collilieux, X., 167–172, 189–193
Coulot, D., 120
Courde, C., 189–193
Crossley, D.J., 12

D
da Silva, A., 153–163
Dach, R., 4, 120, 158, 178, 210
Dawson, J., 190, 223
de La Serve, M., 167–172
de Ligt, H., 175–181
de Matos, A.C.O.C., 153–163
de Witt, A., 195–201
Degnan, J.J., 220
Dehant, V., 14
Del Pino, J., 220
Delva, P., 54, 120, 228
Demirtzoglou, N., 139–145
Denker, H., 54
Dermanis, A., 183
Desai, S.D., 210
Diamantidis, P.-K., 120

Subject Index

© The Author(s) 2024
J. T. Freymueller, L. Sánchez (eds.), *Gravity, Positioning and Reference Frames*,
International Association of Geodesy Symposia 156, https://doi.org/10.1007/978-3-031-63855-8